Probability Distributions for Directional Data on Smooth Manifolds

Probability Distributions for Directional Data on Smooth Manifolds

Ashis SenGupta
Indian Statistical Institute
Kolkata, West Bengal
India

Kunio Shimizu
Center for Training Professors in Statistics
The Institute of Statistical Mathematics
Tachikawa, Tokyo
Japan

WILEY

Published by John Wiley & Sons, Inc., Hoboken, New Jersey.

For general information on our other products and services or for technical support, please contact our Customer Care Department within the United States at (800) 762-2974, outside the United States at (317) 572-3993 or fax (317) 572-4002.

Wiley also publishes its books in a variety of electronic formats. Some content that appears in print may not be available in electronic formats. For more information about Wiley products, visit our web site at www.wiley.com.

Library of Congress Cataloging-in-Publication Data applied for:

Hardback ISBN: 9781119414063

Cover Design: Wiley
Cover Image: © Ashis SenGupta

Set in 9.5/12.5pt STIXTwoText by Straive, Chennai, India

In the fond memory of my respected parents Mr. Shiba Nandan SenGupta and Mrs. Sudha Rani SenGupta.

—Ashis

To my family, who have supported me for so many years.

—Kunio

Contents

Preface

Probability distributions form the indispensable foundational tool for deriving objective inference from data on any scientific research in our stochastic world. With the availability of data on various manifolds – notably torus, cylinder, and hyper-disc, in rapidly emerging areas of applied science, there has recently been a surge of research on this topic. Thus, the need for probability distributions on these manifolds has become crucial. The derivations of such distributions as on these non-Euclidean manifolds have called for innovative techniques and new methods as compared to those existing for the Euclidean manifold. No volume exists to this date on a comprehensive and rigorous treatment of this emerging topic. This book is the maiden attempt to present a systematic approach for the derivations of probability distributions on popular manifolds.

Among the unique features of this book, we feel the compilation of probability distributions on the various manifolds considered should stand out. Next is the development of the generalizations of the methods of derivations. Our approach is to develop the methods ab initio for a simple manifold and let the reader appreciate how these unfold and generalize to more complicated manifolds. So, rather than taking up one distribution at a time, we have tried to present one method at a time and show how it can yield an unified approach to yield many distributions under its ambit. Of course, in certain cases as for the hyper-disc, the approach could be unique and may not amenably follow from that for other manifolds. While producing existing distributions, our approach thus also opens up the possibility of developing many more useful distributions, or even family of distributions, under the same technique. Finally, several new distributions have also been presented as outcomes of these generalizations. We also briefly give real-life examples, which not only attest to the ongoing usefulness but will help as well to visualize other modern day areas of the applications of these novel and important distributions.

The book starts with a brief introduction in Chapter 1, highlighting the fundamental differences in properties, and thus derivations, for distributions on non-Euclidean as compared to those on Euclidean manifolds. Chapters 2–5 present several unified and general methods of deriving distributions on the circle, both for the continuous and discrete random variables, including the cases of asymmetry and multi-modality. In Chapter 6, some more interesting distributions on the circle are presented separately under the miscellaneous collection. Chapters 7 treats univariate and multivariate axial distributions, and Chapter 8 deals with the manifolds on sphere, hemi-sphere, and hyper-sphere in detail. The general approach for the chapters on the circle is also followed in Chapters 9 and 10, where probability distributions for the manifolds on (ring) torus and hyper-torus, and (vertical) cylinder and hyper-cylinder, are taken up, respectively. We conclude with the challenging manifold of disc and hyper-disc in Chapter 11. Each of the Chapters 2–11 has brief and selected overviews of the corresponding statistical inference and real-life applications. Appendix provides some useful mathematical concepts and formulae related to the above probability distributions. The rich collection in references should be quite useful to the readers. While the pool of distributions presented has been created on the basis of mostly unified methodologies for their derivations, the list includes almost, if not, all of the commonly known distributions in this arena.

It is a real pleasure to acknowledge with thanks and appreciation the invaluable suggestions and encouragements made by the evaluators of our proposal. We thank the production team of J. Wiley for their exemplary patience in being with us on this long journey and various inputs they have provided to make our presentation a smooth one. We also thank our many colleagues and students who have helped with their interactions with us toward the preparation of the contents.

This book is intended for graduate students, researchers, and faculty in Mathematical Sciences, who are interested in exploring the rich and fertile area of theoretical research for appreciating and developing further probability distributions on different manifolds. It is also intended for researchers in applied sciences who need to model their data, which are invariably stochastic in nature, on non-Euclidean manifolds with highly flexible families of probability distributions. It can further serve to enhance the all-so-important development of statistical inference for the related data analysis. Finally, we feel that this book is just a stepping stone and envisage significant and novel research motivated by it in the emerging areas of modern data science.

17 February 2026

Ashis SenGupta
Kolkata, India

Kunio Shimizu
Tokyo, Japan

1
Fundamental Concepts

The last couple of decades have witnessed a phenomenal growth in both the extent and nature of real-life scientific problems which encounter nonlinear data. More specifically, such data as being referred to here do not have support on the real line but are mapped on a circle, sphere, torus, cylinder, disc, or their multivariate generalizations. Examples abound from Astrophysics, Bioinformatics, Defense Sciences, Environmental and Ecological Sciences, Geological Sciences, Health Sciences, Molecular Biophysics, and in fact from virtually all areas of applied research. Uncertainty in the physical world thus demands the development of stochastic models, more specifically of probability distributions, for the underlying populations.

In this chapter, we briefly discuss several characteristics and special features which arise in the construction of such distributions as mentioned above. We expose some interesting properties which differ between these distributions and those for the distributions defined on the $\mathbb{R}$ or $\mathbb{R}^k$. Also discussed are the challenges faced and short comings in generalizing the methods presented for the univariate to the multivariate case. Briefly noted at the end are the distinguishing features which call for special considerations needed for developing statistical inference procedures for these distributions.

1.1 Preliminary Notions

The simplest non-Euclidean manifold we start with is that of the circle S, which can be considered to be obtained from the Euclidean manifold $\mathbb{R}^2$ by the embedding approach $\mathbb{R}^2 \to S$. Thus, a point on $(x_1, x_2) \in S$ is defined as (i) $(x_1, x_2)|x_1^2 + x_2^2 = 1$. We will be dealing with points on the circumference of the unit circle. Thus three more representations of this point are available: (ii) the angle $\theta, \theta \in [0, 2\pi)$, (iii) the vector representation in terms of the trigonometric functions, $(\cos\theta, \sin\theta)$, and (iv) a point in the complex plane $\mathbb{C}, e^{i\theta}$, where $i = \sqrt{-1}$. It is usually more convenient to work in terms of θ as compared to the other representations for the case of the circle, torus, and cylinder. However, we will see in the later chapters that the Euclidean coordinates defining the related polar transformations may be quite useful for the cases of hyper-sphere and hyper-disc. The non-Euclidean manifolds, e.g. circle, torus, cylinder, sphere, etc. to be discussed in this book, will be referred to in general, simply as (corresponding) manifolds.

Random variables (rvs) defined on $\mathbb{R}$ will be referred to as linear rvs while those on circle, torus, etc. as circular, toroidal rvs, etc. The support of the probability density function (pdf) $f(\theta)$ of the rv Θ may be taken as $[0, 2\pi)$ or equivalently $[-\pi, \pi)$. Their graphical representations on the real line may differ and tend to be misleading, but their equivalence is obvious from that on the circle. For example, a symmetric pdf $f(\theta)$ with mode at 0 will be U-shaped on $[0, 2\pi)$, but bell-shaped on $[-\pi, \pi)$ when plotted on the real line. Similarly, $f(\theta)$ plots that may appear asymmetric on the real line will emerge as symmetric when plotted on the circle if they are truly so.

The pdf $f(x)$ of a linear rv X may not exist or be available in any analytical form. For example, for X with support on entire $\mathbb{R}$, a pdf is not meaningful. Also, for the stable family of distributions for X, the pdfs of its members, save only three, cannot be presented analytically. In both these cases, however, the pdfs of the "corresponding" (by wrapping) Θ do exist and can be explicitly presented as will be elaborated below.

Just as differential equations, ordinary or partial, are used to study various linear pdfs, the same can be one for circular or toroidal pdfs too (see, e.g. SenGupta et al., 2026). However, we will not pursue that approach here, mainly because it does not usually lead to interesting new distributions on manifolds, but rather reveals some growth behavior of those in general.

1.2 Characteristic Functions and Trigonometric Moments

We briefly discuss here some important features of the characteristic function and (trigonometric) moments of distributions on circle and torus. Such properties corresponding to other manifolds, e.g. sphere, will be taken up in details in Chapter 8.

1.2.1 Characteristic Functions

Characteristic function (c.f.) provides an important tool for the derivation of the pdf of the associated rv Θ. This is so because of its elegant representation in terms of the trigonometric moments of Θ, and in turn the representation of the pdf of Θ in terms of these trigonometric moments for its Fourier series representation.

An important property of the c.f. of Θ, $\phi_p(\Theta)$, is given by

Result 1.1 $\phi_p(\Theta)$ is defined only for $p \in \mathbb{Z}$, the set of all integers.

The result follows from the fact that $\theta \equiv \theta + p \cdot 2\pi,\ p \in \mathbb{Z},\ \forall \theta \in [0, 2\pi)$.

Thus, the c.f. of Θ can be written as

$$\phi_p(\Theta) = E(e^{ip\Theta}) = E\{\cos(p\Theta)\} + iE\{\sin(p\Theta)\} \equiv \alpha_p + i\beta_p.$$

By virtue of the result that any strictly periodic function of Θ can be represented by its Fourier series, the pdf of Θ may thus be written as

$$f(\theta) = \frac{1}{2\pi}\left[1 + 2\sum_{p=1}^{\infty}\{\alpha_p \cos(p\theta) + \beta_p \sin(p\theta)\}\right].$$

Note that for a symmetric $f(\theta)$, $\beta_p = 0,\ \forall p = 1, 2, 3, \ldots$.

Let us obtain a circular rv Θ from a linear rv X by wrapping, that is, $\Theta = X(\bmod 2\pi)$. A very useful and fundamental property relating the c.f. of the generated rv Θ to that of its generating rv X is given by

Result 1.2 $\phi_p(\Theta) = \phi_{t=p}(X),\ p \in \mathbb{Z},\ t \in \mathbb{R}$.

For a symmetric pdf of X, the pdf of the corresponding wrapped rv Θ is then given by

Result 1.3 $$f(\theta) = \phi_p(\Theta) = \frac{1}{2\pi}\left\{1 + 2\sum_{p=1}^{\infty}\phi_{t=p}(X)\cos(p\theta)\right\}.$$

This general approach can yield a wide variety of circular pdfs starting from the c.f.s of linear rvs. The striking fact is that Result 1.3 yields the pdf of Θ in the above simple analytic form even though the pdf of the generating rv X may not be so available. Two important examples of this fact are the wrapped stable and wrapped Linnik families of circular rvs.

1.2.2 Trigonometric Moments

Unlike that for a linear rv X, usual pth power moments $E(\Theta^p)$ are not sensible measures for Θ. Instead, one defines $\gamma_p(\Theta) \equiv (\alpha_p, \beta_p)$ as the trigonometric moment of order p, $p = 1, 2, 3, \ldots$. It is worthwhile to note that, unlike the case of linear pdfs, the (trigonometric) moment γ_p exists for all orders for all circular distributions. However, the descriptive measures need not do so. As an example the population mean direction defined by $\mu = \arctan^*(\beta_1/\alpha_1)$, where "*" denotes the quadrant specific value, does not exist for the circular uniform distribution since $\alpha_1 = \beta_1 = 0$ for it. A measure of circular dispersion is given by $E[1 - \cos(\Theta - \mu)]$.

Bivariate or joint trigonometric moments of order (p, q) of two circular rvs Θ and Φ are defined similarly by using the product moments of the trigonometric functions as $(E\{\cos(p\Theta) \cdot \cos(q\Phi)\}, E\{\cos(p\Theta) \cdot \sin(q\Phi)\}, E\{\sin(p\Theta) \cdot \cos(q\Phi)\}, E\{\sin(p\Theta) \cdot \sin(q\Phi)\})$, $p, q = 1, 2, \ldots$. Any measure of association between the two circular rvs should involve an association measure of their corresponding vectorial representations, $\mathbf{X}(\Theta) = (\cos\Theta, \sin\Theta)'$, $\mathbf{Y}(\Phi) = (\cos\Phi, \sin\Phi)'$, that is, involve all the four components of the joint trigonometric moment. Such a measure was obtained through the canonical correlation between $\mathbf{X}(\Theta)$ and $\mathbf{Y}(\Phi)$ by Jupp and Mardia (1980) and shown to involve the usual product moment correlations of the four pairs of the trigonometric rvs $(\cos\Theta, \cos\Phi)$, $(\cos\Theta, \sin\Phi)$, $(\sin\Theta, \cos\Phi)$, $(\sin\Theta, \sin\Phi)$.

1.3 Derivations of Probability Density Functions

Pdfs of rvs on manifolds possess some unique and interesting features. We discuss a few of those below.

1.3.1 Periodicity, Truncation, and Wrapping

Since Θ is an angle, its pdf $f(\theta)$ must be strictly periodic, that is, $f(\theta) = f(\theta + k \cdot 2\pi)$, $k \in \mathbb{Z}$, $\forall\theta \in [0, 2\pi)$. This requirement is in addition to the usual requirements of non-negativity and integrability to 1 as needed to define a pdf for a linear rv X. Thus, the representation of its pdf in terms of the trigonometric functions and Fourier series as mentioned above follow naturally. This is also true for the joint pdf of two circular rvs Θ and Φ, that is, for the pdf on the torus, for which a bivariate (doubly periodic) Fourier series always exists and can be explicitly written down. Here also the approach of using c.f.s as in the univariate case is rewarding and leads to explicit forms of pdfs on the torus for which the pdfs need not even exist in analytical forms, e.g. bivariate stable and Linnik families of toroidal distributions. Further details will be provided in Chapter 9.

We noted that mere truncation of a linear pdf on $(0, 2\pi)$ will not yield periodicity and also fails to utilize the remaining part of the pdf. However, the method of wrapping, which wraps the truncated pdf as given below, satisfies both these requirements. Let the linear rv X be wrapped as $\Theta = X (\mathrm{mod}\, 2\pi)$. Let X have the pdf $g(x)$. Then the corresponding wrapped pdf $f(\theta)$ of Θ is given by

$$f(\theta) = \sum_{k=-\infty}^{\infty} g(\theta + k \cdot 2\pi), \qquad \theta \in [0, 2\pi).$$

Choice of various distributions for X leads to the corresponding variety of circular wrapped distributions.

Unit distributions may appear to be quite attractive for deriving circular distributions therefrom, since the corresponding rv X supported on [0,1] may be easily transformed to have support on $[0, 2\pi)$. However, here also the periodicity needs to be enforced. Such constructions may nevertheless be fruitful, e.g. see circular beta distribution in Chapter 4.

Another powerful method of constructing distributions on manifolds from those for the Euclidean space is that of projection. For the univariate case, consider the transformation from the real line to the unit circle given by $x = \tan(\theta/2)$. This can be equivalently viewed as the corresponding Möbius transformation from the line to the complex plane. Thus, given the pdf $g(x)$ of X one can derive the pdf of Θ and vice versa. Minh and Farnum (2003) have shown that the linear Cauchy and t-distributions are obtained by this method from the circular uniform and Cartwright distributions. This method also generalizes (stereographic and inverse stereographic projections) to derive distributions on the sphere manifold S^2 from those on $\mathbb{R}^2$. Interestingly, this method faces difficulties in deriving toroidal distributions.

1.3.2 Marginal Distributions and Independence

In the generalizations of univariate to multivariate distributions, two properties are often preferred: (i) marginals and joint distributions should be in the similar families and (ii) joint distributions should include the case of independence of the variables. For example, the bivariate normal distribution possesses both these properties, with the nullity of the parameter ρ, the correlation coefficient in it, yielding independence. However, this is not the case with its counterpart, the bivariate von Mises (BvM) distribution, defined on the torus, as will be seen in Chapter 9. The marginals of the BvM distribution are not univariate von Mises (vM) distributions (unless these are independent) but are of no known standard forms. Independence in BvM distribution however results from the nullity of a matrix of its parameters as will be shown there also. A generalization of vM distribution to the toroidal case which possesses both the properties (i) and (ii) is not yet available.

Toroidal distributions derived from their c.f.s, as in their Fourier series representations, in general will possess property (i). Property (ii) may not be obvious from these representations, but follows from their corresponding c.f.s.

In many cases, c.f. may not be readily available and it may be nontrivial to check whether independence is included in the pdf. An example of this fact is the possibly asymmetric pdf derived by a series of conformal mappings for the hyper-disc given in Chapter 11.

1.4 Statistical Inference and Applications

Special methods, unique to the distributions on manifolds, are often needed for developing related statistical inference procedures.

1.4.1 Estimation and Testing of Hypotheses

Even for the circular case, the normalizing constants can be quite complicated involving series of special functions, and routine maximum-likelihood (ML) methods may be slow in convergence. Here the trigonometric moments method (TMM) are frequently found to be of great help by using the representations in terms of the corresponding Fourier series. In fact, these TMM estimates may well serve as initial values when iterative techniques are called for in the ML method.

For testing of hypotheses, tackling of the nuisance parameter(s) can be a challenging problem. This is more so when the nuisance parameter is the concentration parameter, as for the case of even the vM distribution. Since no invariance or similarity property exists for such a parameter, exact or unconditional tests are often not available to mitigate this nuisance parameter.

1.4.2 Testing Goodness-of-Fit

Indiscriminate use of a distribution for modeling the underlying data can be often grossly misleading with devastating consequences. Thus, the suitability of the distribution used is of prime importance. This is achieved through the Goodness-of-Fit (GoF) tests. However, the GoF tests used for the linear case are not appropriate for the manifolds considered here. This is so due to the requirement of invariance with respect to choice of origin for the data. Simple modifications of the test statistics for linear data have been successfully incorporated to yield the corresponding tests for the circular case, e.g. Watson's U^2 test as a modification of the vM functional test. C.f.-based GoF tests can be used for certain manifolds, e.g. the torus where the c.f. is conveniently available. However, to date there does not seem to exist GoF tests for pdfs on several other manifolds, e.g. the pdfs on the cylinder, hyper-disc, etc. This is an important open problem.

1.4.3 Applications

With the emergence of new technologies, interesting datasets on various manifolds are in the need of proper analysis. In the context of Big Data, data on manifolds is a member of this group due to its rich variety. Thus, for supervised and unsupervised learning with these datasets, derivations of distributions on manifolds have become imperative, e.g. for model-based analysis in Statistical Machine Learning. Herein lies a rich and fertile arena for challenging and useful research with a panorama of datasets in our modern era.

References

Jupp, P. E. and Mardia, K. V. (1980). A general correlation coefficient for directional data and related regression problems, *Biometrika*, 67, 163–173 correction (1981), 68, 738.

Minh, D. L. and Farnum, N. R. (2003). Using bilinear transformations to induce probability distributions, *Communications in Statistics –Theory and Methods*, 32, 1–9.

SenGupta, A., Acharjee, S., and Sundar, S. (2026). Differential equations approach for generating tail behaviors: Novel generalized Cauchy and Laplace conditionals families, Submitted.

2

Preliminary Notions for Circular Distributions

2.1 Circular Distributions

An angular observation θ in $[0, 2\pi)$ has one-to-one correspondence with a point $(\cos\theta, \sin\theta)$ on the unit circle in the real plane (Figure 2.1), and the distribution of an angular random variable (rv) is often called the circular distribution (CD) or the distribution on the unit circle. If Θ denotes an angular or circular random variable which has a distribution "D," we may write $\Theta \sim D$. Similar to the case of a distribution on the real line, a continuous CD is characterized by a probability density function (pdf) and a discrete CD by a probability mass function (pmf).

In this chapter, we present some basic notions for univariate circular or angular distributions. As monographs and reference books for CDs, we mention Mardia (1972), Batschelet (1981), Fisher (1993), Mardia and Jupp (2000), Jammalamadaka and SenGupta (2001), Zar (2010), and Ley and Verdebout (2017).

2.2 Probability Density Function and Probability Mass Function

In this book, we mainly consider the case where a distribution is characterized by a pdf. A circular pdf $f(\theta)$ on the interval $[0, 2\pi)$ satisfies the following basic properties:

(a) $f(\theta) \geq 0$ for all $\theta \in (-\infty, \infty)$;

(b) $\displaystyle\int_0^{2\pi} f(\theta)d\theta = 1$;

(c) $f(\theta) = f(\theta + 2k\pi)$ on $(-\infty, \infty)$ for any integer k (i.e. f is 2π-periodic).

Conversely, any function f satisfying the above conditions (a), (b), and (c) simultaneously is a circular pdf.

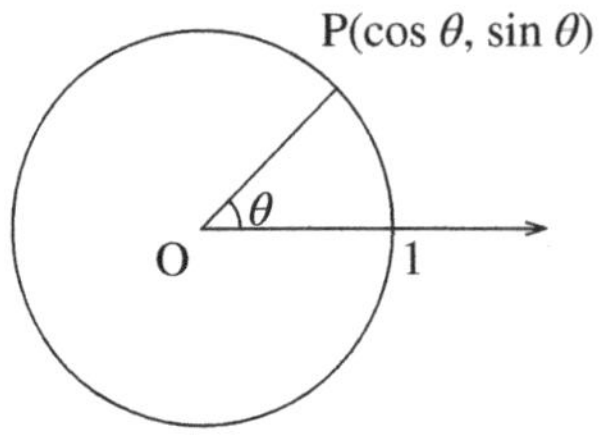

Figure 2.1 Angle from a base line and its corresponding point on the unit circle.

The CD is considered on the interval $[0, 2\pi)$ in the definition above. However, the interval is not restricted to $[0, 2\pi)$ and one may choose, for example $[-\pi, \pi)$ as an interval because pdf is a 2π-periodic function. An important point is the fact that the range is 2π, and henceforth, for all CDs, we will write simply periodic in place of 2π-periodic.

The distribution function $F(\theta)$ of a circular pdf $f(\theta)$ on $[0, 2\pi)$ is defined as follows:

$$F(\theta) = \int_0^{\theta} f(x)dx, \quad \theta \in [0, 2\pi),$$

and

$$F(\theta + 2\pi) - F(\theta) = 1$$

on the whole real line $(-\infty, \infty)$ by the periodic property of $f(\theta)$. It is easy to see that

$$F(0) = 0, \quad F(2\pi) = 1.$$

The distribution function $F(\theta)$ with pdf $f(\theta)$ is a continuous function on $(-\infty, \infty)$ and satisfies

$$\int_{\alpha}^{\beta} f(\theta)d\theta = F(\beta) - F(\alpha)$$

for $-\infty < \alpha \leq \beta < \infty$. A different property from distribution functions on the real line is

$$\lim_{\theta \to -\infty} F(\theta) = -\infty, \quad \lim_{\theta \to \infty} F(\theta) = \infty.$$

Likewise a circular pmf $f(\theta)$ on the interval $[0, 2\pi)$ satisfies the properties:

(a)′ $f(\theta) > 0$ for a countable number of θ $(\theta = \theta_1, \ldots, \theta_m)$, and $f(\theta) = 0$ otherwise;
(b)′ $\sum_{j=1}^{m} f(\theta_j) = 1$;
(c)′ $f(\theta) = f(\theta + 2k\pi)$ on $(-\infty, \infty)$ for any integer k (i.e. f is periodic).

Conversely, any function f satisfying the above conditions (a)′, (b)′, and (c)′ simultaneously is a circular pmf.

2.3 Density Plots

If $f(x)$ is a pdf on the real line, it is usually plotted in the real plane, and if $f(\mathbf{x})$ is a pdf on the real plane, it is plotted in three-dimensional Euclidean space. If $f(\theta)$ is a pdf on the unit circle, it may be plotted in three-dimensional Euclidean space (Figure 2.2a). As from the three-dimensional plot, it may not be easy to visualize the distributional shape, linear plots are also used. Figure 2.2b shows a plot on the interval $[-\pi, \pi)$ by cutting the circle at $(x, y) = (0, -1)$, and Figure 2.2c on $[0, 2\pi)$ by cutting it at $(x, y) = (0, 1)$ and measuring the angle clockwise. These represent the same distribution, though the intervals are different. Recall that pdf is a periodic function on the whole real line. However, these two figures may give us fairly different impressions. The former bell-shaped or inverse bathtub shaped curve has mode at zero and antimode at $-\pi$, whereas the latter bathtub shaped curve has

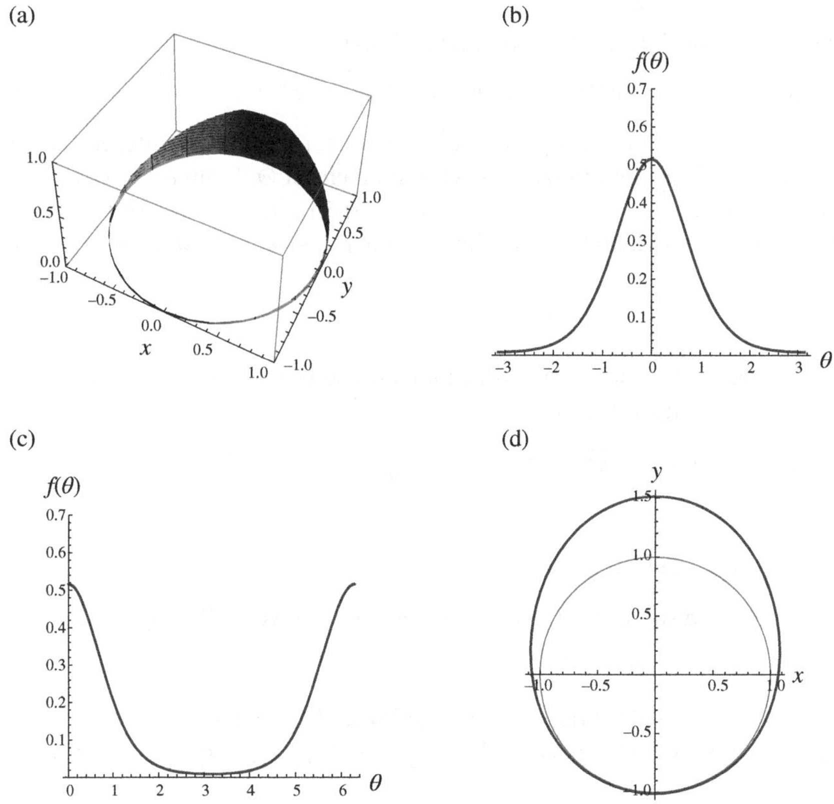

Figure 2.2 Density plots. (a) 3-D plot, (b) linear plot on the interval $[-\pi, \pi)$, (c) linear plot on the interval $[0, 2\pi)$, (d) circular plot.

mode at zero and antimode at π. (The terminologies "mode" and "antimode" will be explained in Section 2.5, but their interpretations are clear in this example.) The area between θ-axis and the curve is one, as in the definition of pdf. Figure 2.2d is a circular plot, and it may appeal to our intuition. However, note that the area between the unit circle and the curve is not necessarily one.

2.4 Characteristic Functions and Trigonometric Moments

For a real number t, the characteristic function of a circular random variable Θ measured anticlockwise on $[0, 2\pi)$ and with pdf $f(\theta)$, is defined by

$$\phi_t = E(e^{it\Theta}) = \int_0^{2\pi} e^{it\theta} f(\theta) d\theta, \quad i = \sqrt{-1}. \tag{2.1}$$

It follows from the periodic property (c) of pdf that

$$\phi_t = E(e^{it\Theta}) = E\{e^{it(\Theta+2k\pi)}\} = e^{2itk\pi} E(e^{it\Theta}) = \phi_t \, e^{2itk\pi}$$

for any integer k. Therefore, $\phi_t = 0$ or $e^{2itk\pi} = 1$, i.e. t must be an integer. For an integer p, we call ϕ_p the pth trigonometric moment of Θ. From $e^{ip\theta} = \cos(p\theta) + i\sin(p\theta)$, it holds that $\phi_p = E\{\cos(p\Theta)\} + iE\{\sin(p\Theta)\}$. $\alpha_p = E\{\cos(p\Theta)\}$ is called the pth cosine moment and $\beta_p = E\{\sin(p\Theta)\}$ the pth sine moment of Θ. It clearly holds that

$$|\phi_p| = |E(e^{ip\Theta})| \leq E(|e^{ip\Theta}|) = 1,$$

and the trigonometric moments, cosine moments, and sine moments always exist. The complex conjugate of ϕ_p is

$$\overline{\phi_p} = \overline{E(e^{ip\Theta})} = E(\overline{e^{ip\Theta}}) = E(e^{-ip\Theta}) = \phi_{-p}.$$

We have

$$\alpha_{-p} = \alpha_p \quad \text{and} \quad \beta_{-p} = -\beta_p.$$

This means that it suffices to treat the case of nonnegative p. We may write

$$\phi_p = \rho_p \, e^{i\mu_p}, \quad p = 0, 1, 2, \dots, \tag{2.2}$$

where $\rho_p = |\phi_p|$ (the absolute value or magnitude of ϕ_p) is nonnegative and $\mu_p = \arg\ \phi_p$ (the argument of ϕ_p) is real. Since $\phi_p = \alpha_p + i\beta_p = \rho_p(\cos\mu_p + i\sin\mu_p)$, α_p, β_p, ρ_p, and μ_p are related by

$$\rho_p = \sqrt{\alpha_p^2 + \beta_p^2} \quad \text{and} \quad \tan\mu_p = \frac{\beta_p}{\alpha_p} \quad (\text{if } \alpha_p \neq 0).$$

The pth central trigonometric moment or pth trigonometric moment about the mean direction μ (Section 2.5) is defined in a way similar to the pth trigonometric moment as

$$E\left\{e^{ip(\Theta-\mu)}\right\} = \overline{\alpha}_p + i\overline{\beta}_p,$$

where $\overline{\alpha}_p = E\{\cos p(\Theta - \mu)\}$ is the pth central cosine moment and $\overline{\beta}_p = E\{\sin p(\Theta - \mu)\}$ is the pth central sine moment. It clearly holds that $\sqrt{\overline{\alpha}_p^2 + \overline{\beta}_p^2} = \rho_p$ and $\arg(\overline{\alpha}_p + i\overline{\beta}_p) = \arg\ \phi_p - p\mu$.

For a symmetric circular pdf, (i) sine moments are zeros, since $\sin\theta$ is an odd function and hence (ii) pth cosine moment, $\alpha_p = \phi_p$.

Let us define $\eta(t) \equiv E\{\exp(t\sin\Theta), \exp(t\cos\Theta)\}$ as the power trigonometric moment generating function (ptmgf). This yields the tth power sine and cosine trigonometric moments using the tth derivative of $\eta(t)$ at $t = 0$. Recall that $\phi_p \equiv E\{\exp(ip\Theta)\}$ yields pth order (regular) trigonometric moments. As an example, using the ptmgf, we conveniently derive the pth power central cosine moment for circular normal distribution (defined in Chapter 3) as, $E\{\cos^p(\Theta - \mu)\} = I_0^{(p)}(\kappa)/I_0(\kappa)$. And, from ϕ_p, the pth central cosine moment is, $E\{\cos p(\theta - \mu)\} = I_p(\kappa)/I_0(\kappa)$ (denoted by $B_p(\kappa)$, which will appear in Section 7.2.2).

The regular and power trigonometric moments can be expressed in terms of each other. For example, expanding both sides of the identity $e^{ip\theta} = (e^{i\theta})^p$, (regular) trigonometric moments $E\{\cos(p\Theta)\}$ and $E\{\sin(p\Theta)\}$ can be expressed in terms of the power trigonometric moments easily. The formulae for the inverse relations are also available – see, e.g. pages 31 and 33 of Gradshteyn and Ryzhik (2007).

2.5 Summary Statistics

For simplicity, suppose that a pdf $f(\theta)$ is a continuous function. Then the mode is the point θ at which the largest value of $f(\theta)$ occurs. The point at which the smallest value occurs is called the antimode. If a distribution has only one peak, i.e. an unique mode, then it is called a unimodal distribution. If the total number of global and local maxima is greater than 1, then the distribution is multimodal.

In (2.2), we consider the frequently used case of $p = 1$. Let us use the following notations:

$$\alpha_1 = \alpha, \quad \beta_1 = \beta, \quad \rho_1 = \rho, \quad \mu_1 = \mu.$$

A complex number can be identified by a vector. Let the magnitude of a complex number (a vector) ϕ_1 be ρ, which is also termed the mean resultant length. If $\rho \neq 0$, then μ is called the mean direction because $\tan\mu = \beta/\alpha$ and μ indicates the direction of a complex number ϕ_1. The quantities ρ_p and μ_p in (2.2) are called the pth mean resultant length and, if $\rho_p \neq 0$, the pth mean direction, respectively.

To clarify these notions further, we consider a discrete CD

$$f(\theta_j) = P(\Theta = \theta_j) = \frac{1}{n}, \quad 0 \leq \theta_1 \leq \cdots \leq \theta_n < 2\pi, \; n \geq 1.$$

Then an angle θ_j has one-to-one correspondence with a vector $(\cos\theta_j, \sin\theta_j)'$ on the unit circle. The resultant vector of $\theta_1, \dots, \theta_n$ is $(C, S)'$, where

$$C = \sum_{j=1}^{n} \cos\theta_j, \quad S = \sum_{j=1}^{n} \sin\theta_j.$$

The first trigonometric moment is

$$E(e^{i\Theta}) = \frac{1}{n}\sum_{j=1}^{n} e^{i\theta_j} = \frac{1}{n}(C + iS)$$

and the first cosine and sine moments are given by

$$\alpha = E(\cos\Theta) = \frac{1}{n}\, C \; (\equiv \overline{C}, \text{ say}), \quad \beta = E(\sin\Theta) = \frac{1}{n}\, S \; (\equiv \overline{S}, \text{ say}).$$

The mean direction μ ($\equiv \overline{\theta}$, say) satisfies

$$\tan\mu = \tan\overline{\theta} = \frac{\beta}{\alpha} = \frac{S}{C} = \frac{\overline{S}}{\overline{C}},$$

which is explicitly expressed by, when $R = \sqrt{C^2 + S^2} > 0$,

$$\overline{\theta} = \begin{cases} \tan^{-1}(S/C), & C > 0,\ S \geq 0, \\ \pi/2, & C = 0,\ S > 0, \\ \tan^{-1}(S/C) + \pi, & C < 0, \\ 3\pi/2, & C = 0,\ S < 0, \\ \tan^{-1}(S/C) + 2\pi, & C > 0,\ S < 0, \end{cases}$$

with $\tan^{-1}(\cdot) \in (-\pi/2, \pi/2)$. If $R = 0$, $\overline{\theta}$ is not defined. The mean resultant length is

$$\rho = \sqrt{\alpha^2 + \beta^2} = \frac{1}{n}\sqrt{C^2 + S^2} = \sqrt{\overline{C}^2 + \overline{S}^2} \; (\equiv \overline{R}, \text{ say}).$$

Note that $\overline{\theta}$ and $\overline{R}$ are the sample mean direction and sample mean resultant length, respectively, when $\theta_1, \dots, \theta_n$, angular observations of size n, are given. Clearly, $0 \leq \overline{R} \leq 1$ since $(\cos\theta_j, \sin\theta_j)'$ are unit vectors. If $\theta_1, \dots, \theta_n$ tend to be clustered at a single point, then $\overline{R}$ will be close to 1, whereas if $\theta_1, \dots, \theta_n$ tend to be dispersed uniformly or clustered with same (similar) frequencies at the n odd (even) equidistant (diametrically opposite) points, then $\overline{R}$ will be close to 0.

Mean direction is equivariant under rotation, whereas mean resultant length is invariant under rotation. To show this, let Θ denote a circular random variable whose mean direction is μ and mean resultant length is ρ, and ψ be a fixed angle. Then the first trigonometric moment of $\Theta - \psi$ is

$$E\left\{e^{i(\Theta-\psi)}\right\} = e^{-i\psi}E(e^{i\Theta}) = e^{-i\psi} \times \rho e^{i\mu} = \rho\ e^{i(\mu-\psi)}.$$

Therefore, the mean direction and mean resultant length of $\Theta - \psi$ are $\mu - \psi$ and ρ, respectively. Replacing ψ by μ in the above equation, we have

$$E\left\{e^{i(\Theta-\mu)}\right\} = \rho, \tag{2.3}$$

which is nonnegative. Thus, the sine moment of $\Theta - \mu$ vanishes, i.e.

$$E\left\{\sin(\Theta - \mu)\right\} = 0.$$

This equation can be viewed as an analogous result to the linear case. Indeed, if X is a random variable on the real line and μ_0 is the mean of X, then the first moment of $X - \mu_0$ clearly vanishes, i.e. $E(X - \mu_0) = 0$. A more general equation than (2.3) is $E\{e^{i(p\Theta-\mu_p)}\} = \rho_p$ for $p = 1, 2, 3, \ldots$, from which it follows that $E\{\sin(p\Theta - \mu_p)\} = 0$.

The circular variance ν of an angular random variable Θ with mean direction μ and mean resultant length ρ is defined by

$$\nu = 1 - \rho = 1 - E\left\{\cos(\Theta - \mu)\right\}.$$

It clearly holds that $0 \le \nu \le 1$. Circular variance has a finite upper bound. This is a big difference between circular and linear variances. There are analogous aspects to the variance of a distribution on the real line. One is that $\nu = 0$ if and only if Θ degenerates to a distribution concentrated at the point μ since $\cos(\theta - \mu)$ is continuous and nonnegative with respect to θ. Another aspect is as follows. In the linear case, minimizing $E\{(X - c)^2\}$ with respect to real number c leads to $E[\{X - E(X)\}^2]$, the variance of X. In the circular case, the square of the distance or chord length between two points on the unit circle, corresponding to angles θ and ξ, is $2\{1 - \cos(\theta - \xi)\}$. If we use

$$V(\alpha) = 1 - E\left\{\cos(\Theta - \alpha)\right\}$$

as a measure of the variability of Θ about any angle α, then

$$V(\alpha) = \nu + 2\rho \sin^2\left(\frac{\mu - \alpha}{2}\right)$$

since

$$\begin{aligned} E\left\{\cos(\Theta - \alpha)\right\} &= E\left\{\cos(\Theta - \mu + \mu - \alpha)\right\} \\ &= E\left\{\cos(\Theta - \mu)\cos(\mu - \alpha) - \sin(\Theta - \mu)\sin(\mu - \alpha)\right\} \\ &= \rho\ \cos(\mu - \alpha). \end{aligned}$$

$V(\alpha)$ has minimum at μ, the mean direction of Θ. Note that it is possible to introduce an alternative circular variance

$$\mathrm{Var}(\Theta) = E\{|e^{i\Theta} - E(e^{i\Theta})|^2\} = 1 - |E(e^{i\Theta})|^2 = 1 - \rho^2, \tag{2.4}$$

where ρ is the mean resultant length of Θ.

The circular standard deviation σ and the circular dispersion δ are defined as

$$\sigma = \{-2\log(1-\nu)\}^{1/2} = \{-2\log\rho\}^{1/2},$$

(and not as $\sqrt{\nu}$), and

$$\delta = \frac{1-\rho_2}{2\rho^2}$$

if $\rho \neq 0$, respectively. For small ν, σ is approximately $\sqrt{2\nu}$. The definition of circular standard deviation is motivated by the fact (Section 4.4.2) that the wrapped normal distribution WN(μ, ρ) (see Chapter 4) obtained from the normal distribution $N(\mu, \sigma^2)$ has the mean resultant length $\rho = \exp(-\sigma^2/2)$. The median direction $\tilde{\mu}$ is defined as a direction ϕ which minimizes

$$E(\pi - |\pi - |\Theta - \phi||),$$

and it satisfies

$$P(\Theta \in [\tilde{\mu} - \pi, \tilde{\mu})) \geq \frac{1}{2}, \quad P(\Theta \in [\tilde{\mu}, \tilde{\mu} + \pi)) \geq \frac{1}{2}.$$

2.6 Symmetry, Skewness, and Kurtosis

A circular random variable Θ or its distribution is said to be symmetric about a given direction μ if its pdf or pmf has the property:

$$f(\mu + \theta) = f(\mu - \theta)$$

for all θ. Otherwise, Θ is not symmetric. If a distribution is symmetric about μ, then it is also symmetric about $\mu \pm \pi$. For a symmetric distribution, the pth central sine moments are zero, i.e. $\overline{\beta}_p = 0$ for $p = 0, \pm 1, \pm 2, \ldots$, and thus the central trigonometric moments are all real. If circular random variables Θ_1 and Θ_2 are independent and identically distributed, then $\Theta_1 - \Theta_2$ is symmetric about zero because the distributions of $\Theta_1 - \Theta_2$ and $\Theta_2 - \Theta_1 = -(\Theta_1 - \Theta_2)$ are the same.

The circular skewness and kurtosis are the measures of asymmetry and peakedness of a CD, respectively. The former and the latter are, respectively, defined by

$$s = \frac{\overline{\beta}_2}{(1-\rho)^{3/2}}, \quad k = \frac{\overline{\alpha}_2 - \rho^4}{(1-\rho)^2}.$$

However, $\overline{\beta}_2$ and $\overline{\alpha}_2$ themselves may be respectively defined as the circular skewness and kurtosis.

Finally for CDs, it may not be meaningful to define any of the popular CN or WN distributions as mesokurtic and then define leptokurtic and platykurtic distributions based on it. Such, if it could be so defined, would have been in the same spirit of adopting normal distribution as mesokurtic for the linear pdfs. This is because kurtosis measure for these distributions is not free of the concentration parameter, unlike the case of the normal distribution with scale parameter.

2.7 Tail Behavior and Points of Inflection

For linear distributions, kurtosis reflects the tail properties. A distribution with larger kurtosis will have larger tails and larger probability for exceeding (falling below) a "large" ("small") value as often subjectively chosen. This in turn yields higher peak of the distribution. However, higher peak or equivalently higher modal density (within the families of symmetric and unimodal CDs) need not imply thicker tail. This fact can be substantiated by Figs 2.5 and 2.4 in Jammalamadaka and SenGupta (2001) corresponding to the symmetric wrapped stable (SWS) (see Chapter 4) and CN (or von Mises) families, respectively, wherefrom it can be seen that for higher modal density SWS has heavier but CN has lighter tails.

For a linear rv with support on $\mathbb{R}$, the tail behavior is quite transparent. However, for CDs, the support is only 2π. More often than not (i.e. when the pdf is highly concentrated), CDs will have tails extending over the entire support as well as nontrivial probabilities thereon. So "long" and "short" tails, as in the linear case, do not arise too often for CDs. Rather, it may be prudent to consider "heavy" and "thin" tails, when comparing the tail behavior for two CDs. Further, for such comparison, we may standardize with respect to the peaks, i.e. constrain the modal densities to be the same, and then explore the concentrations. In Figure 2.3, we display two such peak-matched pdfs of CN and wrapped Cauchy (WC), which will be discussed in Chapter 4. For CN(0,1) the modal density is 0.3419, which gives $\rho = 0.3645$ for WC(0, ρ). It is clear that WC density has heavier tails.

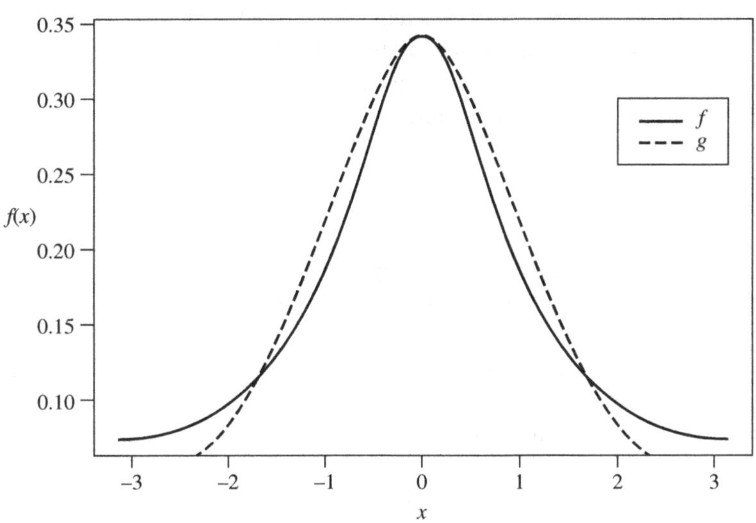

Figure 2.3 Inflection points for peak-matched CN (= g) and WC (= f) pdfs.

Alternatively, we introduce here $\boldsymbol{\mathcal{P}} \equiv$ *Inter Inflection Points Distance*, the circular distance between the two inflection points, as a measure for comparing the tails of unimodal, symmetric circular pdfs within the same family. Note that $\boldsymbol{\mathcal{P}}$ is a monotone function of the modal density, though it may be increasing for one, e.g. WC, while decreasing, e.g. CN, for another family. WC has a smaller $\boldsymbol{\mathcal{P}}$ of 1.1595 rad compared to that of 1.8060 rad for the peak-matched CN(0,1). WC also has higher concentration (around the mean direction) given by its kurtosis of 0.2852 compared to that of 0.2202 for CN(0,1). The measure $\boldsymbol{\mathcal{P}}$ will be exemplified by several other location- and peak-matched distributions in Chapter 4.

2.8 Flexible Family of Distributions

A family of distributions which can encompass variety of Properties – possibly opposite ones, e.g. symmetry and asymmetry, unimodal and multimodal, large and small concentrations, heavy and light tails, by varying their parameter values, are of significant practical importance. This may be met by mixture distributions. However, the aim is to achieve this flexibility with a small number of parameters, thus enhancing such elegant families as with fewer parameters than mixture distributions. These families will in general be termed as flexible families of distributions. It is desirable that the parameters of such distributions be interpretable. Further, for usefulness in practice, it is mandatory that such families be amenable to statistical inference. A judicious introduction of even a single parameter may achieve this to a great extent. For example, the family of multimodal, asymmetric generalized von Mises (AGvM) distributions (Kim and SenGupta, 2013, 2021), a member of the regular exponential family, is derived from the unimodal symmetric family of von Mises distributions by adding a single parameter. The construction of AGvM, along with several other flexible families of distributions, will be discussed in Chapter 5.

References

Batschelet, E. (1981). *Circular Statistics in Biology*. London: Academic Press.

Fisher, N. I. (1993). *Statistical Analysis of Circular Data*. Cambridge: Cambridge University Press.

Gradshteyn, I. S. and Ryzhik, I. M. (2007). *Table of Integrals, Series, and Products*, 7e. Amsterdam: Elsevier.

Jammalamadaka, S. R. and SenGupta, A. (2001). *Topics in Circular Statistics*. Singapore: World Scientific.

Kim, S. and SenGupta, A. (2013). A three-parameter generalized von Mises distribution. *Statistical Papers* 54: 685–693.

Kim, S. and SenGupta, A. (2021). Multimodal exponential families of circular distributions with application to daily peak hours of PM 2.5 level in a large city. *Journal of Applied Statistics*. 48: 3193–3207.

Ley, C. and Verdebout, T. (2017). *Modern Directional Statistics*. Boca Raton, FL: CRC Press.

Mardia, K. V. (1972). *Statistics of Directional Data*. London: Academic Press.

Mardia, K. V. and Jupp, P. E. (2000). *Directional Statistics*. Chichester: Wiley.

Zar, J. H. (2010). *Biostatistical Analysis*, 5e. Upper Saddle River, NJ: Prentice Hall.

3

General Methods of Construction for Circular Distributions

This chapter introduces methods for constructing circular distributions. One can use ad hoc methods based on the definition of a circular probability density function. Systematic approaches include derivations based on characterizations, wrapping, projection, conditioning, linear fractional transformation, Brownian motion, characteristic functions, mixtures, elliptically contoured distributions, transformations, and arithmetic and geometrical methods. Shortcomings and topics for future work are briefly discussed at the end of this chapter.

3.1 Ad Hoc

Any probability density function (pdf) on the real line may be truncated on $[0, 2\pi)$ to yield a weighted distribution as a circular pdf, provided it is periodic. This requirement immediately implies that simply using θ for x in the corresponding pdf will not suffice. Of course, we can use the operator mod 2π, but this will be cumbersome. Trigonometric functions are then the natural choice for the kernel to be periodic. The two fundamental choices as building blocks are given below: a trigonometric polynomial and the exponential of a trigonometric function. For the former, the normalizing constant is trivially obtained, but the pdf does not admit of any nontrivial sufficient statistic and thus usually is not amenable to the development of optimal statistical inference procedures. The reverse scenario occurs for the latter, see e.g. Rukhin (1972).

The continuous uniform distribution is the most basic distribution among distributions on the circle. Its pdf $f(\theta)$ is constant over the circumference of the circle with $\int_0^{2\pi} f(\theta)d\theta = 1$, i.e.

$$f(\theta) = \frac{1}{2\pi}, \quad 0 \leq \theta < 2\pi.$$

Properties of this distribution will be described in Section 4.2.

It can be seen that the distributions mentioned in the following sections include the uniform distribution as a special case.

As an example of a periodic function, we may enhance $\cos\theta$, which takes values in $[-1,1]$ and $1+\cos\theta \geq 0$. It also has only one turning point in $[0, 2\pi)$. Thus, a function

$$f(\theta) = \frac{1}{2K\pi}\sum_{p=1}^{K}\{1+a_p\cos(p\theta)\}, \quad 0 \leq \theta < 2\pi, -1 \leq a_p \leq 1 \quad \forall p,$$

and for some arbitrary K, is a circular pdf since $\int_0^{2\pi} f(\theta)d\theta = 1$. This gives Cartwright distribution and the cardioid distribution for $K = 1$ with $a_1 = 1$ and $-1 \leq a_1 \leq 1$, respectively. However, the Cartwright distribution does not involve any parameter. In Section 4.6.2, we define the cardioid distribution and its generalizations with multiple parameters and study their properties.

As another example, consider a periodic function $\exp(\cos\theta)$, whose values are clearly positive. Thus, a function

$$f(\theta) = c\, e^{\sum_{p=1}^{K} a_p\cos(p\theta)}, \quad 0 \leq \theta < 2\pi,$$

is a circular pdf with normalizing constant $c = \{\int_0^{2\pi} e^{\sum_{p=1}^{K} a_p\cos(p\theta)} d\theta\}^{-1}$. For $K = 1$ with $a_1 > 0$, this gives a von Mises distribution. Section 4.3.3 details properties of the von Mises distribution with an additional parameter. For $K = 2, f(\theta)$ was discussed in Cox (1975). For $K > 2$, it is the multimodal generalized von Mises distribution studied by Maksimov (1967) and also by Kim and SenGupta (2021). We will discuss these and their variants in more details in Chapters 4 and 5.

3.2 Characterizations

Characterization of normal distributions on the real line has been studied by many authors. A summary of this topic can be found in Johnson et al. (1994, pp. 100–110).

One may ask which circular distribution is characterized by properties based on independence, maximization of certain constrained functionals, etc., SenGupta (2004, 2010) extensively gave characterizations based on Shannon's entropy and other entropy measures/functionals to derive distributions on the circle, sphere, torus, cylinder as well as their multivariate generalizations. Mainly based on those papers, this section provides a compact description of characterizations to generate distributions on the circle. The following sections of specific distributions also pick up this topic: The uniform distribution (Section 4.2) is characterized by the independence of the sample mean direction and the sample mean resultant length.

The von Mises distribution (Section 4.3.3) has the maximum entropy property and is also characterized by the maximum likelihood estimator of the mean direction.

3.2.1 Maximizing Entropy

Let $f(\theta)$ and $g(\theta)$ denote two pdfs on the circle. The Kullback–Leibler (KL) divergence from f to g is defined by

$$D_{\mathrm{KL}}(f \parallel g) = -E_f\left\{\log \frac{g(\Theta)}{f(\Theta)}\right\} = \int_0^{2\pi} f(\theta) \log \frac{f(\theta)}{g(\theta)}\, d\theta \ (\geq 0),$$

which is a measure of the difference between two distributions. If $g(\theta)$ is the circular uniform pdf, then the KL divergence is

$$D_{\mathrm{KL}}(f \parallel g) = \log(2\pi) + \int_0^{2\pi} f(\theta) \log f(\theta)\, d\theta.$$

Shannon's entropy of a circular distribution with pdf $f(\theta)$ is defined by

$$H(f) = -\int_0^{2\pi} f(\theta) \log f(\theta)\, d\theta. \tag{3.1}$$

It is easily seen that $H(f) \leq \log(2\pi)$ and the entropy of the uniform distribution attains the upper bound. The problem we consider here is to find the circular distribution for which the entropy is maximized under certain restrictions.

From Theorem 13.2.1 (p. 409) of Kagan et al. (1973), for pdf $f(\theta)$ (> 0) on the circle and integrable functions $h_j(\theta)$ ($j = 1, 2, 3, \ldots$) on the circle satisfying the conditions:

$$\int_0^{2\pi} h_j(\theta) f(\theta) d\theta = g_j \quad (j = 1, 2, 3, \ldots) \tag{3.2}$$

for given constants g_j ($j = 1, 2, 3, \ldots$), the maximum entropy is attained by distributions with pdf

$$f(\theta) = \exp\{a_0 + a_1 h_1(\theta) + a_2 h_2(\theta) + \cdots\}$$

(and only by them), if there exist $a_0, a_1, a_2, \ldots$ such that the above $f(\theta)$ satisfies the conditions (3.2).

The maximizing entropy method yields the uniform distribution (Section 4.2) when one does not impose any restriction on moments and von Mises distribution (Section 4.3.3), the generalized von Mises distribution (Section 5.3.1), and some further generalizations, when such restrictions are invoked.

3.2.2 Minimizing Functionals

A more general functional, which includes Shannon's entropy,

$$\psi(D) = \int_{-\pi}^{\pi} [F(D) + \{\lambda_0 + \lambda_1 \cos\theta + \lambda_2 \sin\theta + \lambda_3 \cos 2\theta + \lambda_4 \sin 2\theta\} D(\theta)] d\theta$$

Table 3.1 Optimal solutions to minimizing the functional $\psi(D)$

F(D)	**Distribution**
D^2	Cardioid
$-D\log D$	von Mises
$-\log D$	Wrapped Cauchy
(Open problem)	Wrapped stable

is minimized (or maximized) to obtain circular distributions with pdf $D(\theta)$. Here $F(D)$ is a particular function of D and λ_j $(j = 0, 1, 2, 3, 4)$ are real constants. Note that we only consider minimization (or maximization) under constraints of first two trigonometric moments of $D(\theta)$. Various choices of $F(D)$ yield various circular distributions. Table 3.1 shows some of distributions obtained by this method: $F(D) = D^2$ yields the cardioid distribution (Section 4.6.2), $F(D) = -D\log D$ the von Mises distribution (Section 4.3.3) and $F(D) = -\log D$ the wrapped Cauchy distribution (Section 4.4.1). However, a simple functional does not exist for the pdf

$$D(\theta) = \frac{\Gamma(\alpha+1)}{2\sqrt{\pi}\,\Gamma(\alpha+1/2)} \cos^{2\alpha}(\theta/2)$$
$$= \frac{\Gamma(\alpha+1)}{2^{\alpha+1}\sqrt{\pi}\,\Gamma(\alpha+1/2)} (1+\cos\theta)^{\alpha}, \quad \alpha > 0, \tag{3.3}$$

except for particular cases such as $\alpha = 1$, which corresponds to the cardioid distribution. For the distribution with the pdf (3.3), see Section 4.6.1. Mitsuyasu et al. (1975) used this circular pdf (3.3), originally proposed by Cartwright (1963), to analyze ocean wave spectra. It is yet to be derived which functional and under which suitable constraints yields the general wrapped stable family, a flexible family of distributions as shown in Sections 4.5 and 5.7.

3.3 Wrapping

Let X be any linear random variable on the real line. The circular wrapped distribution associated with the pdf $f(x)$ of X is derived by defining $\Theta = X \pmod{2\pi}$. It is thus obtained by wrapping $f(x)$ around the circle:

$$g(\theta) = \sum_{m=-\infty}^{\infty} f(\theta + 2m\pi), \quad 0 \le \theta < 2\pi.$$

$g(\theta)$ has the shortcoming that it is not necessarily a periodic function on the whole real line, but it is easy to extend it to a periodic function on the whole real line by considering $g(\theta \pm 2k\pi)$, $k = 0, 1, 2, \ldots$.

For positive integer p, the pth trigonometric moment of Θ is given by

$$\phi_p = E(e^{ip\Theta}) = \int_0^{2\pi} e^{ip\theta} g(\theta) d\theta = \sum_{m=-\infty}^{\infty} \int_{2\pi m}^{2\pi(m+1)} e^{ip(x-2\pi m)} f(x) dx$$

$$= \int_{-\infty}^{\infty} e^{ipx} f(x) dx = E(e^{ipX}) \ (\equiv \alpha_p + i\beta_p, \text{ say})$$

because $e^{-ip2\pi m} = 1$, and ϕ_p coincides with the characteristic function of X.

Based on the above result, it may seem only natural to name the wrapped distribution corresponding to or induced by the linear distribution as the same named circular distribution, e.g. circular Cauchy distribution as the one obtained by wrapping the linear Cauchy distribution (which is popularly termed as wrapped Cauchy distribution), etc. However, it is to be noted that the two characteristic functions defined above coincide only at integer values of their arguments. Thus, $f(x)$ need not uniquely define or characterize $g(\theta)$.

If two random variables X_1 and X_2 are independent, then the wrapped distribution of the sum $X_1 + X_2$ is equal to the distribution of the sum of two wrapped variables $X_1 \pmod{2\pi}$ and $X_2 \pmod{2\pi}$, i.e. $(X_1 + X_2) \pmod{2\pi} \overset{d}{=} X_1 \pmod{2\pi} + X_2 \pmod{2\pi}$, where $\overset{d}{=}$ means that the distributions of both sides are the same. If distributions on the real line have the reproductive property, then the corresponding wrapped distributions also have the reproductive property, and if X is infinitely divisible, then Θ is also infinitely divisible. If the distribution of X is symmetric about 0, then $\beta_p = 0$ and $\phi_p = \alpha_p$. Thus, in this case, the mean direction is 0 and the mean resultant length is ϕ_1.

If $\sum_{p=1}^{\infty} (\alpha_p^2 + \beta_p^2) < \infty$, then the pdf of Θ is expressed by the Fourier representation as

$$g(\theta) = \frac{1}{2\pi} \left[1 + 2 \sum_{p=1}^{\infty} \{\alpha_p \cos(p\theta) + \beta_p \sin(p\theta)\} \right], \quad 0 \le \theta < 2\pi, \quad \text{a.e.}, \tag{3.4}$$

which is a periodic function over $(-\infty, \infty)$. A distribution with location parameter $\mu (0 \le \mu < 2\pi)$ is simply obtained by replacing θ with $\theta - \mu$.

The wrapped Cauchy distribution (Section 4.4.1) is induced by a Cauchy distribution on the real line. Likewise, we can construct the wrapped normal (Section 4.4.2), the wrapped t (Section 4.4.3), the wrapped Laplace (Section 5.8.2), and so on. Skew or asymmetric wrapped distributions are induced by skew distributions on the real line. For example the wrapped skew normal distribution (Section 5.6.2) is induced by the skew normal distribution on the real line. More recently, Jammalamadaka and Kozubowski (2017) showed that many wrapped distributions can be obtained using a general method of wrapping a mixture of linear distributions which coincides with a mixture (with respect to the same mixing distribution) of circular distributions.

Following the univariate approach, one may derive a hyper toroidal distribution $f(\theta_1, \dots, \theta_p)$ of the circular random vector $\boldsymbol{\Theta} = (\Theta_1, \dots, \Theta_p)'$ on the hyper torus $(S^1)^p$ $[\theta_j \in [0, 2\pi),\ j = 1, \dots, p]$ from a multivariate distribution $g(x_1, \dots, x_p)$ of the linear random vector $\boldsymbol{X} = (X_1, \dots, X_p)'$ on $\mathbb{R}^p$ as follows:

$$f(\theta_1, \dots, \theta_p) = \sum_{k_1=-\infty}^{\infty} \cdots \sum_{k_p=-\infty}^{\infty} g(\theta_1 + 2\pi k_1, \dots, \theta_p + 2\pi k_p).$$

While this approach has been very productive for the univariate case, leading to as general a family as the possibly asymmetric wrapped stable family of distributions on the circle, not much success has been had for the multivariate case. In general, no closed form expression for these distributions can be expected. However, an advantage of this method is that the generation of random variables $\boldsymbol{\Theta}$ from the above family $f(\boldsymbol{\theta})$ is easily done if the same is true for the generation of $\boldsymbol{X}$ from $g(\boldsymbol{x})$. Also, the condition characterizing the independence of $\boldsymbol{X}$ suffices for the independence of $\boldsymbol{\Theta}$.

3.4 Projection

Let $(X_1, X_2)'$ be a bivariate random variable on the real plane with joint pdf $g(x_1, x_2)$. Using the polar transformation (Figure 3.1)

$$\begin{cases} X_1 = R\ \cos\Theta, \\ X_2 = R\ \sin\Theta, \end{cases} \tag{3.5}$$

where R and Θ take their values in $[0, \infty)$ and $[0, 2\pi)$, respectively, we transform the pdf of $(X_1, X_2)'$ into that of $(R, \Theta)'$:

$$g(x_1, x_2)dx_1 dx_2 = g(r\cos\theta, r\sin\theta)r\ drd\theta.$$

The marginal distribution of the angular variable Θ with pdf

$$f(\theta) = \int_0^{\infty} g(r\cos\theta, r\sin\theta)r\ dr$$

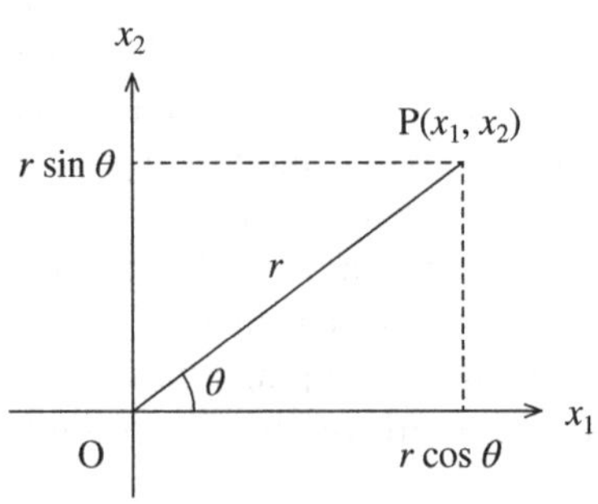

Figure 3.1 Polar transformation in the plane.

is called the projected or offset distribution of $(X_1, X_2)'$. For example if $(X_1, X_2)'$ is distributed as a bivariate normal distribution, then the resulting distribution of Θ is a projected or offset normal distribution. A projected distribution relates to the distribution of the quotient of two random variables (see Section 4.10).

3.5 Conditioning

Let $(X_1, X_2)'$ denote a continuous bivariate random vector on the whole real plane $\mathbb{R}^2$. Consider the polar transformation (3.5). Then the conditional distribution of the angular variable Θ, given $R = r$ yields a distribution on the circle.

As an example, let $(X_1, X_2)'$ be distributed as a bivariate normal distribution $N_2(\boldsymbol{\eta}, I_2)$ with mean vector $\boldsymbol{\eta}$ and identity variance–covariance matrix I_2, and $\boldsymbol{\eta} = \rho\ \mathbf{e}(\tau)$, where $\rho = \| \boldsymbol{\eta} \|$ denotes the length of $\boldsymbol{\eta}$ and $\mathbf{e}(\tau) = (\cos\ \tau, \sin\ \tau)'$ the direction cosine of $\boldsymbol{\eta}$. Then the conditional distribution of Θ given $R = r$ is a von Mises distribution $\text{VM}(\tau, \rho r)$ (see Section 4.3.3.6). Another example is obtained from a bivariate t-distribution on the plane by using the conditioning method. The distribution results in a circular t-distribution (see Section 4.11).

A more general setting can be seen in Jones and Pewsey (2005). Let us consider a bivariate spherically symmetric distribution (Fang et al. 1990) with nonzero location $(m, 0),\ m > 0$, whose random vector is denoted by $(X, Y)'$ and pdf is given by

$$f_s(x, y) \propto g\{(x - m)^2 + y^2\}.$$

The polar random variables R_s and Θ_s are defined by $X = R_s \cos \Theta_s$ and $Y = R_s \sin \Theta_s$. Then the conditional distribution of the angular variable Θ_s, given the length $R_s = r_s \in (0, \infty)$ has pdf

$$g\{(r_s^2 + m^2) - 2mr_s \cos \theta_s\}. \tag{3.6}$$

Alternatively, consider an elliptically symmetric distribution with mean at the origin whose pdf is given by

$$f_e(x, y) \propto g\left(\frac{x^2 - 2\delta xy + y^2}{1 - \delta^2}\right), -1 < \delta < 1.$$

With the polar random variables R_e and Φ, the conditional distribution of the angular variable Φ given $R_e = r_e$ has pdf

$$g\left(\frac{r_e^2\{1 - \delta \sin(2\phi)\}}{1 - \delta^2}\right).$$

This is an axial distribution (Section 7.1). Doubling the angle (see, (7.1)), we have the distribution of $(\Theta_e \equiv 2\Phi)|(R_e = r_e)$, whose pdf is $g(r_e^2(1 - \delta \sin \theta_e)/(1 - \delta^2))$ with mode at $\pi/2$ if $\delta > 0$ and at $-\pi/2$ if $\delta < 0$. If the distribution is recentered at 0, i.e. $\sin\theta_e$ is transformed into $\sin\{\theta_e + \mathrm{sgn}(\delta)\pi/2\} = \mathrm{sgn}(\delta)\cos\theta_e$, the transformed distribution has pdf

$$g\left(\frac{r_e^2(1-|\delta|\cos\theta_e)}{1-\delta^2}\right). \tag{3.7}$$

Formulae (3.6) and (3.7) match when $|\delta| = 2mr_s/(r_s^2+m^2)$, $r_e = |r_s^2 - m^2|/\sqrt{r_s^2+m^2}$ or, equivalently, $m = (r_e/2)d_+(\delta)$, $r_s = (r_e/2)d_-(\delta)$ with $d_\pm(\delta) = (1-|\delta|)^{-1/2} \pm (1+|\delta|)^{-1/2}$.

3.6 Linear Fractional Transformation

A linear fractional, bilinear or Möbius transformation is a conformal mapping of the form

$$w = g(z) = \frac{az+b}{cz+d} \tag{3.8}$$

for $a, b, c, d \in \mathbb{C}$ $(ad - bc \neq 0)$ and $z \in \mathbb{C}$. The condition $ad - bc \neq 0$ is for which w is not constant. For example, if $ad - bc = 0$ and $ad \neq 0$, then

$$\frac{az+b}{cz+d} = \frac{a(z+b/a)}{c(z+d/c)} = \frac{a}{c},$$

independent of z. Equation (3.8) has an inverse transformation

$$z = \frac{dw-b}{-cw+a},$$

which is also a linear fractional transformation. For $c = 0$, since

$$w = \frac{a}{d}\,z + \frac{b}{d},$$

where $d \neq 0$, because $ad \neq 0$ from the condition $ad - bc \neq 0$, (3.8) is decomposed into

$$w_1 = \frac{a}{d}\,z \quad \text{and} \quad w = w_1 + \frac{b}{d}.$$

Similarly, for $c \neq 0$, (3.8) is

$$w = \frac{(bc-ad)/c^2}{z+d/c} + \frac{a}{c}$$

and is decomposed into four transformations

$$w_1 = z + \frac{d}{c}, \quad w_2 = \frac{1}{w_1}, \quad w_3 = \frac{bc-ad}{c^2}\,w_2, \text{and} \quad w = w_3 + \frac{a}{c}.$$

As a result, (3.8) is composed of the following three basic transformations:

$$\text{(a) } w = z + a \text{ (translation), (b) } w = az, \text{ and (c) } w = \frac{1}{z} \text{ (inversion).}$$

Putting $a = re^{i\theta}$ ($i = \sqrt{-1}$, $r = |a|$, $\theta = \arg a$) in (b) above, we can further decompose the transformation into

$$w_1 = rz \text{ (homothety or dilation)} \quad \text{and} \quad w = e^{i\theta} w_1 \text{ (rotation).}$$

The linear fractional transformation carries circles or lines to circles or lines. A general transformation from the unit circle onto itself (Möbius circle transformation) is written as

$$w = e^{i\theta} \frac{z + a}{1 + \overline{a}z}, \qquad |z| = 1, \ |a| \neq 1,$$

where $\overline{a}$ denotes the complex conjugate of a. Assume that $\theta = 0$, since $w = e^{i\theta} w_1$ means rotation, and $|a| < 1$. It is clear that the case $a = 0$ means the identity mapping. The transformation

$$w = \frac{z + a}{1 + \overline{a}z}, \qquad |z| = 1, \ 0 < |a| < 1 \tag{3.9}$$

has two fixed points $z = \pm a/|a|$. Putting $z = e^{i\theta}$ and $a = re^{i\xi}$, we have $w = (e^{i\theta} + re^{i\xi})/(1 + re^{-i\xi + i\theta}) \to e^{i\xi} = a/|a|$ as $|a| = r \to 1$. Thus, a can be interpreted as the parameter that attracts the points z on the unit circle toward $a/|a|$ as $|a|$ goes to 1. It follows from (3.9) that $(\overline{a} + \overline{z})w - (a + z)\overline{w} - a\overline{z} + \overline{a}z = 0$, which means that the three points $a, -z$, and w exist on the same straight line. Figure 3.2a shows the behavior of the transformation w_j in (3.9) for $z = \exp(-i\pi/4)$ and $a_j = 0.3j\exp(i\pi/6)$ for $j = 1, 2, 3$ (Kato et al. 2008). For the case $|a| > 1$, the transformation is composed of two types of transformations:

$$z \mapsto \left(\frac{a}{|a|}\right)\left(\frac{a\overline{z}}{|a|}\right) \quad \text{and} \quad z \mapsto \frac{z + \tilde{a}}{1 + \overline{\tilde{a}}z},$$

i.e.

$$w = \frac{z + a}{1 + \overline{a}z} = \frac{\tilde{z} + \tilde{a}}{1 + \overline{\tilde{a}}\tilde{z}},$$

where $\tilde{z} = (a/|a|)(a\overline{z}/|a|)$ and $\tilde{a} = 1/\overline{a}$ with $|\tilde{a}| < 1$. Figure 3.2b shows the behavior of the transformation w for $z = \exp(-\pi i/6)$ and $a = (5/3)\exp(\pi i/6)$.

By the Möbius circle transformation, a circular distribution can be transformed into another circular distribution. Similarly, a general transformation from $|z| < 1$, the inside of a unit circle, onto $|w| < 1$ is

$$w = e^{i\theta} \frac{z + a}{1 + \overline{a}z}, \qquad |a| < 1.$$

By this transformation, a distribution on the disc can be transformed into another distribution on the disc. Jones (2004) used this transformation to construct the

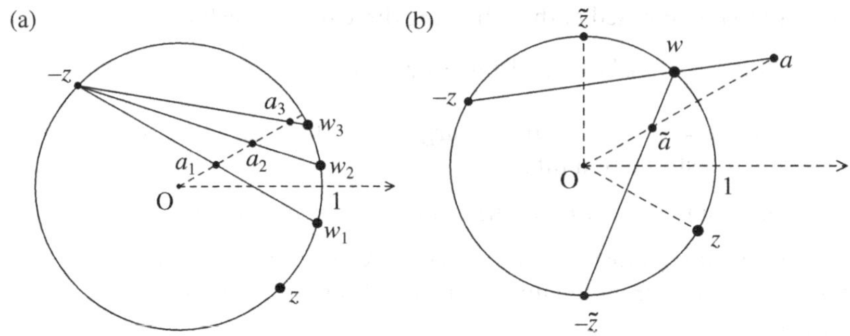

Figure 3.2 (a) Plot of w_j for $z = \exp(-i\pi/4)$ with $a_j = 0.3j\exp(i\pi/6)$ for $j = 1, 2, 3$. (b) Plot of w for $z = \exp(-i\pi/6)$ with $a = (5/3)\exp(i\pi/6)$.

Möbius, a skew, distribution on the disc (Section 11.3.4) from the spherically symmetric beta distribution. The linear fractional transformation from the interval $(-1, 1)$ onto itself is used by Seshadri (1991) to relate a symmetric beta distribution with a distribution analogous to McCullagh's (1989) distribution. This topic is treated in Section 4.7.

The stereographic projection as depicted in Figure 3.3 is another method for obtaining distributions on the circle. A variant of this is the mapping of the point on the x-axis by the tangent at a given point on the circumference of the circle (wherein the north pole is mapped to $+\infty$). Let X be a random variable on the real line. The stereographic projection is expressed by $x = \tan(\theta/2)$ $(-\pi \leq \theta < \pi)$. As an example, if X is distributed as a Cauchy distribution on the real line, then Θ has a circular uniform distribution. However, unlike the method of wrapping which can produce circular distributions even from linear distributions supported only on $\mathbb{R}^+$, for the method of stereographic projection, it is mandatory to have the full support $\mathbb{R}$.

Similarly, the stereographic projection maps a sphere from the North Pole onto a plane. The inverse stereographic projection may be used to produce a spherical distribution from a bivariate distribution on the real plane.

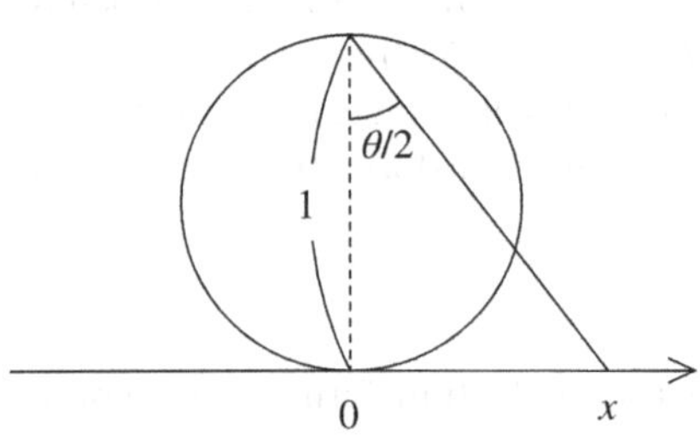

Figure 3.3 Stereographic projection.

3.7 Brownian Motion on the Circle

A p-dimensional Brownian motion taking values in p-dimensional Euclidean space $\mathbb{R}^p$ is a continuous-time stochastic process $\{B_t\}$, $t \geq 0$, that is characterized by

(a) B_t has independent increments, i.e. if $0 \leq t_0 < t_1 < t_2 < \cdots < t_{n-1} < t_n$, then $B(t_0)$, $B(t_2) - B(t_1), \ldots, B(t_n) - B(t_{n-1})$ are independent random variables;
(b) $B_t - B_s$ is distributed as a p-dimensional normal distribution with mean vector $\mathbf{0}$ and variance–covariance matrix $(t-s)I_p$ for $0 \leq s < t$, where I_p denotes the $p \times p$ identity matrix;
(c) $t \to B_t$ is almost surely continuous.

Brownian motion on the real line is used to generate linear distributions. For example, the inverse Gaussian distribution is obtained as the distribution of the waiting time that a Brownian motion with positive drift takes to reach a fixed point level (Chhikara and Folks 1989, p. 27). For Brownian motion on the circle, the wrapped normal distribution (Section 4.4.2) is generated. Stephens (1963) shows this fact as the limiting case of a random walk on the circle with assumption that the successive independent steps are governed by a symmetric density function. Furthermore, the wrapped Cauchy distribution (Section 4.4.1) is given as an exit distribution for Brownian motion in the disc (Durrett 1984). Circular beta models (Section 4.7) are also connected with Brownian motion in p-dimensional Euclidean space as exit distributions (McCullagh 1989). Kato (2009) proposed a bivariate Cauchy model (Section 9.10.1) for a pair of dependent unit vectors by using Brownian motion.

3.8 Characteristic Function-Based Method

Normal distributions in Euclidean space play a central role in statistical inference. However, there may be many situations where normal distributions do not fit real-life data sets even if symmetry is assumed. For instance, a t-distribution may be assumed as a theoretical distribution if its tail is considered heavier than a normal distribution. The t-distribution with n degrees of freedom converges to the standard normal distribution as n tends to infinity. That is, a standard normal distribution is a limit of t-distributions, and in this sense a class of t-distributions with location and scale parameters can be viewed as a family which includes normal distributions. Besides the t-distribution with one degree of freedom is a Cauchy distribution.

Several families are known which naturally include the normal distribution. One such family is the family of α-stable distributions on the real line. Its characteristic function is given by

$$\phi(t) = \begin{cases} \exp\left[-\tau^\alpha |t|^\alpha \left\{1 - i\beta \,\mathrm{sgn}(t) \tan \frac{\pi\alpha}{2}\right\} + i\mu t\right], & \alpha \in (0,1) \cup (1,2], \\ \exp\left[-\tau |t| \left\{1 + i\beta \frac{2}{\pi} \,\mathrm{sgn}(t) \log |t|\right\} + i\mu t\right], & \alpha = 1, \end{cases}$$

where $0 < \alpha \leq 2$, $|\beta| \leq 1$, $\tau \geq 0$, $i = \sqrt{-1}$, μ is a real number, and sgn$(\cdot)$ is a sign function. There is generally no closed form expression for the pdf of an α-stable distribution on the real line. Special cases include a Cauchy distribution with location parameter (mode) μ $(-\infty < \mu < \infty)$ and scale parameter τ $(0 < \tau < \infty)$ whose pdf is

$$g(x) = \frac{\tau}{\pi\{(x-\mu)^2 + \tau^2\}}, \qquad -\infty < x < \infty,$$

for $\alpha = 1$. A normal distribution with mean μ $(-\infty < \mu < \infty)$ and variance $2\tau^2$ $(0 < \tau < \infty)$ is also included in this class. As a result, the α-stable family links Cauchy $(\alpha = 1)$ and normal $(\alpha = 2)$ distributions. Properties of the wrapped α-stable distribution will be given in Sections 4.5.1 and 5.7.1.

Since the α-stable distribution on the real line generally has no closed form expression for the pdf, it is not suitable for obtaining distributions on the circle by the methods of construction based on the explicit form of the pdf on the real line. However, the wrapping method is applicable to the α-stable distribution because, as mentioned in Section 3.3, the pdf of its corresponding wrapped distribution is expressed in terms of the Fourier series.

The above discussion leads to another general method of construction of a circular pdf, when the linear pdf for the associated linear variable is not available. For example, from a random variable X having a symmetric distribution with characteristic function $\phi(p)$, the corresponding circular (wrapped) pdf is obtained from its Fourier series representation with $\alpha_p = \phi(p)$. We will discuss further distributions derived by this method in Chapter 4.

3.9 Mixtures

Let $f_1(\theta)$ and $f_2(\theta)$ denote two circular pdfs on the interval $[0, 2\pi)$. Then a mixture of $f_1(\theta)$ and $f_2(\theta)$ written as

$$f(\theta) = pf_1(\theta) + (1-p)f_2(\theta), \quad 0 < p < 1,$$

yields a pdf on the interval $[0, 2\pi)$, where p maybe known or unknown. As an example, the wrapped Laplace distribution (Section 5.8) is viewed as a mixture of

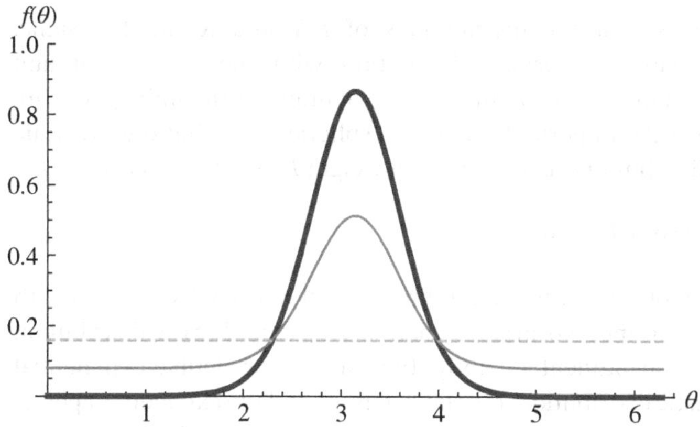

Figure 3.4 Density plots of mixtures. $p = 1$ (thick), $p = 0.5$ (thin), and $p = 0$ (uniform: dotted).

positive and negative wrapped Laplace distributions. Of course, it is possible to consider mixtures of three or more pdfs.

As a particular case, we may choose the circular uniform pdf $f_2(\theta) = 1/(2\pi)$ as a contamination of $f_1(\theta)$. As an example, the case of $f_1(\theta)$ as a symmetric wrapped stable distribution has been introduced by SenGupta and Pal (2001). The mixture distribution is naturally flatter than $f_1(\theta)$ and its tails lie between those of the mixing distributions (see Figure 3.4).

The method of parametric mixtures is another device for constructing circular distributions. Let $(\Theta|\xi)$ be a circular random variable with pdf $h(\theta|\xi)$, where ξ is a parameter. If ξ is a random variable Ξ with pdf $g(\xi;\eta)$ having a (vector) parameter η, then the circular pdf of Θ is given by

$$f(\theta;\eta) = \int h(\theta|\xi)g(\xi;\eta)d\xi.$$

This method also extends to the bivariate and multivariate cases elegantly, see e.g. Ong and SenGupta (2012) and Section 9.8.2.

3.10 Scale Mixtures/Elliptically Contoured Distributions

Let $\boldsymbol{X}$ be a random vector in n-dimensional Euclidean space and $\mathcal{O}(n)$ be the set of orthogonal matrices. Then the distribution of $\boldsymbol{X}$ is called a spherical, spherically symmetric or spherically contoured distribution if it is satisfied that

$$H\boldsymbol{X} \stackrel{\mathrm{d}}{=} \boldsymbol{X} \quad \text{for each} \quad H \in \mathcal{O}(n),$$

where $HX \overset{d}{=} X$ means that the distributions of HX and X are the same. An n-dimensional standard normal distribution with mean vector $\mathbf{0}$ and variance–covariance matrix I_n and a uniform distribution on the unit sphere are spherical distributions. An important subclass of spherical distributions is a scale mixture of normal distributions associated with weight F, whose pdf is given by

$$f(\boldsymbol{x}) = \int_0^\infty \xi(\boldsymbol{x}; \mathbf{0}, a^2 I_n) dF(a),$$

where $\xi(\boldsymbol{x}; \mathbf{0}, a^2 I_n)$ denotes the pdf of an n-dimensional normal distribution with mean vector $\mathbf{0}$ and variance–covariance matrix $a^2 I_n$, and $F(a)$ a distribution function on $(0, \infty)$. The Student t family, the Laplace or double-exponential distribution, the logistic distribution, the stable family, and the exponential power family are constructed by this method. Different weights provide distributions with different tail behavior.

An n-dimensional random vector $\boldsymbol{X}$ is said to have an elliptical, elliptically symmetric, or elliptically contoured distribution with parameters $\boldsymbol{\mu}$ $(n \times 1)$ and Σ $(n \times n)$ if

$$\boldsymbol{X} \overset{d}{=} \boldsymbol{\mu} + A'\boldsymbol{Y},$$

where A is $k \times n$, $A'A = \Sigma$ with $\text{rank}(\Sigma) = k$ and $\boldsymbol{Y}$ $(k \times 1)$ has a spherical distribution. Hereafter, we assume that A is $n \times n$ with $\text{rank}(A) = n$ and $\boldsymbol{X}$ has a density. An n-dimensional normal distribution with mean vector $\boldsymbol{\mu}$ and variance–covariance matrix Σ and a p-dimensional t-distribution with degrees of freedom ν, mean vector $\boldsymbol{\mu}$, and correlation matrix R whose pdf is given by

$$f(\boldsymbol{x}) = \frac{\Gamma((\nu + p)/2)}{(\pi\nu)^{p/2}\Gamma(\nu/2)|R|^{1/2}} \left\{ 1 + \frac{1}{\nu} (\boldsymbol{x} - \boldsymbol{\mu})' R^{-1} (\boldsymbol{x} - \boldsymbol{\mu}) \right\}^{-(\nu+p)/2}$$

belong to the family of elliptical distributions. The wrapping, projection, and conditioning methods are applicable to the construction of circular distributions from elliptical distributions.

3.11 Transformations from Bivariate Distributions

Let $f(x_1, x_2)$ denote a bivariate joint pdf on the real plane. Then the marginal pdf is given by

$$f(x_1) = \int_{-\infty}^\infty f(x_1, x_2) dx_2.$$

Applying a transformation $\theta_1 = 2\tan^{-1} x_1$ $(-\infty < x_1 < \infty, -\pi < \theta_1 < \pi)$ from the whole real line onto the unit circle (inverse stereographic projection, see

Figure 3.3), we have a circular distribution. For example, a bivariate Cauchy distribution with joint pdf

$$f(\boldsymbol{x}) = \frac{1}{2\pi} \frac{|A|^{1/2}}{1 + \{(\boldsymbol{x} - \boldsymbol{\mu})'A(\boldsymbol{x} - \boldsymbol{\mu})\}^{3/2}}, \quad \boldsymbol{x} = (x_1, x_2)' \in \mathbb{R},$$

where A is a 2×2 positive definite matrix and $\boldsymbol{\mu} \in \mathbb{R}$, gives a circular Cauchy pdf (Waine 2001; Mardia 2007).

We can apply this transformation method to scale mixtures/elliptically contoured distributions introduced in Section 3.10 to construct a more general class of circular distributions.

3.12 Construction of Skew or Asymmetric Circular Distributions

Skew circular distributions may be constructed by broadly two methods:

(i) skewing linear or circular symmetric distributions, thus popularly termed as skew-symmetric circular distributions; and
(ii) using the methods of deriving circular distributions presented above now with skew linear distributions.

Since the appearance of Azzalini's (1985) paper, many researchers have extended his derivation of skew-normal distributions to other linear distributions. A variety of circular skew-normal distributions can also be constructed using the wrapping (see, e.g. wrapped skew-normal distributions in Pewsey 2006), projection and conditioning methods mentioned earlier. Detailed properties of such distributions will be presented in Section 5.6.2.

We discuss now a general class of skew distributions which is based on a simple but useful theorem in Azzalini and Capitanio (2003) and Azzalini (2005): If f_0 is a one-dimensional pdf symmetric about zero, and G is a one-dimensional distribution function such that its derivative G' exists and is a pdf symmetric about zero, then

$$f(z) = 2f_0(z)G(w(z)), \quad -\infty < z < \infty,$$

is a pdf for any odd function $w(\cdot)$. Its proof (Azzalini 2005) is as follows: Let Y and X denote independent random variables whose pdfs are f_0 and G'. Then

$$\frac{1}{2} = P\{X - w(Y) \leq 0\} = E[P\{X - w(Y) \leq 0|Y\}] = \int_{\mathbb{R}} G(w(z))f_0(z)dz$$

since both $w(Y)$ and $X - w(Y)$ are symmetric about zero.

A circular version of this theorem is as follows: Suppose that f and g are circular pdfs on $[-\pi, \pi)$ which are symmetric about zero and $G(\theta) = \int_{-\pi}^{\theta} g(t)dt$ is the distribution function of g. If w is an odd function with $-\pi \leq w(\theta) < \pi$ and a periodic function, i.e. $w(\theta) = w(\theta + 2\pi k)$ for all integers k, then

$$f_\mu(\theta) = 2f(\theta - \mu)G(w(\theta - \mu)), \quad -\pi \leq \theta < \pi, -\pi \leq \mu < \pi, \tag{3.10}$$

is a circular pdf. Its proof (Umbach and Jammalamadaka 2009) now follows. By the result of Azzalini (2005) above, f_0 (f_μ with $\mu = 0$ in (3.10)) is a pdf on $[-\pi, \pi)$ since f is a pdf on $[-\pi, \pi)$ symmetric about zero and $w(\theta)$ is an odd function with $-\pi \leq w(\theta) < 0$ for $-\pi \leq \theta < \pi$. Furthermore, f_0 is a periodic function since f and w are periodic functions. Indeed,

$$f_0(\theta + 2\pi k) = f(\theta + 2\pi k)G(w(\theta + 2\pi k)) = f(\theta)G(w(\theta)) = f_0(\theta)$$

for all integers k. We have the result by noting that f_μ is a location change of f_0. Some particular cases of (3.10) will be treated in Section 5.11.

It is possible to construct also a very general class of skew or asymmetric circular distributions by applying the transformation method in Section 3.11 to Dey and Liu's (2005) class of skew multivariate distributions.

3.13 Arithmetic and Geometric Methods

An angle θ measured clockwise or anticlockwise from the base line is a geometric object and may arithmetically take a value in the interval $[0, 2\pi)$ or $[-\pi, \pi)$, which has one-to-one correspondence to a point $P(\cos\theta, \sin\theta)$ on the unit circle in the real plane. Thus, if Θ denotes an angular random variable, then its distribution is also called the distribution on the circle. Since θ ($\in [0, 2\pi)$) corresponds to a unit vector $(\cos\theta, \sin\theta)'$, it also corresponds to a complex number $e^{i\theta} = \cos\theta + i\sin\theta$, a point on the unit circle in the complex plane, where i is the imaginary unit ($i = \sqrt{-1}$). This interpretation has an important role, for instance as shown in the section of wrapped Cauchy distributions (Section 4.4.1).

A point $(x, y)'$ in the plane can be transformed into (r, θ) by the polar transformation, where $r \geq 0$ and $0 \leq \theta < 2\pi$. This idea is used to derive projected distributions (Section 3.4) and conditional distributions (Section 3.5).

A random walk and Brownian motion on the line, in the plane and, more generally, in p-dimensional Euclidean space appeal to our geometric intuition. For example, a problem is proposed by Pearson (1905) (Section 4.2.2). Another example is the wrapped Cauchy distribution, which is derived as an exit distribution of Brownian motion.

The stereographic projection (Section 3.6) is an appealing transformation from circle to line or line to circle. Its geometrical understanding is easy. It is possible to extend it to the case of higher-dimensional space, i.e. from hyper-sphere to hyper-plane.

Another important aspect is the geometric structure of circular distributions. Discussion about the structure of the von Mises distribution (Section 4.3.3) as a member of the regular exponential family can be seen in Jammalamadaka and SenGupta (2001). Geometric terminologies such as statistical curvature, tangent space, and metric tensor are used in this field, see, e.g. SenGupta and Jammalamadaka (2001).

3.14 Shortcomings and Future Work

Methods of construction for circular distributions were introduced in Sections 3.1 to 3.13. Here we summarize remarkable features of the methods with their shortcomings and mention future work.

At the early stage of the study, ad hoc methods (Section 3.1) are the "building blocks." However, one needs some insights to obtain elegant and meaningful forms for the pdfs. Since by definition a pdf on the circle is a periodic function on the whole real line, it is natural to exploit cosine and sine functions to construct circular distributions. For instance, we can review the cardioid distribution (Section 4.6.2) and study its statistically meaningful properties. Next, one may explore which systematic methods are related to the distributions obtained by ad hoc methods. If we use $\cos(k\theta)$ and $\sin(k\theta)$ for $k = 1, 2, 3, \ldots$ and their linear combinations for $0 \leq \theta < 2\pi$, it is possible to derive skew and/or multimodal distributions. Examples are generalized von Mises distributions (Section 5.3.1), generalized t-distributions (Section 5.5.1), and nonnegative trigonometric sums (Section 5.2.3). Characterization is a mathematically interesting branch and, as mentioned in Section 3.2, there are some open problems even in the general wrapped stable family of distributions.

The wrapping method (Section 3.3) is applicable to any continuous distribution on the real line. Examples include wrapped Cauchy distributions (Section 4.4.1), wrapped normal distributions (Section 4.4.2), and more generally wrapped t-distributions (Section 4.4.3). The method is applicable not only to symmetric distributions on the real line but also to skew or asymmetric distributions to get skew distributions on the circle. Wrapped skew-normal (Section 5.6.2), wrapped α-stable (Section 5.7), wrapped Laplace (Section 5.8), and wrapped gamma distributions (Section 5.9) are such examples. By definition, the pdf of a wrapped distribution on $[0, 2\pi)$ is obtained by wrapping a pdf on the real line around the circle. This pdf is not necessarily a periodic function on the whole real line, but it is easy to extend the pdf on $[0, 2\pi)$ to that on the whole real line. Wrapped distributions do not have one-to-one correspondence with distributions on the real line. Since the pdf of a wrapped distribution may be expressed as a Fourier series under a mild condition, the wrapping method is particularly helpful if the characteristic functions or trigonometric moments have simple expressions. For example, wrapped α-stable distributions (Section 5.7) provide a useful class of circular distributions, though the representation of its pdf in terms of an infinite series calls for innovative approaches for related statistical inference.

Projection and conditioning are general methods for deriving circular distributions from a bivariate pdf. It was remarked in Section 3.4 that a projected distribution is related to the distribution of the quotient of two random variables on the real line. Typical examples are projected normal distributions (Section 4.10.2) and, as an extension, projected Pearson type VII distributions (Section 4.10.3), which include projected Cauchy and t-distributions as special cases, but the forms for their pdfs are not simple. It is possible to have projected distributions from skew distributions on the real line. Such an example is a projected lognormal distribution in Section 5.14. Since lognormal distributions are closed with respect to the quotient of two random variables, a projected lognormal distribution is easily obtained. However, it is not clear whether this distribution has nice properties. The von Mises distribution (Section 4.3.3) is derived as the conditional distribution of the angle given the length for a bivariate normal distribution. Analogously, the circular t-distribution (Section 4.11) is derived as the conditional distribution of the angle given the length for a bivariate t-distribution.

Linear fractional transformation (Section 3.6) is an attractive method for generating skew distributions not only on the circle but also on other manifolds such as the (hyper-) sphere, (hyper-) torus, (hyper-) cylinder, and (hyper-) disc.

In Section 3.7, the following were observed. (i) The wrapped normal distribution is derived from Brownian motion on the circle. (ii) The wrapped Cauchy distribution can be viewed as an exit distribution for Brownian motion in the disc. (iii) Circular beta models are connected with Brownian motion in p-dimensional Euclidean space as exit distributions. (iv) A multivariate wrapped Cauchy distribution is obtained from Brownian motion in hyper-sphere. Thus, Brownian motion is one of the sources for obtaining distributions on the circle and other manifolds.

The search of an amenable class of distributions from the simultaneous viewpoints of both distributional properties and statistical inference is of prime importance. The t-distribution on the real line appears as the distribution of the test statistic for the mean of a normal distribution with unknown variance and the t-distribution with location and scale parameters is used as a theoretical model. And, the t-distribution with n degrees of freedom approaches the standard normal distribution as n tends to infinity. A family of t-distributions contains normal distributions as a limit. Literature in directional statistics includes some classes which contain the von Mises distribution, often called the circular normal distribution, and basic distributions such as the circular uniform, cardioid, and wrapped Cauchy distributions. A family of wrapped α-stable distributions (Section 3.8) contains the circular uniform, wrapped Cauchy, and wrapped normal distributions. A mixture of distributions is a general method for constructing circular distributions. Derivation of mathematical properties of a circular mixture distribution (Section 3.9) analytically may be nontrivial except for particular cases, but modern powerful computational techniques may be employed. Scale mixtures and, more generally, elliptically contoured distributions (Section 3.10) can provide meaningful circular distributions by the methods of wrapping (Section 3.3), projection (Section 3.4), conditioning (Section 3.5), and transformations (Section 3.11).

Many researchers are interested in skew or asymmetric circular distributions as well as symmetric ones. Section 3.12 describes some of the recent studies of skew circular distributions. This is one of the developing areas. Finally, both arithmetic and geometric aspects (Section 3.13) are important for a clear understanding of the theory of circular distributions.

References

Azzalini, A. (1985). A class of distributions which includes the normal ones. *Scandinavian Journal of Statistics* 12: 171–178.

Azzalini, A. (2005). The skew-normal distribution and related multivariate families. *Scandinavian Journal of Statistics* 32: 159–188.

Azzalini, A. and Capitanio, A. (2003). Distributions generated by perturbation of symmetry with emphasis on a multivariate skew *t*-distribution. *Journal of the Royal Statistical Society* B65: 367–389.

Cartwright, D. E. (1963). The use of directional spectra in studying the output of a wave recorder on a moving ship. In: *Ocean Wave Spectra*, 203–218. Englewood Cliffs, NJ: U.S. Naval Oceanographic Office, Prentice-Hall.

Chhikara, R. S. and Folks, J. L. (1989). *The Inverse Gaussian Distribution: Theory, Methodology, and Applications*. New York: Dekker.

Cox, D. R. (1975). Discussion of Professor Mardia's paper. *Journal of the Royal Statistical Society* B37: 380–381.

Dey, D. K. and Liu, J. (2005). A new construction for skew multivariate distributions. *Journal of Multivariate Analysis* 95: 323–344.

Durrett, R. (1984). *Brownian Motion and Martingales in Analysis*. Belmont, CA: Wadsworth.

Fang, K. T., Kotz, S., and Ng, K. W. (1990). *Symmetric Multivariate and Related Distributions*. London: Chapman and Hall.

Jammalamadaka, S. R. and Kozubowski, T. J. (2017). A general approach for obtaining wrapped circular distributions via mixtures. *Sankhyā Series A* 79: 133–157.

Jammalamadaka, S. R. and SenGupta, A. (2001). *Topics in Circular Statistics*. Singapore: World Scientific.

Johnson, N. L., Kotz, S., and Balakrishnan, N. (1994). *Continuous Univariate Distributions*, vol. 1, 2e. New York: Wiley.

Jones, M. C. (2004). The Möbius distribution on the disc. *Annals of the Institute of Statistical Mathematics* 56: 733–742.

Jones, M. C. and Pewsey, A. (2005). A family of symmetric distributions on the circle. *Journal of the American Statistical Association* 100: 1422–1428.

Kagan, A. M., Linnik, Y. V., and Rao, C. R. (1973). *Characterization Problems in Mathematical Statistics*. New York: Wiley-Interscience.

Kato, S. (2009). A distribution for a pair of unit vectors generated by Brownian motion. *Bernoulli* 15: 898–921.

Kato, S., Shimizu, K., and Shieh, G. S. (2008). A circular-circular regression model. *Statistica Sinica* 18: 633–645.

Kim, S. and SenGupta, A. (2021). Multimodal exponential families of circular distributions with application to daily peak hours of PM 2.5 level in a large city. *Journal of Applied Statistics*. 48: 3193–3207.

McCullagh, P. (1989). Some statistical properties of the family of continuous univariate distributions. *Journal of the American Statistical Association* 84: 125–129.

Maksimov, V. M. (1967). Necessary and sufficient statistics for a family of shifts of probability distributions on continuous bicompact groups. *Teoriya Verojatnosteĭ i ee Primeneniya* 12: 307–321 (in Russian). English translation: Theory of Probability and its Appllications 12: 267–280.

Mardia, K. V. (2007). On some recent advancements in applied shape analysis and directional statistics, Leeds Annual Statistical Research Workshop 2007, Department of Statistics, University of Leeds. In: *Systems Biology & Statistical Bioinformatics* (ed. S. Barber, P. D. Baxter, and K. V. Mardia), 9–17. Leeds: Leeds University Press.

Mitsuyasu, H., Tasai, F., Suhara, T. et al. (1975). Observations of the directional spectrum of ocean waves using a cloverleaf buoy. *Journal of Physical Oceanography* 5: 750–760.

Ong, S. H. and SenGupta, A. (2012). Bivariate and multivariate circular distributions by mixtures. Journal of the Indian *Statistical Association* Special Golden Jubilee Year Issue, 50: 193–204.

Pearson, K. (1905). The problem of the random walk. *Nature* 72: 294.

Pewsey, A. (2006). Modelling asymmetrically distributed circular data using the wrapped skew-normal distribution. *Environmental and Ecological Statistics* 13: 257–269.

Rukhin, A. L. (1972). Some statistical decisions about distributions on a circle for large samples. *Sankhyā Series A* 34: 243–250.

SenGupta, A. (2004). On the constructions of probability distributions for directional data. *Bulletin of Calcutta Mathematical Society* 96: 139–154.

SenGupta, A. (2010). Probability Distributions for Directional Data. Institute of Mathematical Statistics Lecture Note. *7th International Conference on Multivariate Distributions with Applications*, 8–13 August 2010, Maresias, Sao Paulo, Brazil.

SenGupta, A. and Jammalamadaka, S.R. (2001). Geometry of the von Mises distribution and of tests for its mean direction. *Stochastic Modelling and Applications* 4: 16–37.

SenGupta, A. and Pal, C. (2001). On optimal tests for isotropy against the symmetric wrapped stable-circular uniform mixture family. *Journal of Applied Statistics* 28: 129–143.

Seshadri, V. (1991). A family of distributions related to the McCullagh family. *Statistics & Probability Letters* 12: 373–378.

Stephens, M.A. (1963). Random walk on a circle. *Biometrika* 50: 385–390.

Umbach, D. and Jammalamadaka, S.R. (2009). Building asymmetry into circular distributions. *Statistics & Probability Letters* 79: 659–663.

Waine, B. (2001). Face analysis and the bivariate wrapped Cauchy distribution. PhD thesis. Leeds: University of Leeds.

4

Symmetric Distributions on the Circle

Distributions are roughly divided into two types: symmetric and asymmetric. Symmetric distributions themselves give not only basic models for circular data but also the source of constructing asymmetric distributions. Another classification of distribution is continuous or discrete. In this chapter, we deal with continuous symmetric distributions. Definition of symmetry in the context of circular distributions is given in the first section. The uniform distribution is first introduced since it is the basic distribution on the circle and other distributions include the uniform distribution as a special case. The general methods of construction of circular distributions as detailed in Chapter 3 are then taken up to derive many corresponding distributions and indicate further possibilities. Several appealing properties of these distributions are also presented.

4.1 Preliminary Notions

Definition 4.1 When the support of a pdf $f(\theta)$ of a circular random variable Θ is $[-\pi, \pi)$, the pdf $f(\theta)$ or the distribution of Θ is said to be symmetric about 0 if $f(\theta) = f(-\theta)$ for all θ $(-\pi \leq \theta < \pi)$, i.e. $f(\theta)$ is an even function. If $f(\theta)$ is symmetric about 0, then $f(\theta - \mu)$ on $[-\pi, \pi)$ or $[0, 2\pi)$ is symmetric about μ.

The expression in Section 3.3 can in general be equivalently written as: $\alpha_p + i\beta_p \equiv \rho_p(\cos \mu_p + i \sin \mu_p)$.

If a distribution is symmetric about μ, it is also symmetric about $\mu \pm \pi$. If a distribution is unimodal and symmetric about a mode μ, then $\mu \pm \pi$ is an anti-mode, and μ is the mean direction and the median direction. Moreover, if a circular random variable Θ is symmetric about the mean direction μ, $\mu_p = p\mu$, all the sine moments of $\Theta - \mu$ are zero, i.e. $E[\sin\{p(\Theta - \mu)\}] = 0$ for $p = \cdots, -2, -1, 0, 1, 2, \ldots$, and the Fourier expansion of $f(\theta)$ reduces to

$$f(\theta) = \frac{1}{2\pi}\left[1 + 2\sum_{p=1}^{\infty}\overline{\alpha}_p \cos\{p(\theta - \mu)\}\right], \tag{4.1}$$

where $\overline{\alpha}_p$ denotes the pth cosine moment about μ or the pth cosine moment of $\Theta - \mu$. Since (4.1) is written

$$f(\theta) = \frac{1}{2\pi}\left[1 + 2\sum_{p=1}^{\infty}\left\{\overline{\alpha}_p \cos(p\mu)\cos(p\theta) + \overline{\alpha}_p \sin(p\mu)\sin(p\theta)\right\}\right],$$

the pth cosine moment of Θ is $E\{\cos(p\Theta)\} = \overline{\alpha}_p \cos(p\mu)$ and the pth sine moment of Θ is $E\{\sin(p\Theta)\} = \overline{\alpha}_p \sin(p\mu)$.

4.2 Uniform Distribution

4.2.1 Definition and Properties

In contrast to the uniform distribution on a finite interval of the real line, or unit distributions, the uniform distribution on the unit circle (henceforth, circle) plays a fundamental role in circular statistics. The continuous circular uniform distribution is defined as a distribution that has a constant pdf over the circumference of the unit circle. Thus, its pdf is expressed by

$$f(\theta) = \frac{1}{2\pi}, \quad 0 \le \theta < 2\pi,$$

since the length of the circumference of the unit circle is 2π. The uniform distribution is derived from the maximizing entropy method when one does not ask for the restrictions on moments. In fact, as seen in Section 3.2.1, Shannon's entropy for a circular pdf $f(\theta)$ has an inequality

$$-\int_0^{2\pi} f(\theta)\log f(\theta)d\theta \le \log(2\pi),$$

and the equality holds when $f(\theta)$ is the uniform pdf without any restrictions on moments. Another characterization (Bingham, 1978; Kent et al., 1979) is: When the size of a random sample is fixed, the sample mean direction is stochastically independent of the sample mean resultant length.

The distribution function of the continuous uniform pdf $f(\theta)$ is obviously

$$F(\theta) = \int_0^{\theta} f(x)dx = \frac{\theta}{2\pi}, \quad 0 \le \theta < 2\pi.$$

If Θ is distributed as the continuous circular uniform, the characteristic function or pth trigonometric moment in (2.1) is

$$\phi_p = E(e^{ip\Theta}) = \begin{cases} 1, & p = 0, \\ 0, & p \neq 0. \end{cases}$$

From this, the mean resultant length is zero, and no mean directions or no particular directions exist for the uniform distribution. However, a zero mean-resultant-length or, equivalently, one circular-variance, distribution does not imply the uniform distribution. For example, the mean resultant length of a distribution with pdf

$$f(\theta) = \frac{1}{2\pi}\{1 + 2a\cos(2\theta)\}, \quad |a| \leq \frac{1}{2}, \; 0 \leq \theta < 2\pi \tag{4.2}$$

is zero, since the pth trigonometric moment is given by

$$\int_0^{2\pi} e^{ip\theta} f(\theta) d\theta = \begin{cases} 1, & p = 0, \\ 0, & p = 1, 3, 4, \dots, \\ a, & p = 2. \end{cases}$$

It is obvious that (4.2) with $a \neq 0$ is not a uniform pdf.

The distribution of the sum of two independent circular random variables such that at least one is uniform has a very interesting property. To show this precisely, suppose that a continuous random variable Θ is uniformly distributed with pth trigonometric moment ϕ_p and that another circular random variable Ψ is independent of Θ and has the pth trigonometric moment ψ_p. Then the sum $\Theta + \Psi$ has the pth trigonometric moment

$$E(e^{ip(\Theta+\Psi)}) = \phi_p \psi_p = \begin{cases} 1, & p = 0, \\ 0, & p \neq 0, \end{cases}$$

which is the pth trigonometric moment of the continuous uniform distribution. This means that $\Theta + \Psi$ is continuously uniformly distributed from the uniqueness property of the pth trigonometric moment! It is seen from the above fact that if $\Theta_1, \dots, \Theta_n$ are independent circular random variables and at least one out of these is continuously uniformly distributed, then the sum $\Theta_1 + \cdots + \Theta_n$ is distributed as the continuous uniform. In addition, note that, as in Section A.4 in Appendix A (Central Limit Theorem), if $\Theta_1, \dots, \Theta_n$ are mutually independent and identically distributed continuous circular random variables, then the sum $\Theta_1 + \cdots + \Theta_n$ converges to the continuous uniform distribution as n tends to infinity.

4.2.2 The Problem of the Random Walk

Karl Pearson (1905) proposed a problem:

> "A man starts from a point **O** and walks l yards in a straight line; he then turns through any angle whatever, and walks another l yards in a second

straight line. He repeats this process n times. I require the probability that after these n stretches he is at a distance between r and $r + \delta r$ from his starting point, **O**."

The general solution to this random walk problem was found by Kluyver (1905), which depends upon the theory of Bessel's functions. The solution can be seen in Mardia (1972, Section 4.4), Mardia and Jupp (2000, Section 4.4) and Jammalamadaka and SenGupta (2001, Section 3.2) and it is briefly described here.

First, we begin with a special case $n = 2$. The angle A of a triangle $\triangle$OAT is expressed by Θ which takes its values in $[0, 2\pi)$ with circular uniform distribution (Figure 4.1). Let a_1 and a_2 (constants) denote the lengths of the line segments OA and AT, and R (random variable) the length of OT. Then it follows from the second rule of cosine that

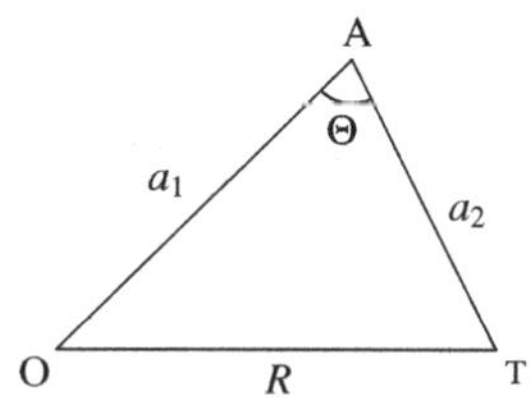

Figure 4.1 Pearson's random walk problem ($n = 2$).

$$R^2 = a_1^2 + a_2^2 - 2a_1 a_2 \cos\Theta.$$

The distribution function of R is, for $0 < r < a_1 + a_2$,

$$\begin{aligned}
F(r) &= P(R \le r) = P(a_1^2 + a_2^2 - 2a_1 a_2 \cos\Theta \le r^2) \\
&= 2P\left(0 \le \Theta \le \cos^{-1}\left(\frac{a_1^2 + a_2^2 - r^2}{2a_1 a_2}\right)\right) \\
&= \frac{1}{\pi}\cos^{-1}\left(\frac{a_1^2 + a_2^2 - r^2}{2a_1 a_2}\right)
\end{aligned}$$

and thus the pdf of R is given by

$$\begin{aligned}
f(r) = \frac{dF(r)}{dr} &= \frac{2r}{\pi\sqrt{4a_1^2 a_2^2 - (a_1^2 + a_2^2 - r^2)^2}} \\
&= \frac{2r}{\pi\sqrt{\{r^2 - (a_1 - a_2)^2\}\{(a_1 + a_2)^2 - r^2\}}},
\end{aligned} \tag{4.3}$$

which reduces to

$$f(r) = \frac{2}{\pi\sqrt{4 - r^2}}, \qquad 0 < r < 2,$$

if $a_1 = a_2 = 1$.

Next, we consider the general case. Let $\Theta_1, \ldots, \Theta_n$ denote a sequence of independent random variables with circular uniform distribution on $[0, 2\pi)$ and $a_1, \ldots, a_n$ the corresponding length of the segments. Then the characteristic function of (C, S), where

$$C = \sum_{j=1}^{n} a_j \cos\Theta_j \quad \text{and} \quad S = \sum_{j=1}^{n} a_j \sin\Theta_j,$$

is

$$\Psi(t) = \prod_{j=1}^{n} J_0(a_j t),$$

with the Bessel function of the first kind and order zero, $J_0(\cdot)$, and the inversion formula leads to the fact that the pdf of the length R is

$$f(r) = r \int_0^\infty t J_0(rt)\Psi(t)dt,$$

which is Kluyver's general solution of the problem of the random walk.

If $n = 2$ as a special case, it is known (Gradshteyn and Ryzhik, 1994 [5th ed.], p. 718; 2007 [7th ed.], p. 686, 6.578.9) that

$$\int_0^\infty J_\nu(ax)J_\nu(bx)J_\nu(cx)x^{1-\nu}dx = \frac{2^{\nu-1}\Delta^{2\nu-1}}{(abc)^\nu\Gamma(\nu+1/2)\Gamma(1/2)}, \qquad \mathrm{Re}\,\nu > -\frac{1}{2},$$

where Δ is the area of a triangle whose sides are a, b, and c, i.e.

$$\Delta = \frac{1}{4}\sqrt{\{c^2-(a-b)^2\}\{(a+b)^2-c^2\}}, \qquad |a-b| < c < a+b.$$

In our problem when $\nu = 0$, we have

$$\int_0^\infty xJ_0(rx)J_0(a_1x)J_0(a_2x)dx = \frac{1}{2\pi\Delta},$$

with

$$\Delta = \frac{1}{4}\sqrt{\{r^2-(a_1-a_2)^2\}\{(a_1+a_2)^2-r^2\}}, \qquad 0 < r < a_1+a_2,$$

and the pdf $f(r)$ of the length R reduces to (4.3), i.e.

$$f(r) = r\int_0^\infty xJ_0(rx)J_0(a_1x)J_0(a_2x)dx = \frac{r}{2\pi\Delta}, \qquad 0 < r < a_1+a_2.$$

4.2.3 Expected Square Displacement

Kareiva and Shigesada (1983) give a formula for expected square displacements with applications to the flight sequences of cabbage white butterfly and movement patterns of pipe-vine swallowtails. Moving with lengths L_j $(1 \le j \le n,\ n \ge 2)$ and turning angles Θ_j $(1 \le j \le n-1)$ between successive moves is as shown in Figure 4.2.

Let L_j be independent and identically distributed random variables with mean $\mu_1 = E(L_j)$ and second moment about zero $\mu_2 = E(L_j^2)$ for all $j = 1, \dots, n$, and Θ_j be independent and identically distributed circular random variables with $c = E(\cos\Theta_j)$ and $s = E(\sin\Theta_j)$ for all $j = 1, \dots, n-1$. Circular uniformity is not necessarily assumed for Θ_j. We suppose that $L_1, \dots, L_n$ and $\Theta_1, \dots, \Theta_{n-1}$ are mutually independent. Moreover, let

$$X_j = L_j\cos\left(\sum_{m=1}^{j-1}\Theta_m\right), \quad Y_j = L_j\sin\left(\sum_{m=1}^{j-1}\Theta_m\right),$$

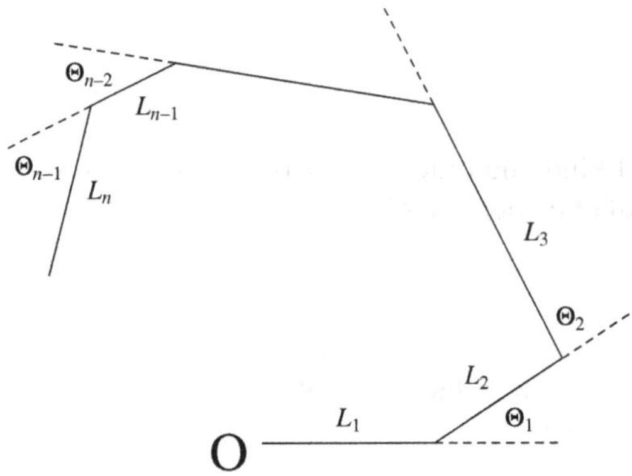

Figure 4.2 Moving with lengths and turning angles.

for $j = 1, \dots, n$. The resultant vector is

$$R_n = \sum_{j=1}^{n} W_j = \left(\sum_{j=1}^{n} X_j, \sum_{j=1}^{n} Y_j \right),$$

where $W_j = (X_j, Y_j)$.

The net squared displacement is calculated by

$$\begin{aligned}
\|R_n\|^2 &= \left(\sum_{j=1}^{n} X_j \right)^2 + \left(\sum_{j=1}^{n} Y_j \right)^2 \\
&= \sum_{j=1}^{n} L_j^2 + 2 \sum_{j>k} L_j L_k \left\{ \cos\left(\sum_{m=1}^{j-1} \Theta_m \right) \cos\left(\sum_{m=1}^{k-1} \Theta_m \right) \right. \\
&\quad \left. + \sin\left(\sum_{m=1}^{j-1} \Theta_m \right) \sin\left(\sum_{m=1}^{k-1} \Theta_m \right) \right\} \\
&= \sum_{j=1}^{n} L_j^2 + 2 \sum_{1 \le k < j \le n} L_j L_k \cos\left(\sum_{m=k}^{j-1} \Theta_m \right) \\
&= \sum_{j=1}^{n} L_j^2 + \sum_{1 \le k < j \le n} L_j L_k \left\{ \exp\left(i \sum_{m=k}^{j-1} \Theta_m \right) + \exp\left(-i \sum_{m=k}^{j-1} \Theta_m \right) \right\}.
\end{aligned}$$

Thus, the expected net squared displacement is

$$\begin{aligned}
E(\|R_n\|^2) &= n\mu_2 + \mu_1^2 \sum_{k=1}^{n-1} \sum_{j=k+1}^{n} \{ (c + is)^{j-k} + (c - is)^{j-k} \} \\
&= n\mu_2 + \mu_1^2 \left[\frac{c + is}{1 - c - is} \left\{ n - \frac{1 - (c + is)^n}{1 - c - is} \right\} \right. \\
&\quad \left. + \frac{c - is}{1 - c + is} \left\{ n - \frac{1 - (c - is)^n}{1 - c + is} \right\} \right].
\end{aligned}$$

In particular, if each Θ_j follows the uniform distribution, then we have $c = s = 0$ and the formula reduces to $E(\|R_n\|^2) = n\mu_2$, and if the distribution of Θ_j is symmetric about zero, then we have $s = 0$ and

$$\begin{aligned} E(\|R_n\|^2) &= n\mu_2 + 2\mu_1^2 \frac{c}{1-c}\left(n - \frac{1-c^n}{1-c}\right) \\ &\approx n\mu_2 + 2n\mu_1^2 \frac{c}{1-c} \quad \text{(for large } n) \end{aligned}$$

since $c^n \to 0\,(n \to \infty)$.

We now take up the general methods for construction of circular distributions detailed in Chapter 3 to provide an array of interesting symmetric distributions on the circle manifold S^1.

4.3 Ad Hoc

4.3.1 Triangular Distribution

A triangular distribution with parameter $\rho\,(-4/\pi^2 \le \rho \le 4/\pi^2)$ has pdf

$$f(\theta) = \frac{1}{8\pi}(4 - \pi^2\rho + 2\pi\rho|\pi - \theta|), \quad 0 \le \theta < 2\pi. \tag{4.4}$$

The distribution is so-called for the shape of the pdf with the graph of a piecewise linear function. The triangular distribution is symmetric about π. It reduces to the uniform distribution if $\rho = 0$, and for $-4/\pi^2 \le \rho < 0$, (4.4) is increasing for $0 < \theta < \pi$ and decreasing for $\pi < \theta < 2\pi$, i.e. the distribution is unimodal with the mode at π. For $0 < \rho \le 4/\pi^2$, (4.4) is decreasing for $0 < \theta < \pi$ and increasing for $\pi < \theta < 2\pi$, i.e. the distribution is unimodal with the mode at zero. It is possible to restrict ρ to be nonnegative if one considers a location parameter.

Let Θ have pdf (4.4). Then the odd cosine moments are

$$E[\cos\{(2p-1)\Theta\}] = \frac{\rho}{(2p-1)^2}, \quad p = 1, 2, 3, \ldots,$$

and the even cosine moments are zero since

$$\int_0^{2\pi} |\pi - \theta| \cos(q\theta)d\theta = \frac{1 - 2\cos(\pi q) + \cos(2\pi q) + \pi q \sin(2\pi q)}{q^2}.$$

All the sine moments are zero since

$$\int_0^{2\pi} |\pi - \theta| \sin(q\theta)d\theta = \frac{\pi q - \pi q \cos(2\pi q) - \sin(\pi q) + \sin(2\pi q)}{q^2}.$$

Thus, for $\rho < 0$, the mean direction is π and the mean resultant length is $-\rho$, and for $\rho > 0$, the mean direction is zero and the mean resultant length is ρ.

4.3.2 Mitsuyasu Distribution

A distribution (Mitsuyasu et al., 1975), originally proposed by Longuet-Higgins et al. (1963),

$$f(\theta) = \frac{2^{2s-1}\Gamma^2(s+1)}{\pi\Gamma(2s+1)} \left|\cos\left\{\frac{1}{2}(\theta-\mu)\right\}\right|^{2s}, \quad 0 \le \theta < 2\pi, \tag{4.5}$$

was also fitted to the above angular distributions. Using $\cos^2(\theta/2) = (1+\cos\theta)/2$ and the duplication formula, (B.2), of $\Gamma(2z)$, the distribution is represented by

$$f(\theta) = \frac{\Gamma(s+1)}{2^{s+1}\sqrt{\pi}\,\Gamma(s+1/2)}\{1+\cos(\theta-\mu)\}^s, \tag{4.6}$$

for $s > -1/2$. This reduces to the Cartwright distribution when $s = n$ for $n = 0, 1, 2, \ldots$.

4.3.3 von Mises Distribution

4.3.3.1 Definition and Properties

The von Mises distribution (von Mises, 1918; Langevin, 1905) with parameters $\mu\,(0 \le \mu < 2\pi)$ and $\kappa\,(\kappa \ge 0)$ has pdf

$$f(\theta) = \frac{1}{2\pi I_0(\kappa)} e^{\kappa\cos(\theta-\mu)}, \quad 0 \le \theta < 2\pi, \tag{4.7}$$

and it is denoted by $\mathrm{VM}(\mu,\kappa)$. The normalizing constant is expressed in terms of $I_0(\cdot)$, the modified Bessel function of the first kind and order zero, with $I_r(\cdot)$ as general order r defined by (B.23). We restrict to $\kappa \ge 0$ because for $\kappa < 0$, $\mathrm{VM}(\mu,\kappa) \equiv \mathrm{VM}(\mu \pm \pi, -\kappa)$ since $I_0(\kappa) = I_0(-\kappa)$.

Though the genesis of the von Mises distribution lies in ad hoc derivation by von Mises as motivated by the deviations in the measurements of atomic weights from integral values, we will see below that it also admits of derivations using multiple techniques, e.g. maximum entropy, mixture, and conditioning described in Chapter 3. Further, it can be identified (SenGupta and Roy, 2011) as a regular dispersion model (for definitions and details, see Jørgensen, 1997) with μ as its location, $1/\kappa$ as its dispersion, and unity as its variance function.

If a random variable Θ has pdf (4.7), the pth trigonometric moment is given by

$$\begin{aligned} E(e^{ip\Theta}) &= \int_0^{2\pi} e^{ip\theta}\frac{1}{2\pi I_0(\kappa)}e^{\kappa\cos(\theta-\mu)}d\theta = \int_0^{2\pi} e^{ip(\theta+\mu)}\frac{1}{2\pi I_0(\kappa)}e^{\kappa\cos\theta}d\theta \\ &= e^{ip\mu}\frac{I_p(\kappa)}{I_0(\kappa)}, \end{aligned} \tag{4.8}$$

since

$$\int_0^{2\pi} \sin(p\theta)e^{\kappa\cos\theta}d\theta = 0, \quad p = 0, \pm1, \pm2, \ldots.$$

Hence, the pth cosine and sine moments are respectively

$$\alpha_p = E\{\cos(p\Theta)\} = \frac{I_p(\kappa)}{I_0(\kappa)}\cos(p\mu), \quad \beta_p = E\{\sin(p\Theta)\} = \frac{I_p(\kappa)}{I_0(\kappa)}\sin(p\mu).$$

From these, the pdf (4.7) of the von Mises distribution is expressible as

$$f(\theta) = \frac{1}{2\pi}\left[1 + 2\sum_{p=1}^{\infty}\frac{I_p(\kappa)}{I_0(\kappa)}\cos\{p(\theta-\mu)\}\right] \tag{4.9}$$

by using the Fourier series (see Eq. (4.1) and Section 3.3). The Fourier series (4.9) is an absolutely convergent series since the pdf of the von Mises distribution satisfies the sufficient condition given in Section 3.3 for absolute convergence of the Fourier series. This means that the von Mises distribution may be regarded as the wrapped distribution corresponding to the distribution on the real line whose characteristic function is $I_{|t|}(\kappa)/I_0(\kappa)$. However, such a distribution is not commonly known. Further, this wrapped representation does not lend itself to be identified as a member of either of the well-known families of wrapped distributions: the wrapped stable or the wrapped Linnik.

Hartman (1976) established that the von Mises distribution $\mathrm{VM}(0,\kappa)$ has the property of infinite divisibility, i.e. $\mathrm{VM}(0,\kappa)$ is a mixture of wrapped normal distributions $\mathrm{WN}(\cdot)$ (mod 2π):

$$\mathrm{VM}(0,\kappa) = \int_0^{\infty}\mathrm{WN}(\lambda)\mu_\kappa(d\lambda),$$

from general results on linear differential equations and from the existence of a distribution μ_κ on $\mathbb{R}^+$ characterized by its Laplace transform $\lambda \to I_{\sqrt{2\lambda}}(\kappa)/I_0(\kappa)$. Kent (1977) used the Hartman and Watson (1974) mixing distribution for the proof of the infinite divisibility. See a note (Gruet, 2000) on hyperbolic von Mises distributions as a related topic.

The mean direction and mean resultant length of $\mathrm{VM}(\mu,\kappa)$ are μ and $\rho = I_1(\kappa)/I_0(\kappa)$, respectively, since $E(e^{i\Theta}) = e^{i\mu}\{I_1(\kappa)/I_0(\kappa)\}$ from (4.8). Thus, the circular variance is given by $\nu = 1-\rho = 1 - I_1(\kappa)/I_0(\kappa)$. If an alternative circular variance (Equation (2.4)) $\mathrm{Var}(\Theta) = 1-\rho^2$ is used, its value is $\mathrm{Var}(\Theta) = 1 - \{I_1(\kappa)/I_0(\kappa)\}^2$. The pth trigonometric moment about the mean direction μ is clearly

$$E\left\{e^{ip(\Theta-\mu)}\right\} = \overline{\alpha}_p + i\overline{\beta}_p = \frac{I_p(\kappa)}{I_0(\kappa)},$$

from (4.8). Hence, the pth central cosine and sine moments are respectively $\overline{\alpha}_p = E[\cos\{p(\Theta-\mu)\}] = I_p(\kappa)/I_0(\kappa)$ and $\overline{\beta}_p = E[\sin\{p(\Theta-\mu)\}] = 0$.

The von Mises distribution $\mathrm{VM}(\mu,\kappa)$ reduces to the circular uniform distribution (Section 4.2) if $\kappa = 0$. The larger κ is, the more the distribution is concentrated around the mean direction. Thus, κ controls concentration of the distribution and is called the concentration parameter. From (B.25), the von Mises distribution degenerates a distribution concentrated at the only one point μ if $\kappa \to \infty$. For $\kappa > 0$, VM(μ, κ) is symmetric about the value $\theta = \mu$, and since

$$f'(\theta) = -\kappa \sin(\theta - \mu) f(\theta),$$

the equation $f'(\theta) = 0$ has its solutions $\theta = \mu,\ \mu + \pi \pmod{2\pi}$ with $f(\mu) > f(\mu + \pi)$, it is unimodal and has mode at μ (antimode at $\mu + \pi$). When $\mu = 0$ and $0 < \kappa < \infty$, the pdf of VM(μ, κ) has two inflection points which satisfy the equation

$$\cos\theta = -\frac{1}{2\kappa} + \sqrt{1 + \frac{1}{4\kappa^2}}, \quad 0 \le \theta < 2\pi.$$

Figure 4.3 shows linear plots of the von Mises distributions with the concentration parameters $\kappa = 1, 5$ on the interval (a) $[-\pi, \pi)$ and (b) $[0, 2\pi)$. The mean direction is assumed to be zero (radian). Figure 4.4 shows circular plots of the von Mises distributions with same parameters.

There is one-to-one correspondence between ρ and κ because $A(\kappa) = I_1(\kappa)/I_0(\kappa)$ is a monotone increasing function of κ with $A(0) = 0$ and $A(\kappa) \to 1$ as $\kappa \to \infty$. In fact, for the pdf $g(\theta)$ of VM$(0, \kappa)$, the Fisher information function leads to

$$E\left[\left\{\frac{\partial}{\partial\kappa} \log g(\Theta)\right\}^2\right] = -E\left\{\frac{\partial^2}{\partial\kappa^2} \log g(\Theta)\right\} = A'(\kappa) > 0.$$

Moreover, it is clear that $A(0) = 0$, since $I_1(0) = 0$ and $I_0(0) = 1$, and we have from l'Hôpital's rule that

$$\ell = \lim_{\kappa\to\infty} \frac{I_1(\kappa)}{I_0(\kappa)} = \lim_{\kappa\to\infty} \frac{I_1'(\kappa)}{I_0'(\kappa)} = \lim_{\kappa\to\infty} \frac{I_0(\kappa) - I_1(\kappa)/\kappa}{I_1(\kappa)} = \frac{1}{\ell},$$

and $\ell = 1$ since $\ell > 0$. The result also follows from an asymptotic expansion given by (B.29).

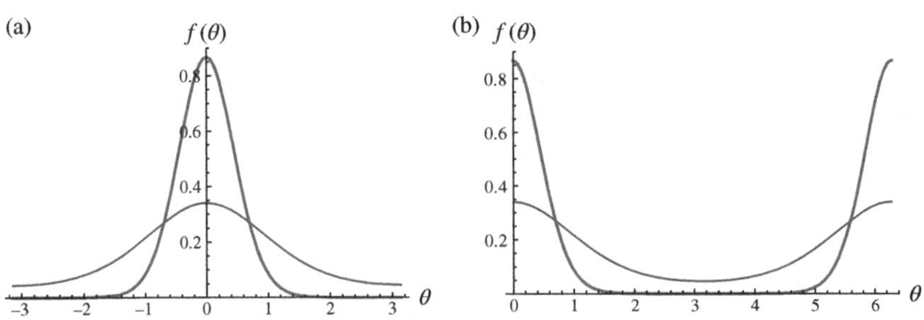

Figure 4.3 Linear plots of von Mises distributions ($\mu = 0$; $\kappa = 1$ (thin), $\kappa = 5$ (thick), (a) $-\pi \le \theta < \pi$, (b) $0 \le \theta < 2\pi$).

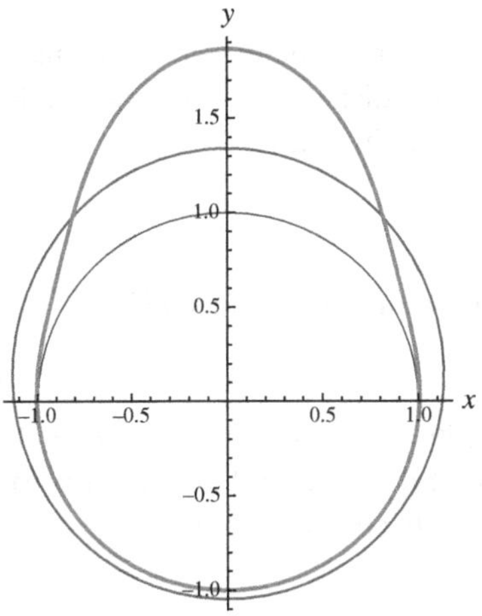

Figure 4.4 Circular plots of von Mises distributions ($\mu = 0$; $\kappa = 1$ (thin), $\kappa = 5$ (thick)).

Since the modified Bessel function of the first kind and order zero has a series expansion given by (B.23), the pdf of $\mathrm{VM}(\mu, \kappa)$ approximates that of a cardioid distribution with parameters μ and $\kappa/2$

$$f(\theta) = \frac{1}{2\pi I_0(\kappa)} e^{\kappa \cos(\theta-\mu)} \approx \frac{1}{2\pi} \{1 + \kappa \cos(\theta - \mu)\}$$

for small κ. For large κ, $\mathrm{VM}(\mu, \kappa)$ can be approximated by a normal distribution. More precisely, if $\Theta \sim \mathrm{VM}(\mu, \kappa)$, then $\sqrt{\kappa}(\Theta - \mu) \stackrel{\mathrm{d}}{\rightarrow} U \sim N(0, 1)$ as $\kappa \to \infty$ from $I_0(\kappa) \sim \exp(\kappa)/\sqrt{2\pi\kappa}$ (B.29).

The distribution function of $\mathrm{VM}(0, \kappa)$ is expressed by the following way. Since the Taylor expansion of the pdf is

$$f(\theta) = \frac{1}{2\pi I_0(\kappa)} \sum_{p=0}^{\infty} \frac{\kappa^p \cos^p \theta}{p!},$$

we have an expression of the distribution function, from term-by-term integration,

$$F(\theta) = \frac{1}{4\pi I_0(\kappa)} \sum_{p=0}^{\infty} \frac{\kappa^p}{p!} B_{\sin^2\theta} \left(\frac{1}{2}, \frac{p+1}{2} \right), \quad 0 \le \theta < 2\pi,$$

where $B_z(\cdot, \cdot)$ denotes the incomplete beta function given by (B.7). An alternative expression of the distribution function is obtained from (4.9), the Fourier expansion of the von Mises pdf, as

$$F(\theta) = \frac{\theta}{2\pi} + \frac{1}{\pi I_0(\kappa)} \sum_{p=1}^{\infty} \frac{I_p(\kappa) \sin(p\theta)}{p}.$$

4.3.3.2 Maximum Entropy

It is known that the normal distribution on the real line is obtained by the maximum entropy method. Likewise, the von Mises distribution on the circle is obtained by the maximum entropy method. Indeed, in (3.2), if

$$\int_0^{2\pi} \cos\theta f(\theta)d\theta = g_1,$$

for a constant g_1, the maximum entropy distribution in this case has pdf

$$f(\theta) = \exp(a_0 + a_1 \cos\theta),$$

for particular constants a_0 and a_1. Since $\int_0^{2\pi} f(\theta)d\theta = 1$, we have $2\pi e^{a_0} I_0(a_1) = 1$, where $I_0(\cdot)$ denotes the modified Bessel function of the first kind and order zero. As a result, the distribution must be a von Mises distribution with pdf

$$f(\theta) = \frac{1}{2\pi I_0(a_1)} e^{a_1 \cos\theta}, \quad 0 \le \theta < 2\pi.$$

The discussion above means that a specified cosine moment gives a von Mises distribution as a maximum entropy distribution. In addition, $g_1 = \int_0^{2\pi} \cos\theta f(\theta) d\theta = I_1(a_1)/I_0(a_1)$.

If, in (3.2), we give the restrictions on the first cosine and sine moments

$$E(\cos\Theta) = \frac{I_1(\kappa)}{I_0(\kappa)} \cos\mu, \; E(\sin\Theta) = \frac{I_1(\kappa)}{I_0(\kappa)} \sin\mu,$$

for $\kappa \ge 0$ and $0 \le \mu < 2\pi$, then the maximum entropy distribution is $\mathrm{VM}(\mu, \kappa)$.

4.3.3.3 Maximum-Likelihood Characterization

The normal distribution $N(\mu, 1)$ on the line is characterized by the sample mean, the maximum-likelihood (ML) estimate of μ. Likewise, the von Mises distribution $\mathrm{VM}(\mu, 1)$ on the circle is characterized by the sample mean direction, the maximum likelihood estimate of μ (von Mises, 1918).

Let $\theta_1, \dots, \theta_n$ $(n \ge 1)$ denote a random sample drawn from $\mathrm{VM}(\mu, 1)$. The log-likelihood function is

$$L(\mu) = -n \log\left\{2\pi I_0(1)\right\} + \sum_{j=1}^{n} \cos(\theta_j - \mu). \tag{4.10}$$

Differentiating (4.10) with respect to μ, we have the likelihood equation

$$\frac{\partial}{\partial\mu} L(\mu) = \sum_{j=1}^{n} \sin(\theta_j - \mu) = 0.$$

The maximum likelihood estimate, $\hat{\mu}$, of μ satisfies $\tan\hat{\mu} = S/C$, where $C = \sum_{j=1}^{n} \cos\theta_j$ and $S = \sum_{j=1}^{n} \sin\theta_j$. We notice that $\hat{\mu}$ is the sample mean direction (Section 2.5). If $R = \sqrt{C^2 + S^2} = 0$, $\hat{\mu}$ is not defined.

Conversely, let $g(\theta - \mu)$ be a population pdf on the circle and assume that the maximam likelihood estimate of μ based on a random sample $\theta_1, \ldots, \theta_n$ $(n \geq 1)$ drawn from the population is the sample mean direction $\hat{\mu}$. Then

$$\sum_{j=1}^{n} \frac{g'(\theta_j - \hat{\mu})}{g(\theta_j - \hat{\mu})} = 0,$$

and

$$\sum_{j=1}^{n} \sin(\theta_j - \hat{\mu}) = 0.$$

Since these two equations are identical for each n, we have

$$\frac{g'(\theta - \hat{\mu})}{g(\theta - \hat{\mu})} = c_0 \sin(\theta - \hat{\mu}),$$

where c_0 is a real constant. Thus, replacing $\hat{\mu}$ by μ, we have

$$g(\theta - \mu) = c_1\, e^{\cos(\theta-\mu)},$$

with $c_1 = 1/\{2\pi I_0(1)\}$, which is the pdf of VM$(\mu, 1)$.

4.3.3.4 Distribution of the Sum

The von Mises distribution is often called a circular normal distribution (Gumbel et al., 1953) because, as shown before, it can be derived from a normal distribution in Euclidean space and its genesis is similar to that of normal distributions. However, the von Mises distribution does not necessarily have good and important properties of normal distributions. One of the great differences between normal and von Mises distributions is the fact that the von Mises distribution does not have the reproductive property.

Let Θ_1 and Θ_2 be independently distributed and have von Mises distributions VM(μ_1, κ_1) and VM(μ_2, κ_2). Then the pdf of the sum $\Theta = \Theta_1 + \Theta_2$ is calculated by the convolution

$$h(\theta) = \frac{1}{4\pi^2 I_0(\kappa_1) I_0(\kappa_2)} \int_0^{2\pi} \exp\left\{\kappa_1 \cos(\xi - \mu_1) + \kappa_2 \cos(\theta - \xi - \mu_2)\right\} d\xi.$$

The exponential part of the integrand is expressible as

$$\begin{aligned}
&\kappa_1 \cos(\xi - \mu_1) + \kappa_2 \cos(\theta - \xi - \mu_2) \\
&= \sqrt{\kappa_1^2 + \kappa_2^2 + 2\kappa_1\kappa_2 \cos\{\theta - (\mu_1 + \mu_2)\}}\, \cos(\xi - \alpha), \\
&\qquad \tan\alpha = \frac{\kappa_1 \sin\mu_1 + \kappa_2 \sin(\theta - \mu_2)}{\kappa_1 \cos\mu_1 + \kappa_2 \cos(\theta - \mu_2)}
\end{aligned}$$

and we have

$$h(\theta) = \frac{1}{2\pi I_0(\kappa_1) I_0(\kappa_2)} I_0\left(\left[\kappa_1^2 + \kappa_2^2 + 2\kappa_1\kappa_2 \cos\left\{\theta - (\mu_1 + \mu_2)\right\}\right]^{1/2}\right).$$

This means that the sum of von Mises random variables is not distributed as a von Mises distribution even if $\kappa_1 = \kappa_2$ and $\mu_1 = \mu_2$.

4.3.3.5 Measures of Similarity or Dissimilarity

If a random variable Θ has VM(μ_1, κ_1) with pdf $f(\theta)$ and if $g(\theta)$ is pdf of VM(μ_2, κ_2), then the Hellinger divergence is given by

$$D_{\mathrm{H}}(f,g) = 2\left\{1 - \frac{I_0\left(\frac{1}{2}\sqrt{\kappa_1^2 + \kappa_2^2 + 2\kappa_1\kappa_2\cos(\mu_1 - \mu_2)}\right)}{\sqrt{I_0(\kappa_1)I_0(\kappa_2)}}\right\},$$

since

$$\begin{aligned}&\int_0^{2\pi} \sqrt{f(\theta)g(\theta)}d\theta \\ &= \frac{1}{2\pi\sqrt{I_0(\kappa_1)I_0(\kappa_2)}}\int_0^{2\pi} \exp\left[\frac{1}{2}\{\kappa_1\cos(\theta - \mu_1) + \kappa_2\cos(\theta - \mu_2)\}\right] d\theta \\ &= \frac{I_0\left(\frac{1}{2}\sqrt{\kappa_1^2 + \kappa_2^2 + 2\kappa_1\kappa_2\cos(\mu_1 - \mu_2)}\right)}{\sqrt{I_0(\kappa_1)I_0(\kappa_2)}}.\end{aligned}$$

The Kullback–Leibler divergence is

$$D_{\mathrm{KL}}(f,g) = \log\left\{\frac{I_0(\kappa_2)}{I_0(\kappa_1)}\right\} + \{\kappa_1 - \kappa_2\cos(\mu_1 - \mu_2)\}\frac{I_1(\kappa_1)}{I_0(\kappa_1)}.$$

In particular, if $\kappa_1 = \kappa_2 (= \kappa)$, the Kullback–Leibler divergence is expressed by

$$D_{\mathrm{KL}}(f,g) = \kappa\{1 - \cos(\mu_1 - \mu_2)\}\frac{I_1(\kappa)}{I_0(\kappa)},$$

in terms of $1 - \cos(\mu_1 - \mu_2)$, a half of the square of the distance or chord length between two points on the unit circle, corresponding to angles μ_1 and μ_2, and $I_1(\kappa)/I_0(\kappa)$, the mean resultant length. If $\mu_1 = \mu_2$ and $\kappa_1 = \kappa_2$, then $D_{\mathrm{H}}(f,g) = D_{\mathrm{KL}}(f,g) = 0$.

Minimum and maximum divergences of $D_{\mathrm{H}}(f(\theta), g(\theta+\xi))$ and $D_{\mathrm{KL}}(f(\theta), g(\theta+\xi))$ for ξ $(0 \le \xi < 2\pi)$ are

$$\begin{aligned}&\min_{\xi} D_{\mathrm{H}}(f(\theta), g(\theta+\xi)) \\ &= \min_{\xi} 2\left\{1 - \frac{I_0\left(\frac{1}{2}\sqrt{\kappa_1^2 + \kappa_2^2 + 2\kappa_1\kappa_2\cos(\xi + \mu_1 - \mu_2)}\right)}{\sqrt{I_0(\kappa_1)I_0(\kappa_2)}}\right\}\end{aligned}$$

$$= 2\left\{1 - \frac{I_0\left(\frac{1}{2}(\kappa_1 + \kappa_2)\right)}{\sqrt{I_0(\kappa_1)I_0(\kappa_2)}}\right\}, \quad \xi = \mu_2 - \mu_1 \,(\mathrm{mod}\, 2\pi),$$

$$\max_{\xi} D_{\mathrm{H}}(f(\theta), g(\theta + \xi))$$

$$= 2\left\{1 - \frac{I_0\left(\frac{1}{2}|\kappa_1 - \kappa_2|\right)}{\sqrt{I_0(\kappa_1)I_0(\kappa_2)}}\right\}, \quad \xi = \mu_2 - \mu_1 + \pi \,(\mathrm{mod}\, 2\pi),$$

$$\min_{\xi} D_{\mathrm{KL}}(f(\theta), g(\theta + \xi))$$

$$= \min_{\xi}\left[\log\left(\frac{I_0(\kappa_2)}{I_0(\kappa_1)}\right) + \{\kappa_1 - \kappa_2\cos(\xi + \mu_1 - \mu_2)\}\frac{I_1(\kappa_1)}{I_0(\kappa_1)}\right]$$

$$= \log\left\{\frac{I_0(\kappa_2)}{I_0(\kappa_1)}\right\} + (\kappa_1 - \kappa_2)\frac{I_1(\kappa_1)}{I_0(\kappa_1)}, \quad \xi = \mu_2 - \mu_1 \,(\mathrm{mod}\, 2\pi),$$

$$\max_{\xi} D_{\mathrm{KL}}(f(\theta), g(\theta + \xi))$$

$$= \log\left\{\frac{I_0(\kappa_2)}{I_0(\kappa_1)}\right\} + (\kappa_1 + \kappa_2)\frac{I_1(\kappa_1)}{I_0(\kappa_1)}, \quad \xi = \mu_2 - \mu_1 + \pi \,(\mathrm{mod}\, 2\pi).$$

4.3.3.6 Derivation by the Conditioning Method

Downs (1966) showed that the conditioning method (Section 3.5) provides a von Mises distribution. Let a random vector $(X_1, X_2)'$ follow a bivariate normal distribution $N_2(\boldsymbol{\eta}, I_2)$ with mean vector $\boldsymbol{\eta} = (\eta_1, \eta_2)'$ and identity variance–covariance matrix I_2. We make the polar transformation $X_1 = R\cos\Theta$, $X_2 = R\sin\Theta$ and $\eta_1 = \rho\cos\tau$, $\eta_2 = \rho\sin\tau$. Then the pdf of $(R, \Theta)'$ is given by

$$g(r, \theta) = \frac{r}{2\pi} e^{-\{r^2 - 2r\rho\cos(\theta - \tau) + \rho^2\}/2}, \quad r \geq 0,\ 0 \leq \theta < 2\pi,$$

and the conditional distribution of Θ given $R = r\,(> 0)$ is a von Mises distribution $\mathrm{VM}(\tau, \rho r)$ with pdf

$$f(\theta | r) = \frac{1}{2\pi I_0(\rho r)} e^{\rho r\cos(\theta - \tau)}, \quad 0 \leq \theta < 2\pi.$$

If we start with $N_2(\boldsymbol{\mu}, I_2/\kappa)$ with mean vector $\boldsymbol{\mu} = (\cos\mu, \sin\mu)'$ and variance–covariance matrix $I_2/\kappa\ (\kappa > 0)$, then the conditional distribution of Θ given $R = 1$ results in $\mathrm{VM}(\mu, \kappa)$ as an angular distribution.

4.3.3.7 Bayesian Perspective

Conjugate distributions mean that prior and posterior distributions belong to the same class of distributions. The von Mises distribution $\mathrm{VM}(\mu, \kappa_0)$ with known $\kappa_0\,(> 0)$ is a member of the curved exponential family and, as shown in Mardia and El-Atoum (1976), it has a conjugate prior.

In fact, let $\theta_1, \dots, \theta_n$ be a random sample drawn from VM(μ, κ_0) with known $\kappa_0\,(> 0)$. Then the likelihood function is

$$p(\theta|\mu) = \frac{1}{\{2\pi I_0(\kappa_0)\}^n} e^{\kappa_0 \sum_{j=1}^{n} \cos(\theta_j - \mu)}, \quad 0 \le \theta_j < 2\pi,\ 0 \le \mu < 2\pi,$$

where $\boldsymbol{\theta} = (\theta_1, \dots, \theta_n)'$. If the prior distribution of μ is

$$\pi(\mu|m, \delta) = \frac{1}{2\pi I_0(\delta)} e^{\delta \cos(\mu - m)}, \quad \delta \ge 0,\ 0 \le m < 2\pi,$$

i.e. the von Mises distribution VM(m, δ) with hyper-parameters m and δ, then the resulting posterior distribution is expressed by

$$p(\mu|\boldsymbol{\theta}) = \frac{p(\boldsymbol{\theta}|\mu)\pi(\mu|m, \delta)}{\int_0^{2\pi} p(\boldsymbol{\theta}|\mu)\pi(\mu|m, \delta)d\mu} = \frac{1}{2\pi I_0(\delta_*)} e^{\delta_* \cos(\mu - \mu_*)},$$

i.e. VM(μ_*, δ_*), where

$$\delta_* = \sqrt{(\kappa_0 R)^2 + \delta^2 + 2\kappa_0\delta \sum_{j=1}^{n} \cos(\theta_j - m)},$$

$$R^2 = C^2 + S^2,\ C = \sum_{j=1}^{n} \cos\theta_j,\ S = \sum_{j=1}^{n} \sin\theta_j,$$

and the posterior mode μ_* depends upon $\boldsymbol{\theta}, m, \delta$ and satisfies

$$\delta_* \cos\mu_* = \kappa_0 C + \delta\cos m,\ \delta_* \sin\mu_* = \kappa_0 S + \delta \sin m.$$

Thus, the posterior distribution is in the same class as the prior distribution.

Let $L(\hat{\mu} - \mu) = 1 - \cos(\hat{\mu} - \mu)$ denote a loss function for the estimator $\hat{\mu}$ of a parameter μ. Then we have the following relationship (Ye and Ohnishi, 2006):

$$\int_0^{2\pi} \{L(\hat{\mu} - \mu) - L(\mu_* - \mu)\}\pi(\mu|\mu_*, \delta_*)d\mu = \frac{I_1(\delta_*)}{I_0(\delta_*)} L(\hat{\mu} - \mu_*) \tag{4.11}$$

since the first central cosine and sine moments of VM(μ_*, δ_*) are respectively $\overline{\alpha}_1 = I_1(\delta_*)/I_0(\delta_*)$ and $\overline{\beta}_1 = 0$. The right side of (4.11) is expressed by $D_{\text{KL}}(\pi(\mu|\mu_*, \delta_*), \pi(\mu|\hat{\mu}, \delta_*))/\delta_*$ in terms of the Kulback–Leibler divergence.

Jeffreys' non-informative prior is defined by

$$p(\xi) \propto |I(\xi)|^{1/2},$$

where $\xi = (\mu, \kappa)'$ and

$$I(\xi) = -nE\left\{\frac{\partial^2 \log f(\Theta|\xi)}{\partial\xi\partial\xi'}\right\},$$

for $\Theta \sim$ VM(μ, κ), is the expected Fisher information matrix about ξ. We can calculate $I(\xi)$ as

$$I(\xi) = n\begin{pmatrix} \kappa A(\kappa) & 0 \\ 0 & A'(\kappa) \end{pmatrix}.$$

Here $A(\kappa) = I_1(\kappa)/I_0(\kappa)$ and $A'(\kappa) = 1 - A(\kappa)/\kappa - A^2(\kappa)$. Thus, we have

$$|I(\xi)|^{1/2} = n\{\kappa A(\kappa)A'(\kappa)\}^{1/2} = n[A(\kappa)\{\kappa - A(\kappa) - \kappa A^2(\kappa)\}]^{1/2},$$

which is independent of μ.

Remark 4.1 Noting the popularity of von Mises or circular normal distribution in applications, and motivated by the success of Box and Cox transformation to near normality for linear data, Johnson (2022) has suggested several interesting and useful transformations to near circular normality for circular data. He has considered computational aspects and theoretical properties associated with these transformations.

4.3.3.8 Mixtures of von Mises Distributions

Mixture distributions can be easily constructed with two or more von Mises components, $\text{VM}(\mu_i, \kappa_i), i = 1, \ldots, k, k \geq 2$. In case μ_is are all equal, say equal to μ, these will lead to symmetric distributions around μ. However, the choice of the mixing proportions can lead to a variety of inflection points and kurtosis values for these distributions.

Bartels (1984) proposed a "reflective" ($\mu_1 = -\mu_2$) bidirectional von Mises mixture distribution given by the pdf

$$\frac{1}{2}\left\{\frac{1}{2\pi I_0(\kappa)} e^{\kappa\cos(\theta-\mu)} + \frac{1}{2\pi I_0(\kappa)} e^{\kappa\cos(\theta+\mu)}\right\}.$$

In general, mixtures of von Mises distributions will lead to multimodal distributions, depending on their parameter configurations. The number and position of modes of a mixture of two von Mises distributions are examined by Mardia and Sutton (1975). Without loss of generality, suppose that one has $\text{VM}(0, \kappa_1)$ with pdf $f_1(\theta) = e^{\kappa_1\cos\theta}/\{2\pi I_0(\kappa_1)\}$ and the other $\text{VM}(\mu, \kappa_2)$ with pdf $f_2(\theta) = e^{\kappa_2\cos(\theta-\mu)}/\{2\pi I_0(\kappa_2)\}$, then the mixture is written

$$f(\theta) = pf_1(\theta) + (1-p)f_2(\theta), \quad 0 \leq \theta < 2\pi, \tag{4.12}$$

where $0 \leq p \leq 1$, $0 \leq \mu \leq \pi$, $\kappa_1 > 0$ and $\kappa_2 > 0$. The modes and antimodes are provided by the solutions of the equation

$$p\kappa_1 f_1(\theta)\sin\theta + (1-p)\kappa_2 f_2(\theta)\sin(\theta-\mu) = 0.$$

Table 4.1 shows a summary of the behavior of (4.12). The quantity $\tilde{\theta}$ in the table is the unique solution of the cubic equation

$$A\cot^3\theta + B\cot^2\theta + C\cot\theta + D = 0, \quad 0 < \theta < \mu < \pi,$$

where $A = \kappa_2\sin^2\mu$, $B = 2\kappa_1\sin\mu - 2\kappa_2\sin(2\mu)$, $C = 2\kappa_2\cos(2\mu) - 3\kappa_1\cos\mu + \kappa_2\cos^2\mu$, and $D = -B/2$.

Table 4.1 Number and position of modes and antimodes (U: Unimodal, B: Bimodal).

Case	μ	Range of p	Type	Modes, M	Antimodes, A
(i)	0	$0 \le p \le 1$	U	0	π
(ii)	π	$p_1 \le p \le p_2$	B	$0, \pi$	$0 < A^* < \pi < A^* < 2\pi$
		$0 \le p < p_1$	U	π	0
		$p_2 < p \le 1$	U	0	π
(iiia)	$0 < \mu < \pi$ $\sin\mu > h(\tilde{\theta})$	$0 \le p \le 1$	U	$0 < M < \mu$	$\pi < A < \mu + \pi$
(iiib)	$0 < \mu < \pi$	$p'_1 \le p \le p'_2$	B	$0 < M < \theta_1 < A < \theta_2 < M \le \mu < \pi < A < \mu + \pi$	
	$\sin\mu \le h(\tilde{\theta})$	$0 \le p < p'_1$	U	$\theta_2 < M < \mu$	$\pi < A < \mu + \pi$
		$p'_2 < p \le 1$	U	$0 < M < \theta_1$	$\pi < A < \mu + \pi$

Case	Notation
(ii)	$p_1 = \{1 + \kappa^* \exp(\kappa_1 + \kappa_2)\}^{-1}$, $p_2 = \{1 + \kappa^* \exp(-\kappa_1 - \kappa_2)\}^{-1}$, $\kappa^* = \{\kappa_1 I_0(\kappa_2)\}/\{\kappa_2 I_0(\kappa_1)\}$, $A^* = \cos^{-1}[\log\{(1-p)/(p\kappa^*)\}/(\kappa_1 + \kappa_2)]$
(iiia)	$h(\theta) = \sin\theta \sin(\theta - \mu)\{\kappa_2 \sin(\theta - \mu) - \kappa_1 \sin\theta\}$; $\tilde{\theta}$ maximizes $h(\theta)\,(0 < \theta < \mu)$
(iiib)	$p'_j = \{1 - \kappa^*/u(\theta_j)\}^{-1}\,(j = 1, 2)$, $u(\theta) = \{\sin(\theta - \mu)/\sin\theta\}\exp\{\kappa_2 \cos(\theta - \mu) - \kappa_1 \cos\theta\}$, θ_1 and θ_2 are the two solutions of the equation $\sin\mu - h(\theta) = 0$ $(0 < \theta < \mu,\ 0 < \theta_1 < \theta_2 < \mu)$

4.3.3.9 Multimodal von Mises Distribution

Let m be an integer greater than or equal to 2. Then, it is possible to obtain a multi-modal von Mises distribution with pdf

$$f(\theta) = \frac{1}{2\pi I_0(\kappa)} e^{\kappa\cos\{m(\theta-\mu)\}}, \quad 0 \le \theta < 2\pi,\ 0 \le \mu < 2\pi/m,\ \kappa \ge 0. \tag{4.13}$$

The numbers of modes and antimodes are m, respectively, and the positions of them are equidistant.

The m-modal von Mises distribution is generated using the method of maximizing entropy (Section 3.2.1). By taking g_1 and g_2 consistent with the expectations of $h_1(\theta) = \cos(m\theta)$ and $h_2(\theta) = \sin(m\theta)$, respectively, i.e. by specifying the mth harmonics of θ, (3.2) yields (4.13).

4.3.3.10 von Mises Distribution Restricted to an Arc

If one restricts the pdf (4.13) to that on the interval $[0, 2\pi/m)$, then the integral value of $f(\theta)$ over $[0, 2\pi/m)$ is $1/m$. Thus, we have a von Mises distribution restricted to part of the circumference with pdf

$$f(\theta) = \frac{m}{2\pi I_0(\kappa)} e^{\kappa \cos\{m(\theta-\mu)\}}, \quad 0 \le \theta < 2\pi/m, \, 0 \le \mu < 2\pi/m, \, \kappa \ge 0.$$

In particular, when $m = 2$, the distribution shows an axial distribution on the semicircle.

4.3.4 Papakonstantinou and Batschelet Distributions

The power cardioid family described in Section 4.6.3 contains the cardioid, von Mises and wrapped Cauchy distributions as special cases. An alternative type of extension of cardioid and von Mises distributions dates back to Papakonstantinou (1979) and Batschelet (1981).

4.3.4.1 Papakonstantinou Distribution

Abe et al. (2009) investigate the Papakonstantinou distribution with pdf

$$f(\theta) = \frac{1}{2\pi\{1 - \kappa J_1(\nu)\}} [1 + \kappa \cos\{(\theta - \mu) + \nu \sin(\theta - \mu)\}], \quad -\pi \le \theta < \pi. \tag{4.14}$$

Here $\mu\,(-\pi \le \mu < \pi)$ is the mean direction, $\kappa\,(0 \le \kappa \le 1)$ is a concentration parameter, $\nu\,(-\infty < \nu < \infty)$ controls both the shape and modality of the distribution, and $J_n(\cdot)$ denotes the Bessel function of the first kind and order n. An integral representation of the Bessel function is (B.18). It is obvious that (4.14) reduces to the cardioid pdf if $\nu = 0$ and the uniform pdf if $\kappa = 0$. The location extension of the distribution proposed by Papakonstantinou (1979, p. 24) has pdf $g(\theta) = \{2\kappa J_1(\nu) - 1\}/(2\pi) + 2\{1 - \kappa J_1(\nu)\}f(\theta)$ different from (4.14) with $|\kappa| \le 1/\{2(1 + |J_1(\nu)|)\}$ and, in order to ensure unimodality, $|\nu| < 1$. This is a mixture (or internal–external division) of the distribution with pdf (4.14) and the uniform distribution.

Under the assumption that $0 < \kappa \le 1$, the distribution with pdf (4.14) is unimodal if and only if $|\nu| \le 1$. Figure 4.5 shows that the distribution is more flat-topped than the corresponding cardioid distribution if $-1 \le \nu < 0$ and more sharply peaked if $0 < \nu \le 1$. A formal definition using curvature for the notion of flat-topped and sharply peaked distributions can be seen later on in this section. For $|\nu| > 1$, (4.14) has $(2m + 1)$ modes if ν satisfies the condition $m\pi < h(\nu) \le (m + 1)\pi\ (m = 1, 2, 3, \ldots)$ and $(2|m| + 1)$ modes if ν satisfies the condition $m\pi \le h(\nu) < (m + 1)\pi\ (m = -1, -2, -3, \ldots)$, where $h(\nu) = \cos^{-1}(-1/\nu) + \nu\sqrt{1 - 1/\nu^2}$. Its proof is in Abe et al. (2009).

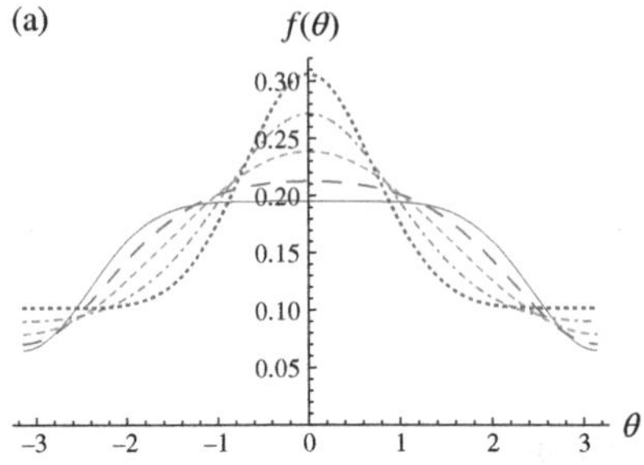

Figure 4.5 Linear plots of the Papakonstantiou distribution for $\nu = -1(0.5)1$, (a) $\kappa = 0.5$ and (b) $\kappa = 1$ with $\nu = -1$ corresponding to the solid line and $\nu = 1$ to the dotted line. The cardioid distribution corresponds to $\nu = 0$.

The function $\cos(\theta + \nu \sin\theta)$ can be represented by the following Fourier series involving Bessel functions of the first kind:

$$\begin{aligned}
&\cos(\theta + \nu\sin\theta)\\
&= -J_1(\nu) + \sum_{s=1}^{\infty}\left[\{J_{2s-2}(\nu) + J_{2s}(\nu)\}\cos\{(2s-1)\theta\} + \{J_{2s-1}(\nu) - J_{2s+1}(\nu)\}\cos(2s\theta)\right]\\
&= -J_1(\nu) + \sum_{s=1}^{\infty}\left[\frac{2(2s-1)}{\nu}J_{2s-1}(\nu)\cos\{(2s-1)\theta\} + 2J'_{2s}(\nu)\cos(2s\theta)\right].
\end{aligned}$$

The first equality follows from $\cos(\theta + \nu\sin\theta) = \cos\theta\cos(\nu\sin\theta) - \sin\theta\sin(\nu\sin\theta)$ and the formulae (B.19) and (B.20), and the second equality from the relationships (B.21) and (B.22). Thus, the pdf (4.14) can be expressed by

$$\begin{aligned}
f(\theta) = \frac{1}{2\pi} &+ \frac{\kappa}{2\pi\{1 - \kappa J_1(\nu)\}}\\
&\times\sum_{s=1}^{\infty}\left[\frac{2(2s-1)}{\nu}J_{2s-1}(\nu)\cos\{(2s-1)(\theta-\mu)\}\right.\\
&\left.+2J'_{2s}(\nu)\cos\{2s(\theta-\mu)\}\right].
\end{aligned} \tag{4.15}$$

Using this expression, when $\mu = 0$ the distribution function of (4.14), for $-\pi \le \theta < \pi$, is given by

$$\begin{aligned}
F(\theta) &= \int_{-\pi}^{\theta} f(t)dt\\
&= \frac{\theta+\pi}{2\pi} + \frac{\kappa}{2\pi\{1-\kappa J_1(\nu)\}}\sum_{s=1}^{\infty}\left[\frac{2J_{2s-1}(\nu)}{\nu}\sin\{(2s-1)\theta\} + \frac{J'_{2s}(\nu)}{s}\sin(2s\theta)\right].
\end{aligned}$$

From (4.15) and the orthogonality of the trigonometric function, for $p = 1, 2, 3, \ldots$ we have the pth cosine moment about the mean direction μ as

$$\overline{\alpha}_p = E[\cos\{p(\Theta - \mu)\}] = \frac{\kappa\{J_{p-1}(\nu) + (-1)^{p-1}J_{p+1}(\nu)\}}{2\{1-\kappa J_1(\nu)\}}.$$

The pth sine moment about the mean direction μ is $\overline{\beta}_p = E[\sin\{p(\Theta - \mu)\}] = 0$. Trivially, the pth trigonometric moment is

$$\phi_p = E(e^{ip\Theta}) = \frac{\kappa\{J_{p-1}(\nu) + (-1)^{p-1}J_{p+1}(\nu)\}}{2\{1 - \kappa J_1(\nu)\}} e^{ip\mu},$$

and the pth cosine and sine moments are

$$\alpha_p = E\{\cos(p\Theta)\} = \frac{\kappa\{J_{p-1}(\nu) + (-1)^{p-1}J_{p+1}(\nu)\}}{2\{1 - \kappa J_1(\nu)\}} \cos(p\mu),$$

and

$$\beta_p = E\{\sin(p\Theta)\} = \frac{\kappa\{J_{p-1}(\nu) + (-1)^{p-1}J_{p+1}(\nu)\}}{2\{1 - \kappa J_1(\nu)\}} \sin(p\mu).$$

The pth mean direction is obtained by $p\mu$ (mod 2π) and the pth mean resultant length $|\overline{\alpha}_p|$.

The (first-order) mean resultant length is $\rho = |\overline{\alpha}_1| = \kappa\{J_0(\nu) + J_2(\nu)\}/[2\{1 - \kappa J_1(\nu)\}]$. For the unimodal case $|\nu| \leq 1$, ρ has range $0 \leq \rho \leq 0.7859$. The lower bound is attained when $\kappa = 0$, independently of ν, and the upper bound when $\kappa = 1$ and $\nu = 1$. Its proof is as follows. Since

$$\frac{\partial \rho}{\partial \kappa} = \frac{J_0(\nu) + J_2(\nu)}{2\{1 - \kappa J_1(\nu)\}^2} > 0,$$

ρ is a monotone increasing function of κ. Therefore,

$$0 \leq \rho = \frac{\kappa\{J_0(\nu) + J_2(\nu)\}}{2\{1 - \kappa J_1(\nu)\}} \leq \frac{J_0(\nu) + J_2(\nu)}{2\{1 - J_1(\nu)\}}.$$

Similarly, $\{J_0(\nu) + J_2(\nu)\}/[2\{1 - J_1(\nu)\}]$ is a monotone increasing function of ν. Thus,

$$0 \leq \rho \leq \frac{J_0(1) + J_2(1)}{2\{1 - J_1(1)\}} = 0.7859.$$

4.3.4.2 Batschelet Distribution

The pdf of the Batschelet distribution is given by

$$f(\theta) = C^{-1} \exp[\kappa \cos\{(\theta - \mu) + \nu \sin(\theta - \mu)\}], \quad -\pi \leq \theta < \pi, \tag{4.16}$$

where $-\pi \leq \mu < \pi$, $\kappa \geq 0$, and $-\infty < \nu < \infty$. Here C is the normalizing constant and is represented by an integral

$$C = \int_{-\pi}^{\pi} \exp\{\kappa \cos(\theta + \nu \sin\theta)\}\, d\theta.$$

The distribution reduces to the von Mises distribution if $\nu = 0$. Modality condition of the Batschelet distribution is the same as that for the Papakonstantinou distribution. Figure 4.6 shows that the distribution is more flat-topped than the corresponding von Mises distribution if $-1 \leq \nu < 0$ and more sharply peaked if $0 < \nu \leq 1$, as in the case of Papakonstantinou distribution.

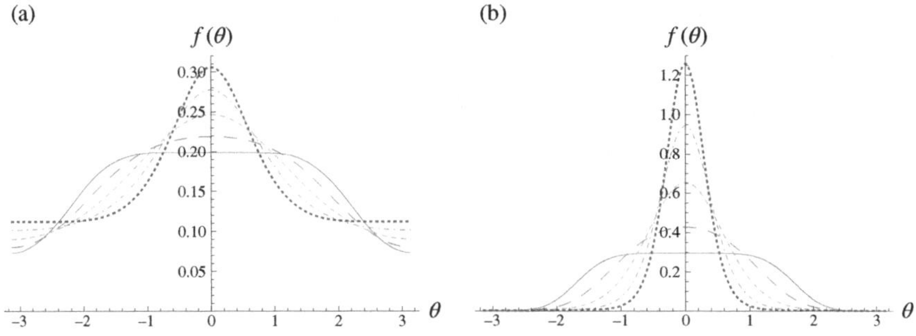

Figure 4.6 Linear plots of Batschelet distribution for $\nu = -1(0.5)1$, (a) $\kappa = 0.5$ and (b) $\kappa = 3$ with $\nu = -1$ corresponding to the solid line and $\nu = 1$ to the dotted line. The von Mises distribution corresponds to $\nu = 0$.

4.3.4.3 A General Form of Symmetric Distributions

A general form of symmetric distributions is proposed by Abe et al. (2013). Let $g(\cos\theta)$ as a function of $\cos\theta\,(-\pi \le \theta < \pi)$ denote a pdf on the circle. Consider a function

$$f_\nu(\theta) = \{c(\nu)\}^{-1} g(\cos(\theta + \nu\sin\theta)), \quad -\pi \le \theta < \pi, \tag{4.17}$$

for $-\infty < \nu < \infty$ with a constant $c(\nu)$ satisfying $\int_{-\pi}^{\pi} f_\nu(\theta)d\theta = 1$, so that $f_\nu(\theta)$ is a pdf on the circle. If $\nu = 0$, then $f_\nu(\theta)$ reduces to the original pdf, i.e. $f_0(\theta) = g(\cos\theta)$, and it is possible to write $f_\nu(\theta) = \{c(\nu)\}^{-1} f_0(\theta + \nu\sin\theta)$ with

$$c(\nu) = \int_{-\pi}^{\pi} f_0(\theta + \nu\sin\theta)d\theta.$$

The distribution with pdf (4.17) is symmetric about zero.

A cardioid distribution with pdf $g(\cos\theta) = (2\pi)^{-1}(1 + \kappa\cos\theta)$, $0 \le \kappa \le 1$, yields the symmetric Papakonstantinou distribution with pdf (4.14) and $\mu = 0$, and a von Mises distribution with pdf $g(\cos\theta) = \{2\pi I_0(\kappa)\}^{-1}\exp(\kappa\cos\theta)$ does the symmetric Batschelet distribution with pdf (4.16) and $\mu = 0$.

Modality can be investigated for (4.17). Assume that the pdf $f_0(\theta) = g(\cos\theta)$ is $C^1[-\pi,\pi)$ and unimodal with $f_0'(0) = 0$ and has a mode at $\theta = 0$ and an antimode at $\theta = -\pi$. Then the distribution with pdf (4.17) is unimodal with a mode at $\theta = 0$ and an antimode at $\theta = -\pi$ if and only if $|\nu| \le 1$. In the case of $|\nu| > 1$, the distribution has $(2|m| + 1)$ modes if ν satisfies $m\pi < h(\nu) \le (m+1)\pi$ for $m = 1, 2, 3, \ldots$ or $m\pi \le h(\nu) < (m+1)\pi$ for $m = -1, -2, -3, \ldots$, where $h(\nu) = \cos^{-1}(-1/\nu) + \nu\sqrt{1 - 1/\nu^2}$. This fact is proved in a similar way to the case of Papakonstantinou distribution.

Under the same assumption about $f_0(\theta) = g(\cos\theta)$ as the above, together with $f_0(\cdot) \in C^2[-\pi,\pi)$ and $f_0(\theta) > 0$ for all $\theta\,(-\pi \le \theta < \pi)$, the distribution with pdf (4.17) has flat-topped and sharply peaked properties in the following sense: Let $-1 \le \nu_1 < \nu_2 \le 1$. Then we have

a) $f_{\nu_1}(0) < f_{\nu_2}(0)$ and $f_{\nu_1}(-\pi) < f_{\nu_2}(-\pi)$, and
b) for the curvature of $f_\nu(\theta)$ at $\theta = 0$, defined by

$$\frac{f_\nu''(0)}{[1 + \{f_\nu'(0)\}^2]^{3/2}} = f_\nu''(0),$$

since $f_\nu'(0) = 0$, it holds that $f_{\nu_1}''(0) > f_{\nu_2}''(0)$, and at $\theta = -\pi, f_{\nu_1}''(-\pi) < f_{\nu_2}''(-\pi)$.

When the above condition, (a) and (b), is fulfilled, f_{ν_1} is said to be more flat-topped than f_{ν_2}, or f_{ν_2} is more sharply peaked than f_{ν_1}. Note that the uniform distribution has a flat pdf, but does not satisfy (a) or (b).

Proof of these facts is as follows.

a) Note that $0 < \theta + \nu_j \sin\theta < \pi\,(j = 1, 2)$ and thus $\theta + \nu_1 \sin\theta < \theta + \nu_2 \sin\theta$ for all θ in the interval $(0, \pi)$. Since $f_0(\theta)$ has a unique mode at $\theta = 0$ and is strictly decreasing on the interval $(0, \pi)$, it holds that

$$c(\nu_1) = 2\int_0^\pi f_0(\theta + \nu_1 \sin\theta)d\theta > 2\int_0^\pi f_0(\theta + \nu_2 \sin\theta)d\theta = c(\nu_2).$$

Hence,

$$f_{\nu_1}(0) = \{c(\nu_1)\}^{-1} f_0(0) < \{c(\nu_2)\}^{-1} f_0(0) = f_{\nu_2}(0).$$

Similarly, we have $f_{\nu_1}(-\pi) < f_{\nu_2}(-\pi)$.

b) It follows from

$$f_\nu''(\theta) = \{c(\nu)\}^{-1}\left\{-\nu \sin\theta f_0'(\theta + \nu\sin\theta) + (1 + \nu\cos\theta)^2 f_0''(\theta + \nu\sin\theta)\right\},$$

and $f_0''(0) < 0$ that

$$f_{\nu_1}''(0) = \{c(\nu_1)\}^{-1}(1 + \nu_1)^2 f_0''(0) > \{c(\nu_2)\}^{-1}(1 + \nu_2)^2 f_0''(0) = f_{\nu_2}''(0).$$

Similarly, we have $f_{\nu_1}''(-\pi) < f_{\nu_2}''(-\pi)$.

4.3.5 A Family of Symmetric, Flat-Topped, and Sharply Peaked Distributions

Since the pdf of the power cardioid family with $\mu = 0$ has the form $g(\cos\theta)$, we can propose a symmetric flat-topped and sharply peaked distribution with pdf

$$f_\nu(\theta) \propto [1 + \rho\cos\{\theta - \mu + \nu\sin(\theta - \mu)\}]^r, \quad -\pi \le \theta < \pi. \tag{4.18}$$

Clearly, the family with pdf (4.18) reduces to the original power cardioid family if $\nu = 0$, the symmetric Papakonstantinou distribution (4.14) if $r = 1$ and $\rho = \kappa$, and the symmetric Batschelet distribution (4.16) if $\rho = \kappa/r$ and $r \to \infty$.

Similar to the fact that the power cardioid family connects the cardioid to von Mises distributions, the distribution with pdf (4.18) connects the symmetric Papakonstantinou to Batschelet distributions. Since the power cardioid family contains the wrapped Cauchy distribution, (4.18) has a flat-topped and sharply peaked wrapped Cauchy pdf

$$f_\nu(\theta) \propto [1 + \rho\cos\{\theta - \mu + \nu\sin(\theta - \mu)\}]^{-1}, \quad -\pi \le \theta < \pi,$$

as a special case. The distribution with pdf (4.18) is more flat-topped than the corresponding power cardioid family if $-1 \leq \nu < 0$, whereas it is more sharply peaked if $0 < \nu \leq 1$.

4.3.6 Geodesic Normal Distribution

The geodesic normal distribution (Coeurjolly and Le Bihan, 2012) has pdf

$$f(\theta) = C^{-1} \exp\left\{-\frac{\gamma}{2} d_G(\mu,\theta)^2\right\}, \quad \mu - \pi \leq \theta < \mu + \pi, -\pi \leq \mu < \pi, \gamma > 0, \tag{4.19}$$

where $d_G(\mu, \theta)$ is the arc length between μ and θ. The normalizing constant is expressed as

$$C = \sqrt{\frac{2\pi}{\gamma}}\{\Phi(\pi\sqrt{\gamma}) - \Phi(-\pi\sqrt{\gamma})\} = \sqrt{\frac{2\pi}{\gamma}}\,\mathrm{erf}\left(\pi\sqrt{\frac{\gamma}{2}}\right),$$

in terms of the distribution function of a standard normal distribution, Φ, and the error function $\mathrm{erf}(z) = (2/\sqrt{\pi})\int_{-\infty}^{z} e^{-t^2} dt$ defined for any complex number z. The case $\gamma = 0$ reduces to the circular uniform distribution. The distribution with pdf (4.19) is actually a truncated normal distribution bounded below and above.

Let Θ denote a circular random variable whose pdf is given by (4.19). Then, the pth trigonometric moment of Θ has an expression

$$E(e^{ip\Theta}) = e^{ip\mu - p^2/(2\gamma)} \frac{\mathrm{Re}\left[\mathrm{erf}(\pi\sqrt{\gamma/2} - ip/\sqrt{2\gamma})\right]}{\mathrm{erf}(\pi\sqrt{\gamma/2})}.$$

Thus, the mean direction of Θ is μ and the mean resultant length of Θ is

$$\rho = e^{-1/(2\gamma)} \frac{\mathrm{Re}\left[\mathrm{erf}(\pi\sqrt{\gamma/2} - i/\sqrt{2\gamma})\right]}{\mathrm{erf}(\pi\sqrt{\gamma/2})}.$$

4.4 Wrapping Method

The wrapping method is a convenient one to derive symmetric circular distributions, when the corresponding symmetric linear distributions have amenable forms. These distributions in general, save a few, e.g. the wrapped Cauchy distribution, admit of only infinite series representation since the summation in the wrapping usually does not yield any elegant single term expression. However, the equivalence between the α_p (since $\beta_p = 0$ in the symmetric case) and the characteristic function $\psi_X(t)|(t = p)$ of the unwrapped random variable X, as shown in Section 3.3, can eliminate the need of obtaining an unwieldy sum especially if the latter has an elegant analytical representation. Note that, while wrapping produces a unique circular distribution from a given linear distribution, various unwrapped linear distributions may yield the same circular distribution, since equality of $\psi_X(t)$ and α_p is needed only at $t = p,\ p \in Z$.

Following the discussions given in Chapter 3, we now present several wrapped distributions below.

4.4.1 Wrapped Cauchy Distribution

The wrapped Cauchy distribution presented below is first derived through the method of wrapping. However similar to the von Mises distribution, as shown in SenGupta and Roy (2011), it too admits of its representation as a dispersion model and also of derivation by several other methods as shown below.

4.4.1.1 Definition and Properties

If a random variable X is distributed as a Cauchy distribution on the real line with 0 median or center of symmetry and scale parameter σ, denoted by $C(0, \sigma)$, whose pdf is given by

$$g(x) = \frac{\sigma}{\pi(x^2 + \sigma^2)}, \quad -\infty < x < \infty,\ 0 < \sigma < \infty,$$

then the characteristic function of X is

$$\phi(t) = E(e^{itX}) = e^{-\sigma|t|}, \quad -\infty < t < \infty.$$

Thus, from the Fourier series representation (3.4), we have the wrapped Cauchy distribution WC$(0, \rho)$ with random variable Θ and pdf

$$f(\theta) = \frac{1}{2\pi}\left\{1 + 2\sum_{p=1}^{\infty}\rho^p\cos(p\theta)\right\}, \quad 0 \le \theta < 2\pi, \tag{4.20}$$

where $\rho = e^{-\sigma}$, since $\alpha_p = E\{\cos(p\Theta)\} = e^{-\sigma p}$, $\beta_p = E\{\sin(p\Theta)\} = 0\,(p = 1, 2, 3, \ldots)$ and

$$\sum_{p=1}^{\infty}(\alpha_p^2 + \beta_p^2) = \sum_{p=1}^{\infty}e^{-2\sigma p} = \frac{1}{e^{2\sigma} - 1} < \infty.$$

This distribution has been studied by Lévy (1939) and Wintner (1947). The mean direction of WC$(0, \rho)$ is 0 (radian) and its mean resultant length is $\rho\,(0 < \rho < 1)$. Equation (4.20) is transformed into

$$\begin{aligned} f(\theta) &= \text{Re}\left[\frac{1}{2\pi}\left(1 + 2\sum_{p=1}^{\infty}\rho^p e^{ip\theta}\right)\right] \\ &= \frac{1 - \rho^2}{2\pi(1 + \rho^2 - 2\rho\cos\theta)}, \quad 0 \le \theta < 2\pi, \end{aligned} \tag{4.21}$$

where Re[z] denotes the real part of a complex number z. The pth trigonometric moment of WC(0, ρ) is $\phi_p = \rho^p$, which coincides with the characteristic function of $C(0, \sigma)$. The wrapped Cauchy distribution with pdf (4.21) is symmetric about zero and unimodal with mode at zero and antimode at π. The distribution function of WC(0, ρ) is given by

$$F(\theta) = \frac{1}{2} + \frac{1}{\pi}\tan^{-1}\left\{\frac{1+\rho}{1-\rho}\tan\left(\frac{\theta}{2}\right)\right\}, \quad -\pi \leq \theta < \pi.$$

Note that this expression is valid for $-\pi \leq \theta < \pi$, i.e. $0 \leq F(\theta) \leq 1$, $F(-\pi) = 0$, $F(\pi) = 1$ and $F(\theta)$ is monotonically increasing. For $0 \leq \theta < 2\pi$, a modification is needed for $F(\theta)$.

Note. The shape of $r = \ell/(1 - \varepsilon\cos\theta)$ with $0 < \varepsilon < 1$ and $\ell > 0$ in polar coordinates represents an ellipse with the eccentricity ε and the semi-latus rectum ℓ. However, the distribution with pdf (4.21) is not called an elliptical distribution.

If ρ is small (σ is large), the shape of WC(0, ρ) pdf is similar to that of the cardioid pdf with parameters $\mu = 0$ and ρ. If $\rho \to 0$, WC(0, ρ) reduces to the uniform distribution on the circle, and if $\rho \to 1$, it does to a degenerate distribution concentrated at $\theta = 0$. A wrapped Cauchy distribution WC(μ, ρ) with mean direction $\mu\,(0 \leq \mu < 2\pi)$ is obtained by setting $\theta - \mu$ instead of θ in (4.20) and (4.21). If Θ is distributed as WC(μ, ρ) and c is a constant, then $c\Theta$ is distributed as WC($c\mu, \rho^{|c|}$) since $E(e^{it\Theta}) = e^{it\mu}\rho^{|t|}$. Wrapped Cauchy distributions have the reproductive property with respect to addition and subtraction because Cauchy distributions on the real line have the reproductive property. In fact, if Θ_1, Θ_2 are independent and $\Theta_1 \sim \mathrm{WC}(\mu_1, \rho_1)$, $\Theta_2 \sim \mathrm{WC}(\mu_2, \rho_2)$, then $\Theta_1 \pm \Theta_2 \sim \mathrm{WC}(\mu_1 \pm \mu_2, \rho_1\rho_2)$. From this, it is easy to see that if $\Theta_j\,(j = 1, \ldots, n)$ are independently distributed as $\mathrm{WC}(\mu_j, \rho_j)$, then $\sum_{j=1}^n \Theta_j \sim \mathrm{WC}(\sum_{j=1}^n \mu_j, \prod_{j=1}^n \rho_j)$, from which it is directly seen that the sum converges to the circular uniform distribution as $n \to \infty$ since $0 < \rho_j < 1\,(j = 1, \ldots, n)$ and $\prod_{j=1}^n \rho_j \to 0$, $n \to \infty$.

4.4.1.2 Derivation by von Mises Mixtures

The wrapped Cauchy distribution is a mixture of von Mises distributions (Mardia, 2007). To show this, let Θ and K be two random variables and suppose that the conditional distribution of Θ given $K = \kappa$, $(\Theta | K = \kappa)$, is VM(0, κ) and K has pdf

$$f(\kappa) = \sqrt{\alpha^2 - 1}\, I_0(\kappa) e^{-\alpha\kappa}, \quad \kappa \geq 0,\ \alpha > 1, \tag{4.22}$$

where $I_0(\cdot)$ stands for the modified Bessel function of the first kind and order zero. It is shown below that (4.22) gives a pdf.

The joint pdf of $(\Theta, K)'$ is

$$f(\theta, \kappa) = \frac{\sqrt{\alpha^2 - 1}}{2\pi} e^{-\kappa(\alpha - \cos\theta)}, \quad 0 \leq \theta < 2\pi,\ \kappa \geq 0, \tag{4.23}$$

and the unconditional pdf of Θ is calculated as

$$f(\theta) = \int_0^\infty f(\theta, \kappa) d\kappa$$
$$= \frac{\sqrt{\alpha^2 - 1}}{2\pi(\alpha - \cos\theta)}, \quad 0 \leq \theta < 2\pi, \ \alpha > 1,$$

which is the wrapped Cauchy $\mathrm{WC}(0, \alpha - \sqrt{\alpha^2 - 1})$ pdf. This means that the wrapped Cauchy distribution is represented by the mixture of von Mises distributions. We have

$$\int_0^\infty f(\kappa) d\kappa = \frac{\sqrt{\alpha^2 - 1}}{2\pi} \int_0^\infty \int_0^{2\pi} e^{\kappa \cos\theta - \alpha\kappa} d\theta d\kappa = 1,$$

i.e. (4.22) is actually a pdf, since $\int_0^{2\pi} f(\theta) d\theta = 1$. Note that the distribution of $(\Theta, K)'$ with pdf (4.23) will appear as Model 1 on the cylinder in Section 10.3.3.

4.4.1.3 Circular Cauchy Distribution

Following McCullagh (1996), we show the brief outlines of another view of the wrapped Cauchy distribution. Let X be a real-valued Cauchy random variable whose pdf is given by

$$f(x) = \frac{|\alpha_2|}{\pi |x - \alpha|^2}, \quad -\infty < x < \infty, \tag{4.24}$$

where $\alpha = \alpha_1 + i\alpha_2$ ($\alpha_1, \alpha_2 \in \mathbb{R}$, $i = \sqrt{-1}$) is a complex number. Here the real part $\alpha_1 = \mathrm{Re}(\alpha)$ is the median or center of symmetry, and the absolute value of the imaginary part, $|\alpha_2| = |\mathrm{Im}(\alpha)|$, is the scale parameter. The parameter space is the whole complex plane and we think of the sample space as the real axis embedded in the parameter space. We simply denote by $X \sim C(\alpha)$. For $\alpha_2 \neq 0$, α and $\overline{\alpha}$ (complex conjugate) give the same Cauchy pdf (4.24). In particular, the standard Cauchy distribution has parameter $\alpha = \pm i$. For $\alpha_2 = 0$, the Cauchy distribution is interpreted as a point mass at α_1.

A random variable Z on the unit circle in the complex plane given by

$$Z = \frac{1 + iX}{1 - iX},$$

has pdf

$$f(z) = \frac{|1 - |\psi|^2|}{2\pi |z - \psi|^2}, \quad |z| = 1, \tag{4.25}$$

where $\psi = (1 + i\alpha)/(1 - i\alpha)$ with $|\psi| \neq 1$. We call the distribution of Z the circular Cauchy (named by McCullagh (1996)) and write $Z \sim C^*(\psi)$.

The wrapped Cauchy distribution is obtained by the distribution of Θ through $Z = \exp(i\Theta)$, where Z is distributed as (4.25). In fact, we have the pdf of the angular variable Θ as

$$f(\theta) = \frac{1-\rho^2}{2\pi\{1+\rho^2-2\rho\cos(\theta-\mu)\}}, \qquad 0 \leq \theta < 2\pi,$$

where $\mu = \arg(\psi)$ and

$$\rho = \begin{cases} |\psi|, & |\psi| < 1, \\ 1/|\psi|, & |\psi| > 1. \end{cases}$$

The circular Cauchy distribution C^* has the following properties:

a) If $Z \sim C^*(\psi)$, then $\beta_0 Z \sim C^*(\beta_0\psi)$ for $\beta_0 = e^{i\alpha} \in \{z \in \mathbb{C}; |z| = 1\}$, $\alpha \in \mathbb{R}$, i.e. the rotation of a circular Cauchy variable is also a circular Cauchy variable.
b) If $Z_1 \sim C^*(\psi_1)$, $Z_2 \sim C^*(\psi_2)$, $|\psi_1| \leq 1$, $|\psi_2| \leq 1$, and Z_1, Z_2 are independent, then $Z_1Z_2 \sim C^*(\psi_1\psi_2)$, i.e. the product of two independent circular Cauchy variables is also a circular Cauchy variable.
c) If $Z \sim C^*(\psi)$, then $(Z+\beta_1)/(1+\overline{\beta_1}Z) \sim C^*((\psi+\beta_1)/(1+\overline{\beta_1}\psi))$ for $\beta_1 \in \mathbb{C}$ and $|\beta_1| \neq 1$. For $|\beta_1| < 1$, the transformation is a straight-line projection through the point β_1 onto the opposite side of the circle (see Figure 3.2a in Section 3.6).

From the property (a), the uniform is unaffected by rotation, i.e. if $Z \sim C^*(0)$, then $\beta_0 Z \sim C^*(0)$. From the properties (a) and (c), if Z is distributed as the uniform distribution $C^*(0)$, then the Möbius circle transformation (Section 3.6) of Z, $(Z+\beta_1)/(1+\overline{\beta_1}Z)$, gives the circular Cauchy distribution $C^*(\beta_1)$ and $\beta_0(Z+\beta_1)/(1+\overline{\beta_1}Z)$ generates the circular Cauchy distribution $C^*(\beta_0\beta_1)$, where $\beta_0 \in \{z \in \mathbb{C}; |z| = 1\}$ and $|\beta_1| \neq 1$.

4.4.1.4 The Poisson Integral

This paragraph follows McCullagh (1996). In real and complex analysis, it is known that the real Poisson kernel is defined by

$$P_r(\theta - t) = \frac{1-r^2}{1+r^2-2r\cos(\theta-t)},$$

for $0 \leq \theta, t < 2\pi$, and $0 \leq r < 1$. Note that $P_r(\theta-t)/(2\pi)$ is the pdf of the wrapped Cauchy distribution WC$(0, r)$. If $z = re^{it}$ (z is on the open unit disc U in the complex plane), we have the relation between $P_r(\theta-t)$ and the complex Poisson kernel defined by $(w+z)/(w-z)$ for $w \in \partial U$ (the boundary of U, i.e. the unit circle) and $z \in U$:

$$P_r(\theta - t) = \mathrm{Re}\left[\frac{e^{it}+z}{e^{it}-z}\right].$$

If g is analytic in a neighborhood of the closed unit disc $\overline{U}$ with $g(0)$ real, then for $z \in U$

$$g(z) = \frac{1}{2\pi} \int_{-\pi}^{\pi} \frac{e^{i\theta} + z}{e^{i\theta} - z} \mathrm{Re}[g(e^{i\theta})] d\theta \tag{4.26}$$

(cf. Rudin, 1987, Theorem 11.10; Simon, 2005, p. 27). In fact, let $g(z) = \sum_{n=0}^{\infty} a_n z^n$ since g is analytic. From

$$\frac{e^{i\theta} + z}{e^{i\theta} - z} = \frac{1 + ze^{-i\theta}}{1 - ze^{-i\theta}} = 1 + 2\sum_{n=1}^{\infty} z^n e^{-in\theta},$$

which is uniformly convergent for $e^{i\theta} \in \partial U$ and $z \in U$, we have

$$\begin{aligned}\frac{1}{2\pi} \int_{-\pi}^{\pi} e^{-in\theta} \mathrm{Re}\left[g(e^{i\theta})\right] d\theta &= \frac{1}{4\pi} \int_{-\pi}^{\pi} e^{-in\theta} \{g(e^{i\theta}) + \overline{g(e^{i\theta})}\} d\theta \\ &= \begin{cases} \mathrm{Re}[a_0], & n = 0, \\ a_n/2, & n = 1, 2, 3, \dots, \end{cases}\end{aligned}$$

since $g(0)$ is assumed to be real. Hence, the right side of (4.26) is actually $a_0 + \sum_{n=1}^{\infty} a_n z^n = g(z)$.

From (4.26), if Y has the circular Cauchy distribution $C^*(\psi)$ with $|\psi| < 1$, then

$$E\{g(Y)\} = g(\psi),$$

provided that g is continuous on $\overline{U}$ and analytic on U. In particular, the kth moments of $C^*(\psi)$ are

$$E(Y^k) = \psi^k,$$

for integer $k = 0, 1, 2, \dots$.

4.4.2 Wrapped Normal Distribution

If X is distributed as a normal distribution $N(0, a^2)$ with pdf

$$f(x) = \frac{1}{\sqrt{2\pi}a} e^{-x^2/(2a^2)}, \quad -\infty < x < \infty,\ a > 0, \tag{4.27}$$

then we have the wrapped normal distribution with pdf

$$f(\theta) = \frac{1}{\sqrt{2\pi}a} \sum_{k=-\infty}^{\infty} \exp\left\{ -\frac{(\theta + 2\pi k)^2}{2a^2} \right\}, \quad 0 \le \theta < 2\pi.$$

Since the characteristic function of X is given by

$$\phi_p = E(e^{ipX}) = \exp(-p^2 a^2/2),$$

the Fourier representation (3.4) leads to

$$f(\theta) = \frac{1}{2\pi} \left\{ 1 + 2\sum_{p=1}^{\infty} \rho^{p^2} \cos(p\theta) \right\}, \qquad 0 \le \theta < 2\pi, \tag{4.28}$$

where $\rho = \exp(-a^2/2)\,(0 < \rho < 1)$ is the mean resultant length. Thus, the variance is $\nu = 1 - \rho = 1 - \exp(-a^2/2)$, from which the standard deviation a of the normal distribution is given by $a = \{-2\log(1-\nu)\}^{1/2}$. That is the reason why the circular standard deviation has been defined by $\{-2\log\rho\}^{1/2}$ in Section 2.5. If we use a theta function given by (B.44), (4.28) is expressed by

$$f(\theta) = \frac{1}{2\pi}\,\vartheta_3(\theta/2, \rho),$$

as studied in Wintner (1933).

The wrapped normal WN$(0, \rho)$ with pdf (4.28) is unimodal and symmetric about zero, and a member of wrapped α-stable distributions in Section 5.7. It reduces to the uniform distribution on the circle if $\rho \to 0\,(a \to \infty)$ and to a degenerate distribution concentrated at $\theta = 0$ if $\rho \to 1\,(a \to 0)$. The wrapped normal distribution is generated from Brownian motion on the circle as the limiting case of a random walk (Stephens, 1963).

The wrapped normal distribution WN(μ, ρ) with mean direction $\mu\,(0 \le \mu < 2\pi)$ is obtained by replacing θ by $\theta - \mu$ in (4.28). The distribution possesses the reproductive property, i.e. if $\Theta_1 \sim \mathrm{WN}(\mu_1, \rho_1)$, $\Theta_2 \sim \mathrm{WN}(\mu_2, \rho_2)$, and Θ_1 and Θ_2 are independent, then $\Theta_1 \pm \Theta_2 \sim \mathrm{WN}(\mu_1 \pm \mu_2, \rho_1\rho_2)$. This follows from the reproductive property of the normal distribution with corresponding mean resultant length

$$e^{-(a_1^2+a_2^2)/2} = e^{-a_1^2/2}e^{-a_2^2/2} = \rho_1\rho_2.$$

Generally, if $\Theta_j\,(j = 1, \ldots, n)$ are independently distributed as WN(μ_j, ρ_j), then the sum $\sum_{j=1}^n \Theta_j$ is distributed as WN$(\sum_{j=1}^n \mu_j, \prod_{j=1}^n \rho_j)$. From this fact, it is easy to see that the sum of WN(μ_j, ρ_j) variables tends to follow the uniform distribution if $n \to \infty$ because $\prod_{j=1}^n \rho_j \to 0$ as $n \to \infty$. The wrapped t-distribution introduced in Section 4.4.3 converges to a wrapped normal distribution as degrees of freedom tend to infinity.

The shape of pdf is quite similar to that of von Mises pdf if a^2, the variance of $N(0, a^2)$, is small. To show this fact, consider a von Mises distribution VM(μ, κ) with large κ. Note that κ is the concentration parameter and large κ means that the distribution tends to be sharply peaked at the mode. The mean resultant length $I_1(\kappa)/I_0(\kappa)$ of VM(μ, κ) has an approximation

$$\frac{I_1(\kappa)}{I_0(\kappa)} \approx 1 - \frac{1}{2\kappa},$$

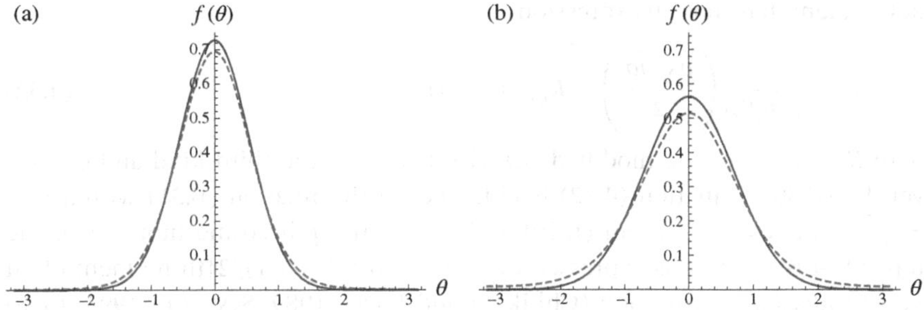

Figure 4.7 von Mises approximation of wrapped normal distribution. (a) WN: $a^2 = 0.3$ (solid), VM: $\kappa = 1/0.3$ (dashed); (b) WN: $a^2 = 0.5$ (solid), VM: $\kappa = 1/0.5$ (dashed).

for large κ from asymptotic expansion given by (B.29). The mean resultant length $\rho = \exp(-a^2/2)$ of WN(0, ρ) is approximated by $\exp(-a^2/2) \approx 1 - a^2/2$ for small a, and we equate $1 - 1/(2\kappa)$ with $1 - a^2/2$. Finally we have $\kappa = 1/a^2$. Figure 4.7 shows von Mises approximation of the wrapped normal distributions WN(0, ρ), $\rho = e^{-a^2/2}$. If $a^2 = 0.3$ and $\kappa = 1/0.3$, then the von Mises approximation is fairly good, but not so good for $a^2 = 0.5$ and $\kappa = 1/0.5$.

4.4.3 Wrapped *t*-Distribution

The wrapped t-distribution has been alluded to in Mardia (1972, p. 57). Here we provide its rigorous derivation. The wrapped t-distribution with n degrees of freedom is obtained by wrapping a t-distribution on the real line with pdf

$$f(x) = \frac{a^n}{\sqrt{n}B(1/2, n/2)}\left(a^2 + \frac{x^2}{n}\right)^{-(n+1)/2}, \quad -\infty < x < \infty,\ 0 < a < \infty \tag{4.29}$$

around the circle. Here $B(\cdot,\cdot)$ is the beta function defined by (B.5) and is expressed as $B(1/2, n/2) = \sqrt{\pi}\,\Gamma(n/2)/\Gamma((n+1)/2)$ in terms of the gamma function, and $n > 0$ is not necessarily integer. The study of the wrapped t-distribution can be seen in the papers by Kato and Shimizu (2004) and Pewsey et al. (2007). The pdf of the wrapped t-distribution is given by

$$g(\theta) = c\sum_{p=-\infty}^{\infty}\left\{1 + \frac{(\theta + 2p\pi)^2}{a^2 n}\right\}^{-(n+1)/2}, \quad 0 \le \theta < 2\pi \tag{4.30}$$

with $c = 1/\{a\sqrt{n}B(1/2, n/2)\}$ and is also expressed as the Fourier series

$$g(\theta) = \frac{1}{2\pi}\left\{1 + 2\sum_{p=1}^{\infty}\alpha_p\cos(p\theta)\right\}, \quad 0 \le \theta < 2\pi. \tag{4.31}$$

The wrapped t-distribution is unimodal and symmetric about zero, the mode, and the pth sine moment β_p is zero. The pth cosine moment α_p in (4.31) with the mean resultant length α_1 has an expression

$$\alpha_p = \frac{2}{\Gamma(n/2)}\left(\frac{p\sqrt{na}}{2}\right)^{n/2} K_{n/2}(p\sqrt{na}), \tag{4.32}$$

where $K_\nu(\cdot)$ denotes the modified Bessel function of the third kind and order ν given by (B.30). Equation (4.32) is obtained by the relation (B.31) with $x = p$, $z = \sqrt{na}$ and $\nu = n/2$. From (B.30) and (B.32), the pth cosine moment of the wrapped t-distribution is expressed in terms of the $\{(n+1)/2\}$th moment of an inverse Gaussian distribution (Chhikara and Folks, 1989; Seshadri, 1993, 1998) $\mathrm{IG}(p\sqrt{na}/2, p^2na^2/2)$:

$$\alpha_p = \frac{2\sqrt{\pi}e^{-\sqrt{na}p}}{\sqrt{na}p\Gamma(n/2)} E\left\{S^{(n+1)/2}\right\}, \quad p = 1, 2, 3, \ldots,$$

where $S \sim \mathrm{IG}(\mu, \lambda)$ with $\mu = p\sqrt{na}/2$ and $\lambda = p^2na^2/2$, whose pdf is

$$h(s) = \sqrt{\frac{\lambda}{2\pi}}\, s^{-3/2} \exp\left\{-\frac{\lambda(s-\mu)^2}{2\mu^2 s}\right\}, \quad s > 0.$$

As $n \to \infty$, the wrapped t-distribution with pdfs (4.30) and (4.31) approaches the wrapped normal distribution $\mathrm{WN}(0, \rho)$ in Section 4.4.2 because (4.29) goes to the normal pdf (4.27).

When the degrees of freedom are odd, i.e. $n = 2m + 1 (m = 0, 1, 2, \ldots)$, from the residue theorem, α_p can be simply expressed by a finite series

$$\begin{aligned}
\alpha_p(n) &= \frac{1}{B(1/2, m+1/2)} \int_{-\infty}^{\infty} \frac{\exp(ip\sqrt{na}z)}{(1+z^2)^{m+1}} dz \\
&= \frac{1}{B(1/2, m+1/2)} (2\pi i)\mathrm{Res}\left[\frac{\exp(ip\sqrt{na}z)}{(1+z^2)^{m+1}}, i\right] \\
&= \frac{2\pi i}{B(1/2, m+1/2)m!} \left[\frac{\exp(ip\sqrt{na}z)}{(z+i)^{m+1}}\right]^{(m)}_{z=i} \\
&= \frac{\pi e^{-p\sqrt{na}}}{B(1/2, m+1/2)2^{2m}m!} \sum_{j=0}^{m} \frac{(2m-j)!}{(m-j)!j!} (2p\sqrt{na})^j.
\end{aligned} \tag{4.33}$$

For $m = 0, 1, 2, 3$, (4.33) reduces to

$$\begin{aligned}
\alpha_p(1) &= e^{-pa}, \\
\alpha_p(3) &= (1 + \sqrt{3}pa)e^{-\sqrt{3}pa}, \\
\alpha_p(5) &= \left\{1 + \sqrt{5}pa + \frac{5}{3}(pa)^2\right\} e^{-\sqrt{5}pa}, \\
\alpha_p(7) &= \left\{1 + \sqrt{7}pa + \frac{14}{5}(pa)^2 + \frac{7\sqrt{7}}{15}(pa)^3\right\} e^{-\sqrt{7}pa}.
\end{aligned}$$

It is clear that if $m = 0$ ($n = 1$), then the distribution is the wrapped t-distribution with 1 degree of freedom, i.e. the wrapped Cauchy distribution WC(0, ρ) with pdf

$$f(\theta) = \frac{1}{2\pi}\left\{1 + 2\sum_{p=1}^{\infty}\rho^p \cos(p\theta)\right\}, \quad \rho = e^{-a},$$

which directly follows from (B.33) or from the characteristic function of the Cauchy distribution as in Section 4.4.1. For $0 < n < 1$, the pdf is more peaked and heavy-tailed than the wrapped Cauchy distribution with the same mean resultant length.

A family of wrapped t-distributions with an additional parameter, mean direction μ ($0 \leq \mu < 2\pi$), is obtained by replacing θ with $\theta - \mu$. The family is flexible in the sense that it links the wrapped Cauchy distribution to the wrapped normal distribution.

4.5 Characteristic Function-Based Method

The method of wrapping as used above is initiated with the existence of an analytical form of the generating (linear) distribution on the real line. The resulting wrapped distribution is useful when the summation can be performed to yield a simplified version, as in the case of wrapped Cauchy and wrapped normal densities. However, there are important families of linear distributions for which the densities need not "exist," i.e. are not available in any closed analytical form. In this section, we show how the corresponding circular distributions, with tractable densities, can be derived for such families based on characteristic functions (c.f.s). The method is based on two main observations: (i) any circular pdf can be represented in terms of the appropriate Fourier series and (ii) the coefficients in the Fourier series are merely the characteristic function of the corresponding linear random variable X, $\psi_X(t)$ at $t = p$, $p \in \mathbb{Z}$. Item (ii) eliminates the need of existence of an analytical form of the pdf of X. Further, the normalizing constant C of such a c.f.-based circular density has an extremely simple form, i.e. $C = 1/(2\pi)$. We consider in particular two families of distributions: stable and Linnik.

4.5.1 Wrapped Stable Family

A random variable X is said to be stable (see, e.g. Samorodnitsky and Taqqu, 1994) if for any positive numbers A and B, and for independent random variables X_1 and X_2 distributed according to the same distribution as X, a linear combination $AX_1 + BX_2$ has the same distribution as $CX + D$ for some positive number C and real number D. For example, a normal distribution with mean μ and variance σ^2, $N(\mu, \sigma^2)$, has this property. Indeed, if X_1 and X_2 are independently distributed as $N(\mu, \sigma^2)$, then the distribution of $AX_1 + BX_2$ is $N((A+B)\mu, (A^2+B^2)\sigma^2)$, and $CX + D$, where $C = (A^2+B^2)^{1/2}$ and $D = (A+B-C)\mu$, follows the same distribution $N((A+B)\mu, (A^2+B^2)\sigma^2)$.

Stable distributions also appear as limit laws of properly normalized sums of independent random variables. This family is highly flexible as it can capture wide varieties of tail behavior – thick-thin, long-short, etc. as well as those of kurtosis – very sharp (high) to very flat (low) peaks. The stable family has thus enjoyed popularity in many different emerging areas of applied sciences.

There is an equivalent definition of the stable distribution using the characteristic function. Details are presented in Section 5.7. Here we deal with symmetric distributions. A random variable X is said to be symmetric (around the location parameter 0) α-stable if there exist parameters $0 < \alpha \leq 2$ and $\tau \geq 0$ such that the characteristic function of X has the form

$$\psi_X(t) = E(e^{itX}) = \exp(-\tau^\alpha |t|^\alpha).$$

The parameter α is termed the index of stability. For example, if $\alpha = 2$, then $E(e^{itX}) = \exp(-\tau^2 t^2)$, which is the characteristic function of $N(0, 2\tau^2)$. If $\alpha = 1$, then $E(e^{itX}) = \exp(-\tau|t|)$, which means that X is distributed as a Cauchy distribution with zero location parameter (median) and scale τ.

The pdfs of α-stable (abbreviated for: stable with index parameter α) random variables are difficult to express in closed forms except for a few special cases, e.g. $\alpha = 1$ and 2. However, the main point here is: contrary to the case of the linear random variables, for circular random variables, the wrapped α-stable distributions are tractable for *all* α. We consider the case of the symmetric α-stable random variable. Noting the characteristic function $\psi_X(t)$ given above, the corresponding trigonometric moments follow as $\phi_p = \exp\{-(p\tau)^\alpha\}$ for $p = 0, 1, 2, \ldots$. By the general Fourier series representation of any circular distribution given in Section 3.3, we finally conclude that the wrapped symmetric α-stable distribution has the pdf

$$\begin{aligned} f(\theta) &= \frac{1}{2\pi}\left[1 + 2\sum_{p=1}^{\infty} e^{-\tau^\alpha p^\alpha} \cos\{p(\theta-\mu)\}\right] \\ &= \frac{1}{2\pi}\left[1 + 2\sum_{p=1}^{\infty} \rho^{p^\alpha} \cos\{p(\theta-\mu)\}\right], \quad 0 \leq \theta, \mu < 2\pi, \end{aligned}$$

$\rho = \exp(-\tau^\alpha)$, as the Fourier series representation. The mean direction here is μ and the mean resultant length is ρ.

4.5.2 Wrapped Linnik Family

Linnik (1963) proposed another family of highly flexible distributions which can capture a variety of inflection points, tail behavior and kurtosis. Neither the stable nor the Linnik families are subfamilies of each other. The Laplace or double exponential distribution plays a pivotal role in the Linnik family as does the normal for the stable family. The Linnik family possesses several appealing properties: (i) the compound geometric-stable family converges to the Linnik family as the success probability $p \to 0$. (ii) The c.f.s $\psi_s(t)$ and $\psi_L(t)$ of the stable and Linnik families respectively are related by $\psi_L(t) = [1 - \log \psi_s(t)]^{-1}$. (iii) The Linnik family is closed under geometric compounding: Let X_i be i.i.d. Linnik and N be geometric, with parameter p, random variables. Assume that N is independent of the X_i's. Then, $p^{1/\alpha} \sum_{i=1}^{N} X_i$ when $N \to \infty$ has the same Linnik distribution as that of the X_i's. For (i) and (ii), see equations (2.3) and (2.4) in Rachev and SenGupta (1993), and for (iii), see Theorem 2.1 in SenGupta et al. (2025).

Define the wrapped Linnik distribution as the distribution corresponding to the c.f. $\psi_{WL(p)}$ given by $\psi_L(p)$, $p \in \mathbb{Z}$. Recall the general Fourier series representation of a circular distribution and that $\alpha_p = \psi_L(p)$, for a symmetric circular distribution. Denoting by μ the location (symmetry) parameter, we then have the following.

Result 4.1 The probability density function of the wrapped Linnik distribution is defined as

$$f(\theta) = \frac{1}{2\pi}\left[1 + 2\sum_{p=1}^{\infty}\{(1 + (\sigma p)^{\alpha})^{-1}\}\cos\{p(\theta - \mu)\}\right],$$

where the parameter space is given by

$$\Omega = \Omega_1 \cup \Omega_2, \quad \Omega_1 = \{(\alpha, \sigma, \mu) : 1 \leq \alpha \leq 2, \sigma \geq 1, 0 \leq \mu < 2\pi\} \quad \text{and}$$
$$\Omega_2 = \{(\alpha, \sigma, \mu) : 1 < \alpha \leq 2, \sigma < 1, 0 \leq \mu < 2\pi\}.$$

Observe the constraints on the parameter space. One should be careful in deriving the wrapped circular distribution from the corresponding linear distribution since it may not exist for some specific range of the parameters.

As mentioned above, Linnik distribution family is highly flexible (see e.g. Erdogan and Ostrovskii, 1998). This was the motivation for deriving the wrapped Linnik family of distributions and the same property extends in a significant way to the corresponding wrapped Linnik family. In fact almost (save for values of stable index parameter quite small, $\alpha^* \leq 0.1$ approximately) uniform tail dominance may be achieved by this family over the wrapped stable family, as can be seen from Figure 4.8. Further properties and discussions including those on convergence of the sum, number of terms needed to compute the pdf, simulation, etc., for this distribution are available from SenGupta and Roy (2023a,b).

Observe that the wrapped Linnik distribution follows as a special case of the wrapped Linnik family with $\alpha = 2$.

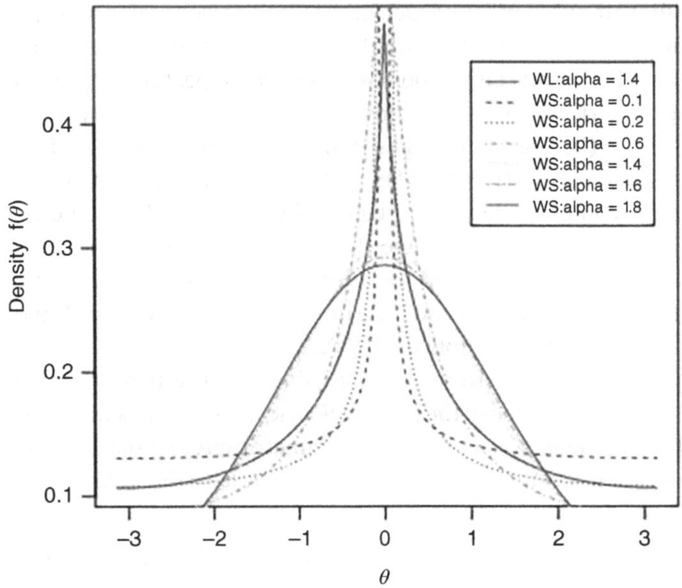

Figure 4.8 Tail dominance of the wrapped symmetric Linnik density with $\alpha = 1.4$ over the wrapped symmetric α^*-stable densities.

4.6 (Möbius) Transformation-Based Method

4.6.1 Cartwright Distribution

Bilinear (Möbius) transformation or stereographic projection of distributions on the real line yields distributions on the circle and vice versa. We exploit this method below to obtain several circular distributions.

Minh and Farnum (2003) showed that this method produces the Cauchy distribution from the circular uniform distribution. They also demonstrated that the Cartwright power-of-cosine distribution on the circle induces a t-distribution on the real line. By virtue of the uniqueness of the transform, the reverse results follow. However, for the purpose of illustration, we provide below the direct derivation for the Cartwright power-of-cosine distribution.

The t-distribution on the real line with m (positive integer) degrees of freedom is given by

$$f(x) = \frac{\Gamma((m+1)/2)}{\Gamma(m/2)\sqrt{\pi m}}\left(1 + \frac{x^2}{m}\right)^{-(m+1)/2}, \qquad -\infty < x < \infty.$$

Apply the (Möbius) transformation,

$$x = u + v\frac{\sin\theta}{1+\cos\theta} = u + v\tan\left(\frac{\theta}{2}\right), \quad -\pi \le \theta < \pi \tag{4.34}$$

with $u = 0$, $v = \sqrt{m}$, $m = 2n + 1\,(n = 0, 1, 2, \ldots)$. It is easily seen that the resulting pdf on the circle is given by

$$g(\theta) = \frac{\Gamma(n+1)}{2^{n+1}\sqrt{\pi}\,\Gamma(n+1/2)}\,(1+\cos\theta)^n, \quad -\pi \le \theta < \pi. \tag{4.35}$$

The distribution with pdf (4.35) is called the Cartwright (1963) power-of-cosine distribution. A special case is the circular uniform distribution when $n = 0\,(m = 1)$. Further properties and characterizations of this distribution are given in Shakil et al. (2019).

Note that (4.35) still represents a pdf for $m = 2s + 1 > 0$ as in (4.6).

4.6.2 Cardioid Distribution

The pdf of a cardioid distribution $C(\mu, \rho)$ with parameters $\mu\,(0 \le \mu < 2\pi)$ and $\rho\,(0 \le \rho \le 1/2)$ is defined by

$$f(\theta) = \frac{1}{2\pi}\left\{1 + 2\rho\cos(\theta - \mu)\right\}, \quad 0 \le \theta < 2\pi. \tag{4.36}$$

This form appears in Jeffreys (1961, p. 328). In the literature, it is often assumed that $|\rho| \le 1/2$, but we restrict ρ to be non-negative to avoid problems of non-identifiability. The name is taken from the fact that the shape of $r = f(\theta)$ in polar coordinates ($x = r\cos\theta$, $y = r\sin\theta$, $r \ge 0$, $0 \le \theta < 2\pi$) resembles a heart (Figure 4.9). The pdf (4.36) is symmetric about μ and unimodal. The distribution with $\mu = 0$ has inflection points at $\theta = \pi/2,\ 3\pi/2$ and the distribution function is

$$F(\theta) = \frac{1}{2\pi}\left(\theta + 2\rho\sin\theta\right), \quad 0 \le \theta < 2\pi.$$

The cardioid distribution with pdf (4.36) reduces to the circular uniform distribution (Section 4.2) if $\rho = 0$. Note that the case of $n = 1$ in (4.35) reduces to a cardioid pdf.

If Θ has pdf (4.36), then the characteristic function or pth trigonometric moment is

$$E(e^{ip\Theta}) = \int_0^{2\pi} e^{ip\theta}\,\frac{1}{2\pi}\left\{1 + 2\rho\cos(\theta-\mu)\right\}d\theta = \begin{cases} 1, & p = 0, \\ \rho\, e^{i\mu}, & p = 1, \\ 0, & p \ge 2, \end{cases}$$

so that the mean direction of Θ is μ and the mean resultant length is ρ. The circular variance is $\nu = 1 - \rho$ and satisfies the inequality $1/2 \le \nu \le 1$. The cardioid distribution never degenerates a distribution concentrated at the only one point

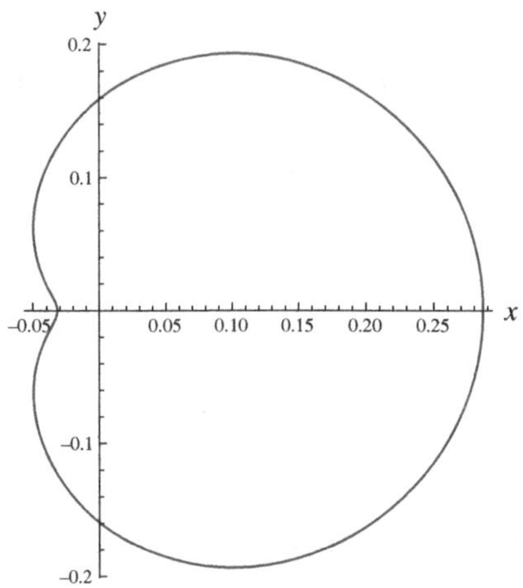

Figure 4.9 Shape of $r = f(\theta)$ in (4.36) with $\mu = 0$ and $\rho = 0.4$ in polar coordinates.

whatever the value of ρ. If Θ_1 is distributed as a cardioid distribution with mean direction μ_1 and mean resultant length ρ_1, and Θ_2 (not necessarily distributed as a cardioid) is independent of Θ_1 and has mean direction μ_2 and mean resultant length ρ_2, then the sum $\Theta_1 + \Theta_2$ is distributed as a cardioid distribution with mean direction $\mu_1 + \mu_2$ and mean resultant length $\rho_1\rho_2$ since the trigonometric moment of $\Theta_1 + \Theta_2$ is

$$E\left\{e^{ip(\Theta_1+\Theta_2)}\right\} = \begin{cases} 1, & p = 0, \\ \rho_1\rho_2 e^{i(\mu_1+\mu_2)}, & p = 1, \\ 0, & p \geq 2. \end{cases}$$

In addition, if Θ_j $(j = 1, \dots, n)$ are mutually independent and each Θ_j is distributed as a cardioid distribution with parameters μ_j and ρ_j, then the sum $\sum_{j=1}^{n} \Theta_j$ is also distributed as a cardioid distribution with parameters $\sum_{j=1}^{n} \mu_j$ and $\prod_{j=1}^{n} \rho_j$. It is easy to see that the distribution of the sum $\sum_{j=1}^{n} \Theta_j$ converges to the circular uniform distribution as n tends to infinity since $\prod_{j=1}^{n} \rho_j \to 0$ as $n \to \infty$ for $0 \leq \rho_j \leq 1/2$ $(j = 1, \dots, n)$.

4.6.3 Power Cardioid Family

4.6.3.1 Definitions and Properties

An extension of the Cartwright power-of-cosine (4.35), cardioid (4.36), and Mitsuyasu (4.5) pdfs is given by

$$g(\theta) = C_{\rho,r}^{-1}\{1 + \rho\cos(\theta - \mu)\}^r, \quad 0 \leq \theta < 2\pi \tag{4.37}$$

with parameters $-1 < \rho < 1$, $-\infty < r < \infty$, $0 \leq \mu < 2\pi$ and the normalizing constant $C_{\rho,r}$. This power cardioid family appears in SenGupta et al. (2021) as the circular marginal pdf of a mixed distribution for linear discrete and circular continuous random variables. If $\rho = \pm 1$ in (4.37), then the distribution reduces to (4.6), with the restriction $r > -1/2$. This distribution also possesses an important and interesting property that unlike most non-exponential family circular distributions, it admits of generalization to the full family of conditionally specified distributions on the torus (SenGupta and Arnold (2025)) as will be shown in Chapter 9.

Though the power cardioid distribution has been enhanced mostly from ad-hoc motivations, we show below (SenGupta, 2015) that it follows from Möbius transformation of a bimodal t-type distribution in the lines of Cartwright power-of-cosine and cardioid distributions and admits of derivation by the conditional method also.

Result 4.2 The power cardioid distribution on the circle is obtained through the Möbius transformation on the probability density function $f(x)$ given by,

$$f(x) = K_{\rho,r}^{-1} \frac{\left\{\left(1 + \frac{x^2}{2r+1}\right) + \rho\left(1 - \frac{x^2}{2r+1}\right)\right\}^r}{\left(1 + \frac{x^2}{2r+1}\right)^{r+1}},$$

where $K_{\rho,r}$ is the normalizing constant.

Proof: Denote (as for the Cartwright distribution) $m = 2r + 1$, $v = \sqrt{m}$ and $w = x/v$. Then the kernel of $f(x)$ reduces to

$$\left\{1 + \rho\left(\frac{1 - w^2}{1 + w^2}\right)\right\}^n \frac{1}{1 + w^2} = (1 + \rho\cos\theta)^n \frac{1}{1 + w^2},$$

since by Möbius transformation, $\theta = 2\arctan(w)$. The desired power cardioid density then follows by multiplying by the Jacobian of the transformation.

The above shows that several circular distributions may be derived from linear ones as special cases: $r = 1$ yields the cardioid distribution from the t distribution with 3 d.f.; $\rho = 1$ yields Cartwright power-of-cosine distribution from a t distribution; $\rho = 0$ and $m = 1$ yield circular uniform distribution from Cauchy distribution. Further, for $r = -1$, and general (not depending on r) $m > 0$, the same proof above yields the wrapped Cauchy distribution from a scaled Cauchy distribution, with scale parameter given by $\{m(1 + \rho)/(1 - \rho)\}^{1/2}$.

The normalizing constant of (4.37) is expressed as

$$C_{\rho,r} = 2\pi\ {}_2F_1\left(-\frac{r}{2}, \frac{1-r}{2}; 1; \rho^2\right).$$

Here ${}_2F_1$ denotes the Gauss hypergeometric function. The hypergeometric series terminates if r is equal to a positive integer, and in that case it also admits of an explicit representation by using De Moivre's theorem ($(\cos\theta + i\sin\theta)^n = \cos(n\theta) + i\sin(n\theta)$ for $n \in \mathbb{Z}$) relating power moments to trigonometric moments of $\cos\Theta$ and $\sin\Theta$. Note that if $r \leq 0$, it is obvious that ${}_2F_1(-r/2, (1-r)/2; 1; \rho^2) > 0$. If $r > 0$, it is not obvious that the value of ${}_2F_1$ is positive, but it follows from a formula

$$ {}_2F_1(a, b; c; z) = (1-z)^{c-a-b}\, {}_2F_1(c-a, c-b; c; z) \tag{4.38} $$

that

$$ {}_2F_1\left(-\frac{r}{2}, \frac{1-r}{2}; 1; \rho^2\right) = (1-\rho^2)^{r+1/2}\, {}_2F_1\left(\frac{r}{2}+1, \frac{r+1}{2}; 1; \rho^2\right) > 0. $$

Special cases include the circular uniform distribution for $r = 0$ as ${}_2F_1(0, 1/2; 1; \rho^2) = 1$, the cardioid distribution for $r = 1$ as ${}_2F_1(-1/2, 0; 1; \rho^2) = 1$, and the wrapped Cauchy distribution for $r = -1$ as ${}_2F_1(1/2, 1; 1; \rho^2) = {}_1F_0(1/2; \rho^2) = (1-\rho^2)^{-1/2}$. The von Mises distribution is obtained if $\rho = \kappa/r\,(\kappa > 0)$ and $r \to \infty$ because ${}_2F_1(-r/2, (1-r)/2; 1; (\kappa/r)^2)$ tends to $I_0(\kappa)$ as r goes to infinity. Modality with $\mu = 0$ is: For $\rho = 0$, the distribution is a uniform distribution. For $\rho \neq 0$, the distribution is unimodal and symmetric about $\theta = 0$ and π: for $0 < \rho < 1\,(-1 < \rho < 0)$, it has a unique mode at $\theta = 0\,(\pi)$ and an antimode at $\pi\,(0)$ if $r > 0$, while if $r < 0$, then it has a unique mode at $\theta = \pi\,(0)$ and an antimode at $0\,(\pi)$.

Parameterization different from (4.37) can be seen in Jones and Pewsey (2005) as

$$ f(\theta) = c_{\kappa,\psi}\{1 + \tanh(\kappa\psi)\cos(\theta-\mu)\}^{1/\psi}, \quad 0 \leq \theta < 2\pi, \tag{4.39} $$

where $\kappa \geq 0$, $-\infty < \psi < \infty$ ($\psi = 0$ by continuity), $0 \leq \mu < 2\pi$ and the normalizing constant of (4.39) is represented by

$$ c_{\kappa,\psi} = \frac{\{\cosh(\kappa\psi)\}^{1/\psi}}{2\pi P_{1/\psi}(\cosh(\kappa\psi))}, $$

with the associated Legendre function $P_{1/\psi}(\cdot)$ given by (B.34). Note that there exists a relationship,

$$ P_{1/\psi}(z) = \left(\frac{z+1}{2}\right)^{1/\psi} {}_2F_1\left(-1/\psi, -1/\psi; 1; \frac{z-1}{z+1}\right), \quad \left|\frac{z-1}{z+1}\right| < 1, $$

between the associated Legendre and Gauss hypergeometric functions, using (B.36) with notation $P_\nu^0(z) = P_\nu(z)$. Further, from (B.13), we have

$$ P_{1/\psi}(z) = z^{1/\psi}\, {}_2F_1\left(-\frac{1}{2\psi}, -\frac{1}{2\psi} + \frac{1}{2}; 1; \left(1-\frac{1}{z}\right)\left(1+\frac{1}{z}\right)\right). $$

Putting $z = \cosh(\kappa\psi)$, we finally have

$$ P_{1/\psi}(\cosh(\kappa\psi)) = \{\cosh(\kappa\psi)\}^{1/\psi}\, {}_2F_1\left(-\frac{1}{2\psi}, -\frac{1}{2\psi} + \frac{1}{2}; 1; \tanh^2(\kappa\psi)\right). $$

For a random variable Θ with pdf (4.37) and $\mu = 0$, the qth trigonometric moment of Θ, $\phi_q = E(e^{iq\Theta})$, for positive integer q, is equal to the qth cosine moment $\phi_q = E\{\cos(q\Theta)\} = \alpha_q$ as the qth sine moment of Θ is zero by symmetry of the pdf. For a positive integer r,

$$\alpha_q = \begin{cases} C_{\rho,r}^{-1} 2\pi \left(\frac{\rho}{2}\right)^q \sum_{k=0}^{[(r-q)/2]} \binom{r}{k} \binom{r-k}{q+k} \left(\frac{\rho}{2}\right)^{2k}, & r \geq q, \\ 0, & r < q, \end{cases} \tag{4.40}$$

where $[z]$ denotes the greatest integer less than or equal to z. For the others,

$$\begin{aligned} \alpha_q &= \frac{(-r)_q (-\xi)^q}{q!} \frac{{}_2F_1(-r, q-r; q+1; \xi^2)}{{}_2F_1(-r, -r; 1; \xi^2)} \\ &= \frac{(-r)_q (-\rho)^q}{2^r q! (1+\sqrt{1-\rho^2})^{q-r}} \frac{{}_2F_1(-r, q-r; q+1; \xi^2)}{{}_2F_1(-r/2, (1-r)/2; 1; \rho^2)} \end{aligned} \tag{4.41}$$

with $\xi = \rho/(1+\sqrt{1-\rho^2})$. Here $(\cdot)_q$ denotes Pochhammer's symbol. A proof of the expressions for α_q above has been given in SenGupta et al. (2021). In particular, for $r = 1$ (cardioid distribution), $\alpha_1 = C_{\rho,1}^{-1} 2\pi(\rho/2) = \rho/2$ as $C_{\rho,1} = 2\pi$ and $\alpha_q = 0$ for $q > 1$, and for $r = -1$ (wrapped Cauchy distribution), $\alpha_q = \{-\rho/(1+\sqrt{1-\rho^2})\}^q = \tanh^q(\kappa/2)$.

Note that (4.41) actually reduces to α_q $(r \geq q)$ in (4.40) using formulae (4.38) and ${}_2F_1(2a, 2a+1-c; c; z) = (1+z)^{-2a}\, {}_2F_1(a, a+1/2; c; 4z/(1+z)^2)$ if r is a positive integer.

4.6.3.2 Derivation by the Conditional Method

Bivariate spherically symmetric distributions with random vector $(X, Y)'$ and pdf $f_s(x, y) \propto g\{(x-m)^2 + y^2\}$ lead to circular distributions with pdf proportional to $g\{(r_s^2 + m^2) - 2mr_s \cos\theta_s\}$ as the conditional distribution of Θ_s given $R_s = r_s$, where Θ_s and R_s are the polar random variables of X and Y. As in Jones and Pewsey (2005), the power cardioid family is covered by three special choices for g.

First $g(z) = e^{-z/2}$ yields a von Mises distribution with pdf $f(\theta_s) \propto \exp(\kappa\cos\theta_s)$ in (4.39), where $0 < \kappa = mr_s < \infty$. Second the bivariate t-distribution with

$$g(z) = \left(1 + \frac{z}{\nu}\right)^{-\nu/2-1}, \quad \nu > 0,$$

gives f in (4.39) with $-1 < \psi < 0$ with $0 \leq \tanh(\kappa\psi) = \nu + m^2 + r_s^2 \leq 1$. Third consider the bivariate symmetric beta distribution with

$$g(z) = (1-z)^{b-3/2} I(0 < z < 1), \quad b > \frac{1}{2}.$$

For $b > 3/2$, we obtain f in (4.39) with $\psi > 0$ after conditioning on $R_s = c_s$ $(1-m)$ for $0 \leq c_s \leq 1$ with $0 \leq \tanh(\kappa\psi) = 2mc_s/\{1 + m - c_s^2(1-m)\} \leq 1$, and for $1/2 < b < 3/2$, f with $-1 < 1/\psi < 0$, i.e. $-\infty < \psi < -1$.

4.7 Circular Beta Distributions

Section 3.6 shows that bilinear (Möbius) transformation or (inverse) stereographic projection of distributions on the real line yield distributions on the circle. Consider here the Möbius transformation from line to line. This idea is used by Seshadri (1991). Let a random variable U follow a symmetric beta distribution with pdf

$$h(u) = \frac{(1-u^2)^{\lambda/2-1}}{2^{\lambda-1}B(\lambda/2, \lambda/2)}, \qquad -1 < u < 1,$$

where $\lambda > 0$ and $B(\cdot,\cdot)$ denotes the beta function. When we apply the transformation

$$X = \left(U - \frac{z}{2-z}\right) \bigg/ \left(\frac{z}{2-z}U - 1\right),$$

for each $z \in (-1, 1)$, which maps the interval $(-1, 1)$ onto itself, the resulting distribution of X has pdf

$$g_1(x) = \frac{2(1-z)^{\lambda/2}(1-x^2)^{\lambda/2-1}}{B(\lambda/2, \lambda/2)(2-z-zx)^{\lambda}}, \qquad -1 < x < 1. \tag{4.42}$$

This is a special case of the generalized beta pdf (4.44). In fact, (4.42) is obtained by putting $\alpha = \beta = \lambda/2$ and $\gamma = \lambda$ in (4.44) and using ${}_1F_0(\lambda/2; -z) = (1+z)^{-\lambda/2}$. Leipnik's (1947) or McCullagh's (1989) distribution $H(\eta, \nu)$ with pdf

$$g_2(x) = \frac{(1-x^2)^{\nu-1/2}}{(1-2\eta x+\eta^2)^{\nu}B(1/2, \nu+1/2)}, \qquad -1 < x < 1 \tag{4.43}$$

for $\nu > -1/2$ and $|\eta| \le 1$ is also obtained from (4.44) with $\alpha = \beta = \nu + 1/2$, $\gamma = \nu$ and $z = -4\eta/(1+\eta)^2$ using (B.6) and (B.15).

The related distribution (McCullagh, 1989) to $H(\eta, \nu)$ with pdf

$$g_3(x') = \frac{(1-x'^2)^{\nu-1/2}(1-\eta^2)}{(1-2\eta x'+\eta^2)^{\nu+1}B(1/2, \nu+1/2)}, \qquad -1 < x' < 1,$$

is denoted by $H'(\eta, \nu)$. The random variables X and X' which follow $H(\eta, \nu)$ and $H'(\eta, \nu)$ respectively are related via the transformation $X' - \eta = (X - \eta)(\eta^2 - 1)/(1 - 2\eta X + \eta^2)$ which each $-1 < \eta < 1$ maps the interval $(-1, 1)$ onto itself. See Section 8.4.10 for connection between the distribution $H'(\eta, \nu)$ and Brownian motion.

Transforming $x = \cos\theta$ in the equation

$$\frac{1}{2^{\alpha+\beta-1}B(\alpha, \beta)} \int_{-1}^{1} (1+x)^{\alpha-1}(1-x)^{\beta-1}dx = 1 \quad (\alpha > 0,\ \beta > 0),$$

we get a circular pdf

$$f(\theta) = \frac{1}{2^{\alpha+\beta}B(\alpha, \beta)}(1+\cos\theta)^{\alpha-1/2}(1-\cos\theta)^{\beta-1/2}, \quad 0 \le \theta < 2\pi,$$

which reduces to the uniform distribution if $\alpha = \beta = 1/2$. The first cosine moment of the distribution is expressible as $\int_0^{2\pi} \cos\theta f(\theta)d\theta = (\alpha - \beta)/(\alpha + \beta)$. Likewise Equations (4.42) and (4.43) lead to circular distributions with pdfs

$$f_1(\theta) = \frac{(1-z)^{\lambda/2}(1-\cos^2\theta)^{\lambda/2-1/2}}{B(\lambda/2,\lambda/2)(2-z-z\cos\theta)^{\lambda}}, \quad 0 \le \theta < 2\pi,$$

and

$$f_2(\theta) = \frac{(1-\cos^2\theta)^{\nu}}{2(1-2\eta\cos\theta+\eta^2)^{\nu}B(1/2,\nu+1/2)}, \quad 0 \le \theta < 2\pi,$$

respectively. More generally, equation (4.44) yields a circular pdf

$$f(\theta) = \frac{1}{2^{\alpha+\beta-\gamma}B(\alpha,\beta)\,{}_2F_1(\gamma,\alpha;\alpha+\beta;-z)} \times \frac{(1+\cos\theta)^{\alpha-1/2}(1-\cos\theta)^{\beta-1/2}}{(2+z+z\cos\theta)^{\gamma}}, \quad 0 \le \theta < 2\pi,$$

whose first cosine moment is given by

$$\int_0^{2\pi} \cos\theta f(\theta)d\theta = \frac{2\alpha}{\alpha+\beta}\,\frac{{}_2F_1(\gamma,\alpha+1;\alpha+\beta+1;-z)}{{}_2F_1(\gamma,\alpha;\alpha+\beta;-z)} - 1.$$

Finally, we note that circular beta models on the interval $(-1,1)$ may be viewed as models on the one-dimensional unit disc. Distributions on the p-dimensional ($p \ge 2$) disc will be treated in Section 11.5.

4.8 Circular Transformed Beta Distributions

The Cartwright power-of-cosine or Minh–Farnum distribution is induced by the transformation of a beta variable, as observed in Jones and Pewsey (2005, p. 1423).

We start with a generalized beta distribution (Johnson et al., 1994, p. 253) with pdf

$$g(x) = \frac{1}{2^{\alpha+\beta-\gamma-1}B(\alpha,\beta)\,{}_2F_1(\gamma,\alpha;\alpha+\beta;-z)}\,\frac{(1+x)^{\alpha-1}(1-x)^{\beta-1}}{(2+z+zx)^{\gamma}}, \quad -1 < x < 1, \tag{4.44}$$

for $\alpha > 0$ and $\beta > 0$, where ${}_2F_1$ denotes the Gauss hypergeometric function, then the transformation $x = \sin(\theta/2)$ with $\alpha = \beta = \delta + 1/2\,(\delta > -1/2)$ yields a circular distribution with pdf

$$f(\theta) = \frac{\Gamma(\delta+1)}{2^{\delta+1-\gamma}\sqrt{\pi}\,{}_2F_1(\gamma,\delta+1/2;2\delta+1;-z)\Gamma(\delta+1/2)} \times \frac{(1+\cos\theta)^{\delta}}{\{2+z+z\sin(\theta/2)\}^{\gamma}}, \quad -\pi \le \theta < \pi \tag{4.45}$$

using the duplication formula of gamma function, (B.2). The distribution with $\gamma = 0$ or $z = 0$ in (4.45) gives the Cartwright power-of-cosine (Minh–Farnum) and Mitsuyasu distributions as special cases.

4.9 A Circular *t*-Distribution and Its Extension

4.9.1 A Circular *t*-Distribution Induced by Inverse Stereographic Projection

As shown in Section 4.6.1, the Cartwright power-of-cosine or Minh–Fahnum distribution whose pdf is expressed by (4.35) is induced by stereographic projection. However, the pdf (4.35) lacks continuity, in the sense that the pdf of the t-distribution with $m (= 2n + 1)$ degrees of freedom tends to the standard normal pdf as $m \to \infty$, while the distribution with pdf (4.35) goes to a degenerate distribution concentrated at the point $\theta = 0$ when $n \to \infty$.

To resolve this problem, Abe et al. (2010) employed a modified stereographic projection (4.34) in Section 4.6.1 with $u = 0$ and $v > 0$ independent of n. The resulting pdf is

$$f(\theta) = \frac{v\{1 + \tan^2(\theta/2)\}}{2\sqrt{m}\, B(1/2, m/2)\{1 + v^2\tan^2(\theta/2)/m\}^{(m+1)/2}}, \tag{4.46}$$

where $B(\cdot, \cdot)$ denotes the beta function. An alternative representation of (4.46) is

$$f(\theta) = \frac{v_0(1 + \cos\theta)^n}{B(1/2, n + 1/2)(1 + v_0^2)^{n+1}(1 - z\cos\theta)^{n+1}}, \tag{4.47}$$

where $m = 2n + 1$, $v_0 = v/\sqrt{m}$ and $z = (v_0^2 - 1)/(v_0^2 + 1)$. The distribution with pdf (4.46) or (4.47) may be viewed as a t-distribution on the circle. As $m \to \infty$, (4.46) tends to

$$f(\theta) = \frac{v}{2\sqrt{2\pi}} \left\{1 + \tan^2\left(\frac{\theta}{2}\right)\right\} \exp\left\{-\frac{v^2}{2}\tan^2\left(\frac{\theta}{2}\right)\right\}, \tag{4.48}$$

without going to a degenerate distribution. Indeed, (4.48) is the pdf of the distribution when the modified stereographic projection described above is applied to the standard normal distribution on the real line. Thus, the transformation with $v > 0$ independent of n has continuity from t to the normal distribution on the circle. The pdf (4.48) is unimodal if $v \geq \sqrt{2}$ and is bimodal if $v < \sqrt{2}$.

4.9.2 An Extension

The form of the pdf (4.47) gives a hint about its extension. In fact, we consider a pdf

$$f(\theta; \mu, \rho_1, \rho_2, r) = C^{-1}_{\rho_1,\rho_2,r} \frac{\{1 + \rho_1\cos(\theta - \mu)\}^r}{\{1 - \rho_2\cos(\theta - \mu)\}^{r+1}}, \quad -\pi \leq \theta < \pi \tag{4.49}$$

where $-\pi \leq \mu < \pi$ is a location parameter, $-1 \leq \rho_1, \rho_2 \leq 1$, $-\infty < r < \infty$, and $C_{\rho_1,\rho_2,r}$ is the normalizing constant. It holds that $f(\theta; \mu, -\rho_1, -\rho_2, r) = f(\theta; \mu + \pi, \rho_1, \rho_2, r)$ if $-\pi \leq \mu < 0$, $= f(\theta; \mu - \pi, \rho_1, \rho_2, r)$ if $0 \leq \mu < \pi$. The distribution is clearly symmetric about μ. The normalizing constant is obtained as, after the use of transformation $\cos\theta = 1 - 2t$,

$$C_{\rho_1,\rho_2,r} = \int_{-\pi}^{\pi} \frac{(1 + \rho_1 \cos\theta)^r}{(1 - \rho_2 \cos\theta)^{r+1}} d\theta$$
$$= \frac{2(1 + \rho_1)^r}{(1 - \rho_2)^{r+1}} \int_0^1 \frac{(1 - z_1 t)^r}{(1 - z_2 t)^{r+1}} t^{-1/2}(1 - t)^{-1/2} dt,$$

where $z_1 = 2\rho_1/(1 + \rho_1)$ and $z_2 = -2\rho_2/(1 - \rho_2)$. The last integral is expressible in terms of Appell's function F_1 and the Gauss hypergeometric function ${}_2F_1$ as

$$\int_0^1 \frac{(1 - z_1 t)^r}{(1 - z_2 t)^{r+1}} t^{-1/2}(1 - t)^{-1/2} dt = \pi F_1\left(\frac{1}{2}; -r, 1 + r; 1; z_1, z_2\right)$$
$$= \pi(1 - z_2)^{-1/2} \, {}_2F_1\left(\frac{1}{2}, -r; 1; \frac{z_1 - z_2}{1 - z_2}\right), \quad (4.50)$$

due to the relation (B.39) between Gauss hypergeometric (B.12) and Appell's (B.38) functions. It is obvious that (4.49) reduces to the power cardioid pdf (4.37) if $\rho_2 = 0$. Abe et al. (2010) illustrate the flexibility of the pdf for $\mu = 0$ and various combinations of other three parameters and give other properties such as modality and trigonometric moments.

Additionally, using a formula (15.3.16) in Abramowitz and Stegun (1972),

$${}_2F_1(a, b; 2b; z) = \left(1 - \frac{1}{2}z\right)^{-a} {}_2F_1\left(\frac{1}{2}a, \frac{1}{2} + \frac{1}{2}a; b + \frac{1}{2}; \left(\frac{z}{2 - z}\right)^2\right),$$

the expression of ${}_2F_1$ in (4.50) is transformed into

$${}_2F_1\left(\frac{1}{2}, -r; 1; \frac{z_1 - z_2}{1 - z_2}\right) = \left\{\frac{2 - (z_1 + z_2)}{2(1 - z_2)}\right\}^r$$
$${}_2F_1\left(-\frac{r}{2}, \frac{1 - r}{2}; 1; \left\{\frac{z_1 - z_2}{2 - (z_1 + z_2)}\right\}^2\right),$$

since ${}_2F_1(a, b; 2b; z) = {}_2F_1(b, a; 2b; z)$. Finally, we have

$$C_{\rho_1,\rho_2,r} = \frac{2\pi(1 + \rho_1\rho_2)^r}{(1 - \rho_2^2)^{r+1/2}} \, {}_2F_1\left(-\frac{r}{2}, \frac{1 - r}{2}; 1; \left(\frac{\rho_1 + \rho_2}{1 + \rho_1\rho_2}\right)^2\right),$$

in terms of ρ_1 and ρ_2 instead of z_1 and z_2. If $\rho_2 = 0$, then $C_{\rho_1,\rho_2,r}$ reduces to $C_{\rho_1,r}$, the normalizing constant of the corresponding power cardioid pdf.

4.10 Projection Method

4.10.1 Rao Distribution

In the book by Rao (1973), p. 178, a bimodal distribution with density

$$f(\theta) = \frac{\sqrt{1-\rho^2}}{2\pi\{1-\rho\sin(2\theta)\}}, \quad 0 \le \theta < 2\pi, \tag{4.51}$$

is introduced. The distribution is obtained as follows.

Let a random vector $(X_1, X_2)'$ be distributed as a bivariate normal distribution with means zero, unit variances, and correlation coefficient $\rho(-1 < \rho < 1)$. Then the quotient $W = X_1/X_2$ has pdf

$$f(w) = \frac{\sqrt{1-\rho^2}}{\pi}\frac{1}{w^2 - 2\rho w + 1}, \quad -\infty < w < \infty,$$

and the pdf of $\Theta = \tan^{-1}(X_1/X_2)$ is given by

$$f(\theta) = \frac{\sqrt{1-\rho^2}}{\pi\{1-\rho\sin(2\theta)\}}, \quad -\frac{\pi}{2} < \theta < \frac{\pi}{2}, \tag{4.52}$$

where $\tan^{-1}$ denotes the inverse function of the tangent function, $\tan x$, for $-\pi/2 < x < \pi/2$. This extends the fact that if X_1 and X_2 are independently distributed random variables each having the standard normal distribution, then X_1/X_2 has the standard Cauchy distribution and $\Theta = \tan^{-1}(X_1/X_2)$ has the uniform distribution on the interval $(-\pi/2, \pi/2)$. We extend the pdf (4.52) to a function on $[-\pi, \pi)$ (this is possible because (4.52) is a periodic function), shift it to a function on $[0, 2\pi)$, and divide by 2 to get a pdf on $[0, 2\pi)$. We finally have (4.51).

The above argument illustrates the fact that the distribution of the quotient of two random variables on the real line is related to a projected distribution. In fact, the distribution with density (4.51) is a projected normal distribution in Section 4.10.2.

4.10.2 Projected Normal Distribution

The projected normal distribution $\mathrm{PN}(\boldsymbol{\eta}, \Sigma)$ is constructed from a bivariate normal distribution $N_2(\boldsymbol{\eta}, \Sigma)$ with mean vector $\boldsymbol{\eta} (\in \mathbb{R}^2)$ and variance–covariance matrix $\Sigma (> 0$: positive definite) as the marginal distribution of the angular variable after the polar transformation.

First let us derive the pdf of $\mathrm{PN}(\boldsymbol{\eta}, \Sigma)$ under the assumption $\boldsymbol{\eta} = \mathbf{0}$. Since the joint pdf of $N_2(\mathbf{0}, \Sigma)$ is

$$\phi(x_1, x_2; \mathbf{0}, \Sigma) = \frac{1}{2\pi|\Sigma|^{1/2}} \exp\left(-\frac{1}{2}\boldsymbol{x}'\Sigma^{-1}\boldsymbol{x}\right), \quad \boldsymbol{x} = (x_1, x_2)',$$

the polar transformation (3.5) yields the joint pdf of $(R, \Theta)'$

$$h(r,\theta) = \frac{1}{2\pi|\Sigma|^{1/2}} \exp\left(-\frac{r^2}{2}\boldsymbol{t}'\Sigma^{-1}\boldsymbol{t}\right) r, \quad \boldsymbol{t} = (\cos\theta, \sin\theta)'.$$

Therefore, the marginal distribution of Θ is obtained by

$$f(\theta) = \int_0^\infty h(r,\theta)dr = \frac{1}{2\pi|\Sigma|^{1/2}(\boldsymbol{t}'\Sigma^{-1}\boldsymbol{t})}. \tag{4.53}$$

An alternative expression of (4.53) is provided by

$$f(\theta) = \frac{\sqrt{1-b^2}}{2\pi[1 - b\cos\{2(\theta - \theta_0/2)\}]}, \quad 0 \le \theta < 2\pi, \tag{4.54}$$

where

$$\begin{aligned} b &= \frac{(\mathrm{tr}\Sigma - 2|\Sigma|^{1/2})^{1/2}(\mathrm{tr}\Sigma + 2|\Sigma|^{1/2})^{1/2}}{\mathrm{tr}\Sigma} \\ &= \frac{1}{\sigma_1^2 + \sigma_2^2}\sqrt{(\sigma_1^2 - \sigma_2^2)^2 + 4\sigma_1^2\sigma_2^2\rho^2}, \\ \tan\theta_0 &= \frac{2\sigma_1\sigma_2\rho}{\sigma_1^2 - \sigma_2^2}. \end{aligned}$$

It is clear that the circular density $f(\theta)$ given in (4.54) is symmetric around $\theta_0/2$. Equation (4.54) has a property $f(\theta) = f(\theta + \pi)$, $0 \le \theta < \pi$. This means that $\mathrm{PN}(\mathbf{0}, \Sigma)$ is a bimodal distribution. If Θ is distributed as $\mathrm{PN}(\mathbf{0}, \Sigma)$ with pdf (4.54), then 2Θ has a wrapped Cauchy distribution $\mathrm{WC}(\mu, \rho)$ if the range is extended to $[0, 2\pi)$ (this is possible because (4.54) is a periodic function), where

$$\rho^2 = \frac{\mathrm{tr}\Sigma - 2|\Sigma|^{1/2}}{\mathrm{tr}\Sigma + 2|\Sigma|^{1/2}}, \quad \tan\mu = \frac{2\sigma_1\sigma_2\rho}{\sigma_1^2 - \sigma_2^2}, \quad \Sigma = \begin{pmatrix} \sigma_1^2 & \sigma_1\sigma_2\rho \\ \sigma_1\sigma_2\rho & \sigma_2^2 \end{pmatrix}.$$

If $\sigma_1 = \sigma_2$ in (4.54), then

$$f(\theta) = \frac{\sqrt{1-\rho^2}}{2\pi\{1 - \rho\sin(2\theta)\}}, \quad 0 \le \theta < 2\pi, \tag{4.55}$$

which has already appeared as (4.51), and the distribution with pdf (4.55) reduces to the circular uniform distribution if $\rho = 0$. The distribution with pdf (4.53) or (4.54) might be called a central projected normal distribution since it is derived from $N_2(\mathbf{0}, \Sigma)$. However, this name is not necessarily appropriate because distributions different from $N_2(\mathbf{0}, \Sigma)$ can lead to the same projected distribution. Let us explain this fact in the paragraph below.

A random vector $\boldsymbol{X}\,(n \times 1)$ is said to have a spherical distribution if it satisfies the condition $\boldsymbol{X} \stackrel{\mathrm{d}}{=} \Gamma\boldsymbol{X}$ for all $n \times n$ orthogonal matrices Γ, where $\stackrel{\mathrm{d}}{=}$ stands for the left and right distributions are the same. The uniform distribution on the unit sphere and the standard normal distribution $N_n(\mathbf{0}, I_n)$ in n-dimensional Euclidean space belong to a family of spherical distributions. Another example is a scale mixture of normal distributions whose density function is given by

$$g_n(x) = \int_0^\infty \phi_n\left(\frac{x}{\sigma}\right) dG(\sigma),$$

where ϕ_n denotes the pdf of the n-dimensional standard normal distribution and $G(\sigma)$ represents a distribution function as a weight. Now let $\boldsymbol{X}$ be spherical and suppose that $P(\boldsymbol{X} = \boldsymbol{0}) = 0$. Consider a function $t(\boldsymbol{X})$. Then it is known (Fang et al., 1990, Theorem 2.22) that $t(\boldsymbol{X}) \stackrel{d}{=} t(\boldsymbol{Z})$, $\boldsymbol{Z} \sim N_n(\boldsymbol{0}, I_n)$ provided that $t(a\boldsymbol{X}) \stackrel{d}{=} t(\boldsymbol{X})$ for all $a > 0$. If $\boldsymbol{Y}\,(2 \times 1)$ is spherical, $P(\boldsymbol{Y} = \boldsymbol{0}) = 0$ and $\Sigma\,(2 \times 2)$ is positive definite, then $t(\boldsymbol{Y}) = \Sigma^{1/2}\boldsymbol{Y}/\|\Sigma^{1/2}\boldsymbol{Y}\|$, the direction cosine of $\boldsymbol{X} = \Sigma^{1/2}\boldsymbol{Y}$, satisfies $t(a\boldsymbol{Y}) \stackrel{d}{=} t(\boldsymbol{Y})$ for all $a > 0$. Thus, $t(\boldsymbol{Y})$ has the same distribution as $t(\boldsymbol{Z})$ when $\boldsymbol{Z} \sim N_2(\boldsymbol{0}, I_2)$. The direction cosine of $\boldsymbol{X}$ has the same distribution as the direction cosine of $\Sigma^{1/2}\boldsymbol{Z} \sim N_2(\boldsymbol{0}, \Sigma)$. Since a projected distribution of $\boldsymbol{X} = (X_1, X_2)'$ is the marginal distribution of Θ, whose pdf $f(\theta)$ is obtained from $T = \tan\Theta$, a function of the direction cosine $(\cos\Theta, \sin\Theta)'$, where $X_1 = R\cos\Theta$, $X_2 = R\sin\Theta$, the direction cosine of a spherical $\boldsymbol{X}$ has a central projected normal distribution.

Next we derive the pdf of $\mathrm{PN}(\boldsymbol{\eta}, \Sigma)$ for a general $\boldsymbol{\eta} \in \mathbb{R}^2$. The joint pdf of $N_2(\boldsymbol{\eta}, \Sigma)$ is

$$\begin{aligned}
\phi(x_1, x_2; \boldsymbol{\eta}, \Sigma) &= \frac{1}{2\pi|\Sigma|^{1/2}} \exp\left\{-\frac{1}{2}(\boldsymbol{x}-\boldsymbol{\eta})'\Sigma^{-1}(\boldsymbol{x}-\boldsymbol{\eta})\right\} \\
&= \frac{1}{2\pi\sigma_1\sigma_2\sqrt{1-\rho^2}} \\
&\quad \times \exp\left[-\frac{1}{2(1-\rho^2)}\left\{\left(\frac{x_1-\eta_1}{\sigma_1}\right)^2\right.\right. \\
&\quad \left.\left. - 2\rho\left(\frac{x_1-\eta_1}{\sigma_1}\right)\left(\frac{x_2-\eta_2}{\sigma_2}\right) + \left(\frac{x_2-\eta_2}{\sigma_2}\right)^2\right\}\right], \\
&\boldsymbol{x} = (x_1, x_2)', -\infty < x_1, x_2 < \infty,
\end{aligned}$$

where

$$\begin{aligned}
&\boldsymbol{\eta} = (\eta_1, \eta_2)', -\infty < \eta_1, \eta_2 < \infty, \\
&\Sigma = \begin{pmatrix} \sigma_1^2 & \sigma_1\sigma_2\rho \\ \sigma_1\sigma_2\rho & \sigma_2^2 \end{pmatrix}, \ 0 < \sigma_1, \sigma_2 < \infty, -1 < \rho < 1,
\end{aligned}$$

η_j denotes the mean of X_j, σ_j^2 the variance of $X_j\ (j = 1, 2)$, and ρ the correlation coefficient between X_1 and X_2. Then we have the pdf of $\mathrm{PN}(\boldsymbol{\eta}, \Sigma)$

$$f(\theta) = \frac{1}{\boldsymbol{t}'\Sigma^{-1}\boldsymbol{t}}\left\{\phi(\eta_1, \eta_2; \boldsymbol{0}, \Sigma) + \frac{D(\theta)}{|\Sigma|^{1/2}}\Phi(D(\theta))\,\phi\left(\frac{\boldsymbol{\eta}\wedge\boldsymbol{t}}{|\Sigma|^{1/2}(\boldsymbol{t}'\Sigma^{-1}\boldsymbol{t})^{1/2}}\right)\right\} \tag{4.56}$$

by using an equality

$$d^2 \int_0^\infty r e^{-d^2(r^2-2br)/2}\, dr = 1 + \sqrt{2\pi} bd\, e^{b^2d^2/2} \Phi(bd),$$

for $0 < d < \infty, -\infty < b < \infty$. Here $\phi(\cdot)$ denotes the pdf of the univariate standard normal distribution $N(0, 1)$, $\Phi(\cdot)$ the distribution function of $N(0, 1)$, i.e.

$$\phi(x) = \frac{1}{\sqrt{2\pi}} e^{-x^2/2}, \ \Phi(x) = \int_{-\infty}^{x} \phi(s)\, ds, \quad -\infty < x < \infty,$$

respectively, and

$$D(\theta) = \frac{\boldsymbol{\eta}'\Sigma^{-1}\boldsymbol{t}}{(\boldsymbol{t}'\Sigma^{-1}\boldsymbol{t})^{1/2}}, \ \boldsymbol{t} = (\cos\theta, \sin\theta)', \ \boldsymbol{\eta} \wedge \boldsymbol{t} = \eta_1 \sin\theta - \eta_2 \cos\theta. \tag{4.57}$$

If $\boldsymbol{\eta} = \boldsymbol{0}$ as a special case in (4.56), then we have (4.53). As another special case, letting $\boldsymbol{\eta} = (\eta_1, 0)'$ and $\Sigma = I_2$, we have

$$f(\theta) = \frac{1}{\sqrt{2\pi}} \phi(\eta_1) + \eta_1 \cos\theta\, \phi(\eta_1 \sin\theta)\Phi(\eta_1 \cos\theta).$$

As noted in Mardia (1972, pp. 52–53), this pdf appears in the context of calculating the power of a test (cf. Klotz, 1964).

4.10.3 Projected Pearson Type VII Distribution

Consider a scale mixture of bivariate normals $N_2(\boldsymbol{\eta}, \sigma^2\Sigma)$ with weight function $G(\sigma)$. Its pdf is

$$g(\boldsymbol{x}) = \int_0^\infty \frac{1}{2\pi|\sigma^2\Sigma|^{1/2}} \exp\left\{-\frac{1}{2}(\boldsymbol{x}-\boldsymbol{\eta})'(\sigma^2\Sigma)^{-1}(\boldsymbol{x}-\boldsymbol{\eta})\right\} dG(\sigma).$$

From (4.56), the corresponding projected distribution has pdf

$$\begin{aligned} f(\theta) \ &= \int_{\sigma=0}^{\infty} \frac{\sigma^2}{\boldsymbol{t}'\Sigma^{-1}\boldsymbol{t}} \{\phi(\eta_1, \eta_2; \boldsymbol{0}, \sigma^2\Sigma) \\ &\quad + \frac{D(\theta)}{\sigma^3|\Sigma|^{1/2}} \Phi\left(\frac{1}{\sigma}D(\theta)\right) \phi\left(\frac{\boldsymbol{\eta}\wedge\boldsymbol{t}}{\sigma|\Sigma|^{1/2}(\boldsymbol{t}'\Sigma^{-1}\boldsymbol{t})^{1/2}}\right)\right\} dG(\sigma), \end{aligned} \tag{4.58}$$

where $D(\theta) = \boldsymbol{\eta}'\Sigma^{-1}\boldsymbol{t}/(\boldsymbol{t}'\Sigma^{-1}\boldsymbol{t})^{1/2}$ and $\boldsymbol{t}, \boldsymbol{\eta} \wedge \boldsymbol{t}$ are given in (4.57).

If $G(\sigma)$ in (4.58) is given by

$$\frac{dG(\sigma)}{d\sigma} = \frac{2^{1-\nu}A^{2\nu}}{\Gamma(\nu)} \sigma^{-1-2\nu} \exp\left(-\frac{A^2}{2\sigma^2}\right), \quad 0 < A < \infty, 0 < \nu < \infty, \tag{4.59}$$

i.e. the distribution of $1/\sigma^2$ is a gamma distribution with shape parameter ν and scale parameter $2/A^2$, the resulting projected distribution is called the projected Pearson type VII distribution because it is derived from the bivariate Pearson type VII distribution with pdf

$$g(\boldsymbol{x}) = \frac{\nu A^{2\nu}}{\pi|\Sigma|^{1/2}} \left\{A^2 + (\boldsymbol{x}-\boldsymbol{\eta})'\Sigma^{-1}(\boldsymbol{x}-\boldsymbol{\eta})\right\}^{-(\nu+1)}.$$

Equation (4.58) with (4.59) is calculated by

$$\begin{aligned}
&\int_0^\infty \frac{\sigma^2}{\boldsymbol{t}'\Sigma^{-1}\boldsymbol{t}} \frac{D(\theta)}{\sigma^3|\Sigma|^{1/2}} \Phi\left(\frac{1}{\sigma}D(\theta)\right) \phi\left(\frac{\boldsymbol{\eta}\wedge\boldsymbol{t}}{\sigma|\Sigma|^{1/2}(\boldsymbol{t}'\Sigma^{-1}\boldsymbol{t})^{1/2}}\right) dG(\sigma) \\
&= \frac{|\Sigma|^{-1/2}D(\theta)2^{1-\nu}A^{2\nu}}{(\boldsymbol{t}'\Sigma^{-1}\boldsymbol{t})\Gamma(\nu)\sqrt{2\pi}} \int_0^\infty \sigma^{-2-2\nu}\left\{\frac{1}{2}+\frac{D(\theta)}{\sqrt{2\pi}\sigma}\,{}_1F_1\left(\frac{1}{2};\frac{3}{2};-\frac{(D(\theta))^2}{2\sigma^2}\right)\right\} \\
&\quad \times \exp\left[-\frac{1}{2\sigma^2}\left\{|\Sigma|^{-1}(\boldsymbol{t}'\Sigma^{-1}\boldsymbol{t})^{-1}(\boldsymbol{\eta}\wedge\boldsymbol{t})^2+A^2\right\}\right] d\sigma \\
&= \frac{|\Sigma|^{-1/2}D(\theta)}{(\boldsymbol{t}'\Sigma^{-1}\boldsymbol{t})\Gamma(\nu)}\left\{1+\frac{1}{A^2}|\Sigma|^{-1}(\boldsymbol{t}'\Sigma^{-1}\boldsymbol{t})^{-1}(\boldsymbol{\eta}\wedge\boldsymbol{t})^2\right\}^{-\nu} \\
&\quad \times \left[\frac{\Gamma(\nu+1/2)}{2\sqrt{\pi}A}\left\{1+\frac{1}{A^2}|\Sigma|^{-1}(\boldsymbol{t}'\Sigma^{-1}\boldsymbol{t})^{-1}(\boldsymbol{\eta}\wedge\boldsymbol{t})^2\right\}^{-1/2}\right. \\
&\quad + \frac{D(\theta)\Gamma(\nu+1)}{\pi A^2}\left\{1+\frac{1}{A^2}|\Sigma|^{-1}(\boldsymbol{t}'\Sigma^{-1}\boldsymbol{t})^{-1}(\boldsymbol{\eta}\wedge\boldsymbol{t})^2\right\}^{-1} \\
&\quad \left.\times\,{}_2F_1\left(\frac{1}{2},\nu+1;\frac{3}{2};-\frac{(D(\theta))^2}{|\Sigma|^{-1}(\boldsymbol{t}'\Sigma^{-1}\boldsymbol{t})^{-1}(\boldsymbol{\eta}\wedge\boldsymbol{t})^2+A^2}\right)\right]
\end{aligned}$$

provided that $\{D(\theta)\}^2\left\{|\Sigma|^{-1}(\boldsymbol{t}'\Sigma^{-1}\boldsymbol{t})^{-1}(\boldsymbol{\eta}\wedge\boldsymbol{t})^2+A^2\right\}^{-1}<1$, since

$$\int_0^\infty \frac{\sigma^2}{\boldsymbol{t}'\Sigma^{-1}\boldsymbol{t}}\phi(\eta_1,\eta_2;\boldsymbol{0},\sigma^2\Sigma)dG(\sigma) = \frac{1}{2\pi|\Sigma|^{1/2}(\boldsymbol{t}'\Sigma^{-1}\boldsymbol{t})}\left(1+\frac{1}{A^2}\boldsymbol{\eta}'\Sigma^{-1}\boldsymbol{\eta}\right)^{-\nu}$$

and

$$\Phi(z) = \frac{1}{2}+\frac{z}{\sqrt{2\pi}}\,{}_1F_1\left(\frac{1}{2};\frac{3}{2};-\frac{z^2}{2}\right),$$

where ${}_1F_1$ denotes the confluent hypergeometric function and ${}_2F_1$ the Gauss hypergeometric function defined by (B.9) and (B.12) respectively.

Setting $\nu = n/2$, $A^2 = n$ as a special case, we have, under the assumption $\{D(\theta)\}^2 < |\Sigma|^{-1}(\boldsymbol{t}'\Sigma^{-1}\boldsymbol{t})^{-1}(\boldsymbol{\eta}\wedge\boldsymbol{t})^2+n$, the projected t-distribution with n degrees of freedom

$$\begin{aligned}
g(\theta) &= \frac{1}{(\boldsymbol{t}'\Sigma^{-1}\boldsymbol{t})}\left[\frac{1}{2\pi|\Sigma|^{1/2}}\left(1+\frac{1}{n}\boldsymbol{\eta}'\Sigma^{-1}\boldsymbol{\eta}\right)^{-n/2}\right. \\
&\quad + \frac{|\Sigma|^{-1/2}D(\theta)\Gamma((n+1)/2)}{2\sqrt{2\pi n}\Gamma(n/2)}\left\{1+\frac{1}{n}|\Sigma|^{-1}(\boldsymbol{t}'\Sigma^{-1}\boldsymbol{t})^{-1}(\boldsymbol{\eta}\wedge\boldsymbol{t})^2\right\}^{-(n+1)/2} \\
&\quad + \frac{|\Sigma|^{-1/2}\{D(\theta)\}^2}{2\pi}\left\{1+\frac{1}{n}|\Sigma|^{-1}(\boldsymbol{t}'\Sigma^{-1}\boldsymbol{t})^{-1}(\boldsymbol{\eta}\wedge\boldsymbol{t})^2\right\}^{-(n+2)/2} \\
&\quad \left.\times\,{}_2F_1\left(\frac{1}{2},\frac{n}{2}+1;\frac{3}{2};-\{D(\theta)\}^2\left\{|\Sigma|^{-1}(\boldsymbol{t}'\Sigma^{-1}\boldsymbol{t})^{-1}(\boldsymbol{\eta}\wedge\boldsymbol{t})^2+n\right\}^{-1}\right)\right]. \quad (4.60)
\end{aligned}$$

If $n = 1$, then (4.60) represents the projected Cauchy distribution. If $n \to \infty$, (4.60) converges to (4.56), the pdf of the projected normal distribution $\mathrm{PN}(\boldsymbol{\eta},\Sigma)$, from Stirling's formula (B.3) and

$$
{}_2F_1\left(\tfrac{1}{2}, \tfrac{n}{2}+1; \tfrac{3}{2}; -\{D(\theta)\}^2\left\{|\Sigma|^{-1}(t'\Sigma^{-1}t)^{-1}(\eta \wedge t)^2 + n\right\}^{-1}\right)
$$
$$
\to {}_1F_1\left(\tfrac{1}{2}; \tfrac{3}{2}; -\tfrac{\{D(\theta)\}^2}{2}\right).
$$

If $\boldsymbol{\eta} = \mathbf{0}$, then (4.58) gives the same pdf (4.54) regardless of choice among $G(\sigma)$. The reason is shown in Section 4.10.2. It is clear that the projected Cauchy distribution of a bivariate standard Cauchy distribution with pdf

$$
f(x_1, x_2) = \frac{1}{2\pi(1 + x_1^2 + x_2^2)^{3/2}}, \qquad -\infty < x_1 < \infty, -\infty < x_2 < \infty,
$$

is the circular uniform distribution. However, it is not true that if X_1, X_2 are independent random variables each having the univariate standard Cauchy distribution with pdf

$$
f(x) = \frac{1}{\pi(1 + x)^2}, \qquad -\infty < x < \infty,
$$

then $\Theta = \tan^{-1}(X_2/X_1)$ is distributed as the circular uniform distribution. Actually, the pdf of $Y = X_2/X_1$ is given by

$$
g(y) = \frac{1}{\pi^2(y^2 - 1)} \log y^2, \quad -\infty < y < \infty, \ y \neq 0
$$

and $\Theta = \tan^{-1} Y$ has pdf

$$
h(\theta) = -\frac{\log(\tan^2\theta)}{\pi^2 \cos(2\theta)}, \quad -\frac{\pi}{2} < \theta < \frac{\pi}{2}, \ \theta \neq 0,
$$

where $h(\theta)$ takes its value in the sense of limit when $y = \pm 1$ $(\theta = \pm\pi/4)$.

4.11 Mixtures

Many symmetric circular distributions may be derived through the method of mixtures, some of which are presented below.

Another variant of a circular t-distribution compared to the wrapped t-distribution presented earlier now follows. Shimizu and Iida (2002) started with a scale mixture of bivariate normal distribution $N_2(\boldsymbol{\eta}, \sigma^2 I_2)$ with pdf

$$
f(\boldsymbol{y}) = \int_0^\infty \frac{1}{2\pi\sigma^2} \exp\left\{-\frac{1}{2\sigma^2}(\boldsymbol{y} - \boldsymbol{\eta})'(\boldsymbol{y} - \boldsymbol{\eta})\right\} dG(\sigma),
$$

whose random variable is denoted by $\boldsymbol{Y}$, where $G(\sigma)$ denotes a distribution function on $[0, \infty)$. After polar transformation $\boldsymbol{Y} = R\,\mathbf{e}(T)$, $\boldsymbol{\eta} = \rho\,\mathbf{e}(\tau)$, where $R = \|\boldsymbol{Y}\|$ denotes the length of $\boldsymbol{Y}$, $\mathbf{e}(T) = (\cos T, \sin T)'$ the direction cosine of $\boldsymbol{Y}$, $\rho = \|\boldsymbol{\eta}\|$ (> 0) and $\mathbf{e}(\tau) = (\cos\tau, \sin\tau)'$ with $-\pi \le \tau < \pi$, the conditional pdf of T given $R = r$

$$
f(t|r) = \frac{f(r, t)}{f(r)}, \qquad -\pi \le t < \pi, \tag{4.61}
$$

provides a circular pdf, where $f(r,t)$ is the joint pdf of $(R, T)'$

$$f(r,t) = \int_0^\infty \frac{1}{2\pi\sigma^2} \exp\left[-\frac{1}{2\sigma^2}\left\{r^2+\rho^2-2\rho r\cos(t-\tau)\right\}\right] r\,dG(\sigma),$$

and $f(r)$ is the marginal pdf of R

$$f(r) = r\int_0^\infty \frac{1}{\sigma^2}\exp\left(-\frac{r^2+\rho^2}{2\sigma^2}\right) I_0\left(\frac{\rho r}{\sigma^2}\right) dG(\sigma).$$

The family with pdf (4.61) forms a class given by the conditioning method of the scale mixture of normals with weight function $G(\sigma)$. Choice of the weight leads to various circular distributions with different tail behavior. A spherical Laplace distribution (Section 8.4.8.2) is an example. If $G(\sigma) = 0\,(0 \le \sigma < 1),\ 1\,(1 \le \sigma < \infty)$, then (4.61) obviously reduces to a von Mises distribution $\mathrm{VM}(\tau, \rho r)$.

Consider a bivariate t- or more generally Pearson type VII distribution generated by the above scale mixture with the same weight as (4.59). The joint pdf of $(R, T)'$ is

$$f_P(r,t) = r\nu(\pi A^2)^{-1}\left\{1+\frac{r^2+\rho^2-2\rho r\cos(t-\tau)}{A^2}\right\}^{-\nu-1},$$

and, from (4.61), the conditional pdf of t given $R = r$ is

$$f_P(t|r) = \frac{1}{2\pi}\left[{}_2F_1\left(\frac{\nu}{2}+\frac{1}{2},\frac{\nu}{2}+1;1;\left\{\frac{2}{A^2}\kappa(A^2,\rho|r)\right\}^2\right)\right]^{-1}$$
$$\times\left\{1-\frac{2}{A^2}\kappa(A^2,\rho|r)\cos(t-\tau)\right\}^{-\nu-1}, \quad -\pi \le t < \pi, \tag{4.62}$$

where ${}_2F_1$ denotes the Gauss hypergeometric function (B.12) and

$$\kappa(A^2,\rho|r) = \frac{\rho r}{1+(r^2+\rho^2)/A^2}.$$

The distribution with pdf (4.62) is a Pearson type VII distribution on the circle with parameters ν, A and ρ, whose mean direction and mean resultant length are τ and

$$\frac{1}{2}(\nu+1)\frac{2}{A^2}\kappa(A^2,\rho,r)\frac{{}_2F_1\left(\nu/2+1,\nu/2+3/2;2;\left\{2\kappa(A^2,\rho|r)/A^2\right\}^2\right)}{{}_2F_1\left(\nu/2+1/2,\nu/2+1;1;\left\{2\kappa(A^2,\rho|r)/A^2\right\}^2\right)},$$

as a member of the power cardioid family.

A distribution with parameters $\nu = n/2$, $A^2 = n$ (n a positive number) in (4.62) may be called the t-distribution with n degrees of freedom on the circle. Its pdf is

$$f_t(\theta) = \frac{\left\{1-(2/n)\kappa_n(\rho|r)\cos(\theta-\tau)\right\}^{-n/2-1}}{2\pi\,{}_2F_1\left(n/4+1/2,n/4+1;1;\left\{2\kappa_n(\rho|r)/n\right\}^2\right)}, \quad -\pi \le \theta < \pi,$$

with mean direction τ and concentration parameter

$$\kappa_n(\rho|r) = \frac{\rho r}{1+(\rho^2+r^2)/n}.$$

The t-distribution with n degrees of freedom on the real line converges to the standard normal distribution as n tends to infinity. Likewise the t-distribution with n degrees of freedom on the circle converges to a von Mises distribution VM$(\tau, \rho r)$ as n tends to infinity.

Mixtures (additive) of circular distributions can be constructed as for the linear case. It follows trivially that wrapping and such mixing commute – see e.g., wrapped stable-circular uniform mixture family of distributions suggested by SenGupta and Pal (2001).

4.12 Statistical Inference

The parameters of von Mises distribution can be easily computed by the method of maximum likelihood (ML). However, this is rarely the case of other distributions, mainly either because of their cumbersome normalizing constants or their representation in terms of infinite (Fourier) series. Of notable exception is the wrapped Cauchy, since it admits of a simple representation aligned with the linear Cauchy distribution – this fact was nicely exploited by Kent and Tyler (1988) to provide the ML estimators of its parameters.

The estimation problem for the wrapped stable family using the method of trigonometric moments (MTM) appears to have been first enhanced by SenGupta (1996). The MTM resulted in strikingly elegant forms of the estimators, in particular for the index parameter α. This also enhanced the MTM for estimating the parameters of the wrapped Linnik family of distributions. Recall that the index parameter α is preserved in the wrapping method. Thus, the above approach also provided a solution to the long-standing problem of obtaining an universal estimator of α, i.e. one which can be used for its entire parameter space $\Omega : \alpha \in [0, 2)$, in contrast to the Hill estimator defined only for $\alpha \geq 1$, in the stable family. This method was also conveniently used for estimating α of the Linnik family through its wrapping to the wrapped Linnik family. The asymptotic distributions, efficiencies, and CAN property of these estimators have been established – see, SenGupta and Roy (2019, 2023a,b). It is clear that the MTM approach can be used conveniently for all circular distributions which admit of tractable coefficients in their Fourier series representations.

Unlike the linear case, testing for uniformity is a fundamental step in the circular case and the classical Rayleigh's test is known to be optimal against the von Mises alternative. This test was also shown (SenGupta and Pal, 2001) to be locally most powerful (location) invariant against the entire wrapped stable – circular uniform mixture family, whatever be the value of α. This result is somewhat surprising when we recall that the nuisance parameter α is neither a location nor a scale parameter.

A fundamental problem in testing of hypotheses for even basic symmetric circular distributions arises when the nuisance parameter is not the location parameter, e.g. κ in von Mises, α in wrapped Linnik and wrapped stable, or ρ in the power cardioid distributions. The lack of invariance and sufficiency for these models lead to difficulties in deriving universally efficient and optimal tests. However, recently (SenGupta and Kulkarni, 2020) such a test was derived for testing the homogeneity of mean directions of several independent von Mises distributions with the same concentration parameter κ (similar to the ANOVA for normal distributions) by a novel technique based on integrated likelihood function.

Mardia and Jupp (2000) and Jammalamadaka and SenGupta (2001) have introduced the test for reflective symmetry by Schach (1969) against rotational alternatives and the tests for ℓ-fold symmetry by Jupp and Spurr (1983). More recently, Ley and Verdebout (2014) proposed locally and asymptotically optimal (in the maximin sense) tests for reflective symmetry about a fixed central direction against k-sine skewed alternatives (Section 5.11),

$$f^k_{\mu,\lambda}(\theta) = f_0(\theta - \mu)\left[1 + \lambda \sin\{k(\theta - \mu)\}\right], \quad -\pi \le \theta < \pi,$$

with location parameter $-\pi \le \mu < \pi$ and skewness parameter $-1 < \lambda < 1$, where f_0 is a base unimodal and symmetric density. The test statistics are based on the sine moments. When $f_0(\theta)$ is uniform with $k = 1$, the sine-skewed model corresponds to a cardioid distribution, and the test statistic coincides with the Rayleigh test statistic for testing the null hypothesis of uniformity against cardioid alternatives. When $k = 2$, the test statistic corresponds to the test by Pewsey (2004) based on the sample second moment about a known direction.

Observe that wrapped distributions are easy to simulate since they are obtained by simply wrapping (mod 2π) the simulated values from their corresponding unwrapped linear distributions. This facilitates various studies regarding their properties, comparisons, and statistical inference.

4.13 Applications

There are numerous applications of von Mises distribution documented in the books on circular statistics as well as in journals. It is when this distribution is not adequate then one explores other distributions. We give only a few such examples below.

Bartels (1984) was motivated by his study of *Dictyostelium* slugs to propose the special bidirectional reflective mixture of von Mises distributions as given earlier. In Mooney et al. (2003), mixtures of two von Mises distributions were fitted to monthly sudden infant death syndrome (SIDS) data. The SIDS data were also analyzed by Mooney et al. (2006) using circular regression, Jones and Pewsey

(2012) using inverse Batschelet distribution (Section 5.4) and Abe et al. (2013) using Papakonstantinou–Batschelet power cardioid distribution (Section 4.3.4).

Pewsey et al. (2007) illustrated the wrapped t-modeling of a dataset, consisting of cross-bed measurements from Himalayan molasse (a kind of sandstone) in Pakistan. An illustrative example is presented in Abe et al. (2010) for an analysis of the same grouped dataset of 714 observations as was considered in Jones and Pewsey (2005). The data consist of the vanishing angles of non-migratory British mallard ducks taken from Table 1 of Mardia and Jupp (2000). The results obtained from fitting the proposed distribution with pdf (4.49) as a full model and the power cardioid family of distributions by Jones and Pewsey (2005) and the modified distribution with pdf (4.46) or (4.47) submodels are presented. The AIC values 1743.44, 1744.72, and 1743.82, respectively, identify the full family as providing an optimal model among the three models, but the difference is small. The corresponding p-values 0.082, 0.047, and 0.120 provide more evidence of the superior fit of the modified distribution with pdf (4.46). Likelihood-based inference for the Batschelet distribution is studied by Pewsey et al. (2011) with an illustrative example of cross-bed measurements from Himalayan molasse (a type of sandstone).

Fisher and Lee (1992) and Gill and Hangartner (2010) considered the following circular regression. The von Mises distributions have mean directions μ_j ($j = 1, \ldots, n$) and equal concentration parameter κ with pdf

$$f(\theta_j) = \frac{1}{2\pi I_0(\kappa)} \exp\{\kappa \cos(\theta_j - \mu_j)\}, \quad 0 \le \theta_j < 2\pi,\ 0 \le \mu_j < 2\pi,\ \kappa > 0,$$

and μ's are related to the vector of covariates $\boldsymbol{x}_j$ in the form $\mu_j = \mu + g(\boldsymbol{\beta}'\boldsymbol{x}_j)$ with a link function, g, and a vector of regression coefficients, $\boldsymbol{\beta}$.

Gill and Hangartner (2010) give an analysis of domestic terrorism data (data in political science) in the United States using the circular regression with a link function $g(t) = 2\tan^{-1}(t)$. Potential explanatory variables include the nature of the target, the number of attackers, the level of violence inflicted, the motivation for the attack, a count of deaths and injuries to targeted individuals and to attackers, the type of attacker, any property damage, and the number of pages in the newspaper describing the event. The outcome variable is the day of the year. Central research question (Gill and Hangartner, 2010) is whether there is a cyclical pattern within the 150 years of the data. The result shows that there is evidence of a within-year cyclical trend, and there is some evidence of an across-year trend.

Wrapped Linnik distribution has been used (SenGupta and Roy, 2023a) to adequately model time of traffic accidents (in Srinagar, India) and the well-known Jander's ants data, for both of which neither von Mises nor any member of the wrapped stable family gave an acceptable fit.

Circular distributions have been used in an inverse way for problems with analyzing real-line data. For fitting the long-tailed stable and Linnik distributions to financial data, e.g. the log-returns of the gold price data which were collected per ounce in US dollars over the years 1980–2008, the index parameters were conveniently estimated through their wrapped versions (SenGupta and Roy, 2019, 2023b). Also the possibly bimodal distribution which was presented to derive the power cardioid distribution through the Möbius transformation has been enhanced (SenGupta, 2015) as a competitor to stable and Linnik distributions for modeling financial data, and also employed in bicolor image analysis, e.g. chest X-rays for COVID-19 patients (SenGupta and Dinda, 2025).

References

Abe, T., Pewsey, A. and Shimizu, K. (2009). On Papakonstantinou's extension of the cardioid distribution, *Statistics & Probability Letters*, 79, 2138–2147.

Abe, T., Pewsey, A. and Shimizu, K. (2013). Extending circular distributions through transformation of argument, *Annals of the Institute of Statistical Mathematics*, 65, 833–858.

Abe, T., Shimizu, K. and Pewsey, A. (2010). Symmetric unimodal models for directional data motivated by inverse stereographic projection, *Journal of the Japan Statistical Society*, 40, 45–61.

Abramowitz, M. and Stegun, I. A. Eds. (1972). *Handbook of Mathematical Functions with Formulas, Graphs, and Mathematical Tables*, Dover, New York.

Bartels, R. (1984). Estimation in a bidirectional mixture of von Mises distributions, *Biometrics*, 40, 777–784.

Batschelet, E. (1981). *Circular Statistics in Biology*, Academic Press, London.

Bingham, M. S. (1978). A characterization of the uniform distribution on the circle in the analysis of directional data, *Journal of Applied Probability*, 15, 852–857.

Cartwright, D. E. (1963). The use of directional spectra in studying the output of a wave recorder on a moving ship, *Proceedings of Conference on Ocean Wave Spectra*, sponsored by The U.S. Naval Oceanographic Office and Division of Earth Sciences, National Academy of Sciences, National Research Council. Prentice-Hall, Englewood Cliffs, NJ, 203–218.

Chhikara, R. J. and Folks, J. L. (1989). *The Inverse Gaussian Distribution: Theory, Methodology, and Applications*, Dekker, New York.

Coeurjolly, J. F. and Le Bihan, N. (2012). Geodesic normal distribution on the circle, *Metrika*, 75, 977–995.

Downs, T. D. (1966). Some relationships among the von Mises distributions of different dimensions, *Biometrika*, 53, 269–272.

Erdogan, M. B. and Ostrovskii, I. V. (1998). Analytic and asymptotic properties of generalized Linnik probability densities, *Journal of Mathematical Analysis and Applications*, 217, 555–578.

Fang, K. T., Kotz, S. and Ng, K. W. (1990). *Symmetric Multivariate and Related Distributions*, Chapman and Hall, London.

Fisher, N. I. and Lee, A. J. (1992). Regression models for an angular response, *Biometrics*, 48, 665–677.

Gill, J. and Hangartner, D. (2010). Circular data in political science and how to handle it, *Political Analysis*, 18, 316–336.

Gradshteyn, I. S. and Ryzhik, I. M. (1994). *Table of Integrals, Series, and Products*, Fifth Edition, Academic Press, London.

Gradshteyn, I. S. and Ryzhik, I. M. (2007). *Table of Integrals, Series, and Products*, Seventh Edition, Elsevier, Amsterdam.

Gruet, J. C. (2000). A note on hyperbolic von Mises distributions, *Bernoulli*, 6, 1007–1020.

Gumbel, E. J., Greenwood, J. A. and Durand, D. (1953). The circular normal distribution: Theory and tables, *Journal of the American Statistical Association*, 48, 131–152.

Hartman, P. (1976). Completely monotone families of solutions of n-th order linear differential equations and infinitely divisible distributions, *Annali della Scuola Normale Superiore di Pisa - Classe di Scienze* (4), 111, 267–287.

Hartman, P. and Watson, G. S. (1974). "Normal" distribution functions on spheres and the modified Bessel functions, *The Annals of Probability*, 2, 593–607.

Jammalamadaka, S. R. and SenGupta, A. (2001). *Topics in Circular Statistics*, World Scientific, Singapore.

Jeffreys, H. (1961). *Theory of Probability*, Third Edition, Clarendon Press, Oxford.

Johnson, N. L., Kotz, S. and Balakrishnan, N. (1994). *Continuous Univariate Distributions*, Volume 2, Second Edition, Wiley, New York.

Johnson, R. A. (2022). Transformations to improve the approximation by a von Mises distribution, In *Directional Statistics for Innovative Applications: A Bicentennial Tribute to Florence Nightingale*, (Editors: SenGupta, A. and Arnold, B. C.), Springer Nature, Singapore, 147–158.

Jones, M. C. and Pewsey, A. (2005). A family of symmetric distributions on the circle, *Journal of the American Statistical Association*, 100, 1422–1428.

Jones, M. C. and Pewsey, A. (2012). Inverse Batschelet distributions for circular data, *Biometrics*, 68, 183–193.

Jørgensen, B. (1997). *The Theory of Dispersion Models*, Chapman and Hall, London.

Jupp, P. E. and Spurr, B. D. (1983). Sobolev tests for symmetry of directional data, *The Annals of Statistics*, 11, 1225–1231.

Kareiva, P. M. and Shigesada, N. (1983). Analyzing insect movement as a correlated random walk, *Oecologia*, 56, 234–238.

Kato, S. and Shimizu, K. (2004). A further study of t-distributions on spheres, *Research Report KSTS/RR-04/012*, Department of Mathematics, Keio University.

Kent, J. T. (1977). The infinite divisibility of the von Mises-Fisher distribution for all values of the parameter in all dimensions, *Proceedings of the London Mathematical Society*, s3-35, 359–384.

Kent, J. T. and Tyler, D. E. (1988). Maximum likelihood estimation for the wrapped Cauchy dstribution, *Journal of Applied Statistics*, 15, 247–254. Also in *50 Years of Journal of Applied Statistics: Directional Statistics: Advances in Theory and Emerging Applications*, Editor: A. SenGupta. Taylor and Francis.

Kent, J. T., Mardia, K. V. and Rao, J. S. (1979). A characterization of the uniform distribution on the circle, *The Annals of Statistics*, 7, 882–889.

Klotz, J. (1964). Small sample power of the bivariate sign tests of Blumen and Hodges, *The Annals of Mathematical Statistics*, 35, 1576–1582.

Kluyver, J. C. (1905). A local probability problem, *Proceedings of the Koninklijke Nederlandse Akademie van Wetenschappen (Ned. Akad. Wet. Proc.)*, A8, 341–350.

Langevin, M. P. (1905). Magnétisme et théorie des électrons, *Annales de Chimie et de Physique*, série 8, tome 5, 70–127.

Leipnik, R. B. (1947). Distribution of the serial correlation coefficient in a circularly correlated universe, *The Annals of Mathematical Statistics*, 18, 80–87.

Lévy, P. (1939). L'addition des variables aléatoires définies sur une circonférence, *Bulletin de la Société Mathématique de France*, 67, 1–41.

Ley, C. and Verdebout, T. (2014). Simple optimal tests for circular reflective symmetry about a specified median direction, *Statistica Sinica*, 24, 1319–1339.

Linnik, Y. V. (1963). Linear forms and statistical criteria, I, II. *Selected Translations in Mathematical Statistics and Probability*, 3, 41–90.

Longuet-Higgins, M. S., Cartwright, D. E. and Smith, N. D. (1963). Observations of the directional spectrum of sea waves using the motions of a floating buoy, *Proceedings of Conference on Ocean Wave Spectra*, sponsored by The U.S. Naval Oceanographic Office and Division of Earth Sciences, National Academy of Sciences, National Research Council. Prentice-Hall, Englewood Cllifs, 111–132.

Mardia, K. V. (1972). *Statistics of Directional Data*, Academic Press, London.

Mardia, K. V. (2007). On some recent advancements in applied shape analysis and directional statistics, Leeds Annual Statistical Research Workshop 2007, Department of Statistics, University of Leeds, *Systems Biology & Statistical Bioinformatics*, S. Barber, P. D. Baxter and K. V. Mardia (Eds.), Leeds University Press, Leeds, 9–17.

Mardia, K. V. and El-Atoum, S. A. M. (1976). Bayesian inference for the von Mises-Fisher distribution, *Biometrika*, 63, 203–206.

Mardia, K. V. and Jupp, P. E. (2000). *Directional Statistics*, Wiley, Chichester.

Mardia, K. V. and Sutton, T. W. (1975). On the modes of a mixture of two von Mises distributions, *Biometrika*, 62, 699–701.

McCullagh, P. (1989). Some statistical properties of a family of continuous univariate distributions, *Journal of the American Statistical Association*, 84, 125–129.

McCullagh, P. (1996). Möbius transformation and Cauchy parameter estimation, *The Annals of Statistics*, 24, 787–808.

Minh, D. L. and Farnum, N. R. (2003). Using bilinear transformations to induce probability distributions, *Communications in Statistics - Theory and Methods*, 32, 1–9.

Mitsuyasu, H., Tasai, F., Suhara, T., Mizuno, S., Ohkusu, M., Honda, T. and Rikiishi, K. (1975). Observations of the directional spectrum of ocean waves using a cloverleaf buoy, *Journal of Physical Oceanography*, 5, 750–760.

Mooney, J. A., Helms, P. J. and Jolliffe, I. T. (2003). Fitting mixtures of von Mises distributions: A case study involving sudden infant death syndrome, *Computational Statistics & Data Analysis*, 41, 505–513.

Mooney, J. A., Jolliffe, I. T. and Helms, P. J. (2006). Modelling seasonally varying data: A case study for Sudden Infant Death Syndrome (SIDS), *Journal of Applied Statistics*, 33, 535–547.

Papakonstantinou, V. (1979). *Beiträge zur zirkulären Statistik*, Ph.D. dissertation, University of Zurich, Switzerland.

Pearson, K. (1905). The problem of the random walk, *Nature*, 72, 294.

Pewsey, A. (2004). The large-sample joint distribution of key circular statistics, *Metrika*, 60, 25–32.

Pewsey, A., Lewis, T. and Jones, M. C. (2007). The wrapped *t* family of circular distributions, *Australian & New Zealand Journal of Statistics*, 49, 79–91.

Pewsey, A., Shimizu, K. and de la Cruz, R. (2011). On an extension of the von Mises distribution due to Batschelet, *Journal of Applied Statistics*, 38, 1073–1085. Also in *50 Years of Journal of Applied Statistics: Directional Statistics: Advances in Theory and Emerging Applications*, Editor: A. SenGupta. Taylor and Francis.

Rachev, S. T. and SenGupta, A. (1993). Laplace-Weibull mixtures for modeling price changes, *Management Science*, 39, 1029–1038.

Rao, C. R. (1973). *Linear Statistical Inference and Its Applications*, Second Edition, Wiley, New York.

Rudin, W. (1987). *Real and Complex Analysis*, McGraw-Hill, New York.

Samorodnitsky, G. and Taqqu, M. S. (1994). *Stable Non-Gaussian Random Processes: Stochastic Models with Infinite Variance*, Chapman and Hall, New York.

Schach, S. (1969). Nonparametric symmetry tests for circular distributions, *Biometrika*, 56, 571–577.

SenGupta, A. (1996). Analysis of directional data in agricultural research using DDSTAP (Invited Paper), *Proceedings of the Golden Jubilee International Conference of the Indian Society of Agricultural Statistics*, 43–59.

SenGupta, A. (2015). On an efficient universal estimator of the index parameter of a symmetric stable distribution through circular statistics (Invited Lecture), *International Symposium in Statistics*, St. John's, Canada, 6–8 July 2015.

SenGupta, A. and Arnold, B.C. (2025). Circular and toroidal families of discrete distributions. *Sankhya A*, https://doi.org/10.1007/s13171-025-00424-7.

SenGupta, A. and Dinda, S. (2025). On a new bimodal distribution: A novel approach for efficient reconstruction of bicolor images, Submitted.

SenGupta, A. and Kulkarni, H. V. (2020). Universal and efficient tests for homogeneity of mean directions of circular populations, *Statistica Sinica*, 30, 1995–2021.

SenGupta, A. and Pal, C. (2001). On optimal tests for isotropy against the symmetric wrapped stable-circular uniform mixture family, *Journal of Applied Statistics*, 28, 129–143. Also in *50 Years of Journal of Applied Statistics: Directional Statistics: Advances in Theory and Emerging Applications*, Editor: A. SenGupta. Taylor and Francis.

SenGupta, A. and Roy, R. (2011). Testing homogeneity of mean directions for circular dispersion models, *Calcutta Statistical Association Bulletin: Special 7-th Triennial Proceedings Volume*, 63, 249–252.

SenGupta, A. and Roy, M. (2019). On a universal and simple estimator of the index parameter of a stable distribution using circular statistics (Invited Paper), *Financial Statistics and Data Analytics, Special Issue, Journal of Risk and Financial Management*, 12 (171), 1–28.

SenGupta, A. and Roy, M. (2023a). A characteristic function based circular distribution family and its goodness of fit: The flexible wrapped Linnik family, *Journal of Applied Statistics*, https://doi.org/10.1080/02664763.2023.2283689. Also in *50 Years of Journal of Applied Statistics: Directional Statistics: Advances in Theory and Emerging Applications*, Editor: A. SenGupta. Taylor and Francis, 1–18.

SenGupta, A. and Roy, M. (2023b). Circular-statistics-based estimators and tests for the index parameter α of distributions for high-volatility financial markets (Invited Paper), *Journal of Risk and Financial Management*, 16 (9), 405, https://doi.org/10.3390/jrfm16090405. Also in *Financial Data Analytics and Statistical Learning*, Special Issue Reprint, Editors: S. Liu, T. Ma and S. H. Ong, MDPI, 63–76.

SenGupta, A., Ozalp, M. and Yildirak, K. (2025). Compound geometric-asymmetric heavy tailed collective risk models with applications to COVID-19 regime, Submitted.

SenGupta, A., Shimizu, K., Ong, S. H. and Das, R. (2021). A flexible family of mixed distributions for discrete linear and continuous circular random variables, *Methodology and Applications of Statistics: A Volume in Honor of C.R. Rao on the Occasion of his 100th Birthday*, Arnold, B. C., Balakrishnan, N. and Coelho, C. A. (Eds.), Springer International Publishing, 365–381.

Seshadri, V. (1991). A family of distributions related to the McCullagh family, *Statistics & Probability Letters*, 12, 373–378.

Seshadri, V. (1993). *The Inverse Gaussian Distribution: A Case Study in Exponential Families*, Clarendon Press, Oxford.

Seshadri, V. (1998). *The Inverse Gaussian Distribution: Statistical Theory and Applications*, Lecture Notes in Statistics 137, Springer, New York.

Shakil, M., Ahsanullah, M. and Golam Kibria, B. M. (2019). On the Cartwright power-of-cosine circular distribution CPC(μ, ψ) - Some distributional properties and characterizations, *International Journal of Computational and Theoretical Statistics*, 6, 73–88.

Shimizu, K. and Iida, K. (2002). Pearson type VII distributions on spheres, *Communications in Statistics - Theory and Methods*, 31, 513–526.

Simon, B. (2005). *Orthogonal Polynomials on the Unit Circle*, Part 1: Classical Theory, American Mathematical Society, Providence, RI.

Stephens, M. A. (1963). Random walk on a circle, *Biometrika*, 50, 385–390.

von Mises, R. (1918). Über die "Ganzzahligkeit" der Atomgewichte und verwandte Fragen, *Physikalische Zeitschrift*, 19, 490–500.

Wintner, A. (1933). On the stable distribution laws, *American Journal of Mathematics*, 55, 335–339.

Wintner, A. (1947). On the shape of the angular case of Cauchy's distribution curves, *The Annals of Mathematical Statistics*, 18, 589–593.

Ye, X. and Ohnishi, T. (2006). Empirical Bayes estimation in von Mises distribution, *Proceedings of the Institute of Statistical Mathematics*, 54, 177–190 (in Japanese).

5

Asymmetric and Multimodal Distributions on the Circle

5.1 Preliminary Notions

In Chapter 4, several datasets were discussed which were fitted well by symmetric and unimodal distributions. But circular data are neither symmetric nor unimodal in many real-life situations. In this chapter, we consider asymmetric and multimodal probability distributions which attempt to fill this gap. In some cases, the resulting distributions include symmetric and unimodal distributions as special cases.

Wrapping, conditioning, and projection are fruitful methods for generating circular distributions from linear univariate or bivariate distributions, and thus we expect that application of wrapping, conditioning, and projection to asymmetric distributions such as gamma, lognormal, and skew-normal distributions leads to asymmetric circular distributions. On the other hand, since there are many methods for generating asymmetric distributions in linear distributions, the ideas may be applicable to the case of circular distributions. For example, in the linear case, we may transform a normal distribution into the corresponding lognormal distribution to make an asymmetric distribution by using an exponential function. However, in the circular case, it is not appropriate to apply the exponential function to a circular random variable, though it is possible to apply it to a circular density function as a method. A Möbius circle transformation is applicable to a circular random variable to get an asymmetric circular distribution from the corresponding symmetric circular distribution. Skew-normal distributions (cf. Azzalini and Capitanio, 2014) are well known and the skewing method is applicable to circular distributions. Transformation of scale as another method for generating asymmetric distributions has application to the circular case.

An alternative approach for obtaining asymmetric circular distributions is to increase the degree of the harmonic or the degree of the trigonometric polynomial in the generating symmetric distribution by, e.g. judiciously including higher-order terms such as $\cos(p\theta)$, $\sin(p\theta)$, $p \geq 2$. This may additionally lead to multimodal circular distributions, which is very desirable for a model to be highly flexible. This approach is particularly appealing when applied to generating exponential family distributions because of its closure property as well as of its amenability to optimal statistical inference (see, e.g. Beran, 1979).

We present below a rich variety of asymmetric circular and multimodal distributions, some of which are highly flexible for capturing various shapes in terms of skewness, points of inflection, high and low kurtosis (sharply peaked and flat-topped), thick and thin tails, multiple modes, etc., which are useful for modeling real-life data.

5.2 Ad Hoc

5.2.1 Asymmetric Triangular Circular Distribution

An asymmetric triangular circular distribution can be seen in Mardia (1972, p. 52). It has pdf

$$f(\theta) = \frac{1}{2\pi}\{1 + \rho(\theta - \pi)\}, \quad 0 \leq \theta < 2\pi, 0 < \rho \leq 1/\pi.$$

For $\rho = 0$, this distribution reduces to the circular uniform distribution.

5.2.2 Asymmetric Mitsuyasu Distribution

Based on the angular nature of the random variable, Mitsuyasu et al. (1975, equation (6)) proposed an angular distribution as a partial (four terms) Fourier sum, with the coefficients being suitably chosen to ensure positivity of the function and also to correspond to their directional wave spectra data. These distributions are generally asymmetric. We next observe that such partial sums of any order can be constructed with weights satisfying some general properties to ensure the positivity of the sum.

5.2.3 Fourier Series or Trigonometric Sums of Fixed Order

Further circular distributions based on nonnegative trigonometric sums or truncated Fourier series distributions on the circle are enhanced by Fernández-Durán (2004) using Fejér's (1916) result. A truncated Fourier series

$$f(\theta) = \frac{1}{2\pi}\left[1 + 2\sum_{k=1}^{n}\{a_k\cos(k\theta) + b_k\sin(k\theta)\}\right] \tag{5.1}$$

gives a pdf provided that $\sum_{k=0}^{n} |c_k|^2 = 1/(2\pi)$ and $a_k - ib_k = 2\sum_{v=0}^{n-k} c_{v+k}\bar{c}_v$ for $k = 1, \dots, n$. The uniform distribution (Section 4.2) is obtained by putting $a_k = b_k = 0$ for $k = 1, \dots, n$. For $n = 1$, Eq. (5.1) reduces to the cardioid distribution (Section 4.6).

The pth cosine moment of (5.1) is

$$\alpha_p = \int_0^{2\pi} \cos(p\theta) f(\theta) d\theta = a_p,$$

and the pth sine moment is

$$\beta_p = \int_0^{2\pi} \sin(p\theta) f(\theta) d\theta = b_p.$$

From the first trigonometric moment $\phi_1 = a_1 + ib_1$, it is easy to determine the circular mean μ and the mean resultant length ρ by putting $\phi_1 = \rho e^{i\mu}$, i.e. $\rho = \sqrt{a_1^2 + b_1^2}$ and $\cos\mu = a_1/\sqrt{a_1^2 + b_1^2}$, $\sin\mu = b_1/\sqrt{a_1^2 + b_1^2}$.

5.3 Exponential Family-Based Generalizations

The pivotal role and popularity of von Mises distribution in the arena of symmetric circular distributions are mainly attributable to its membership in the exponential family, which admits of sufficient statistics and optimal inference – for further motivations see, e.g. Beran (1979). This has enhanced asymmetric and multimodal generalizations of von Mises distribution while retaining their membership in the exponential family. We present next several such generalizations, including those by Kim and SenGupta (2013, 2021) which are highly flexible as these are members of the exponential family as well as are also capable of modeling asymmetry and multimodality simultaneously with a very small number of parameters.

5.3.1 Generalized von Mises Distributions

5.3.1.1 Definition and Properties

The generalized von Mises distribution (GvM), a special case of (5.4) with $n = 2$, is defined by pdf

$$f(\theta) = C^{-1} \exp[\kappa_1 \cos(\theta - \mu_1) + \kappa_2 \cos\{2(\theta - \mu_2)\}], \quad 0 \le \theta < 2\pi \tag{5.2}$$

for $\kappa_1, \kappa_2 \ge 0$, $0 \le \mu_1 < 2\pi$, $0 \le \mu_2 < \pi$ with normalizing constant

$$C = \int_0^{2\pi} \exp[\kappa_1 \cos(\theta - \mu_1) + \kappa_2 \cos\{2(\theta - \mu_2)\}] d\theta$$

and is denoted by $\mathrm{GvM}(\mu_1, \mu_2, \kappa_1, \kappa_2)$. This distribution has been mentioned or suggested by Maksimov (1967), Cox (Discussion of Mardia's (1975) paper) and Rukhin (1972), and studied by Yfantis and Borgman (1982) and Gatto and Jammalamadaka (2007).

The Fourier series expansion of $\exp\{\kappa_1 \cos(\theta - \mu_1)\}$ is

$$e^{\kappa_1 \cos(\theta-\mu_1)} = I_0(\kappa_1) + 2\sum_{n=1}^{\infty} I_n(\kappa_1)\cos\{n(\theta - \mu_1)\},$$

where $I_\nu(\cdot)$ is the modified Bessel function of the first kind and order ν. Similarly, we have

$$e^{\kappa_2 \cos\{2(\theta-\mu_2)\}} = I_0(\kappa_2) + 2\sum_{n=1}^{\infty} I_n(\kappa_2)\cos\{2n(\theta - \mu_2)\}$$

since

$$\int_0^{2\pi} \cos(m\theta)\, e^{\kappa\cos(2\theta)} d\theta = \begin{cases} 2\pi I_n(\kappa), & m = 2n, \\ 0, & m = 2n+1, \end{cases}$$

for $n = 0, 1, 2, \ldots$. Thus, the normalizing constant is expressed by

$$C = 2\pi \left[I_0(\kappa_1)I_0(\kappa_2) + 2\sum_{n=1}^{\infty} I_{2n}(\kappa_1)I_n(\kappa_2)\cos\{2n(\mu_2 - \mu_1)\} \right],$$

which follows from

$$\int_0^{2\pi} \cos\{j(\theta - \mu_1)\}\cos\{2k(\theta - \mu_2)\} d\theta = \begin{cases} 0, & j \neq 2k, \\ \pi\cos(2k(\mu_2 - \mu_1)), & j = 2k, \end{cases}$$

for $j, k = 1, 2, 3, \ldots$.

A three-parameter distribution with $\mu_2 = 0$ has been suggested by Rattihalli and SenGupta (2002), whose normalizing constant can be expressed in terms of the weighted sum of independent noncentral χ^2 variables. If $\kappa_2 = 0$, (5.2) reduces to the von Mises pdf with parameters μ_1 and κ_1, and if $\kappa_1 = 0$, (5.2) reduces to a von Mises pdf of the form

$$f(\theta) = \frac{1}{2\pi I_0(\kappa_2)} e^{\kappa_2\cos\{2(\theta-\mu_2)\}}, \quad 0 \le \theta < 2\pi,\ 0 \le \mu_2 < \pi,\ \kappa_2 > 0.$$

If $\kappa_1 = \kappa_2 = 0$, (5.2) reduces to the circular uniform pdf.

5.3.1.2 The Modes and Antimodes

The pdf (5.2) gives symmetric and unimodal, symmetric and bimodal, asymmetric and unimodal, or asymmetric and bimodal distributions for various combinations of the parameters. Figure 5.1 shows linear plots of symmetric and unimodal (von Mises), symmetric and bimodal (von Mises), and asymmetric and bimodal (generalized von Mises) distributions.

The modes and antimodes of (5.2) are provided by the solutions of the equation

$$\kappa_1 \sin(\theta - \mu_1) + 2\kappa_2 \sin\{2(\theta - \mu_2)\} = 0.$$

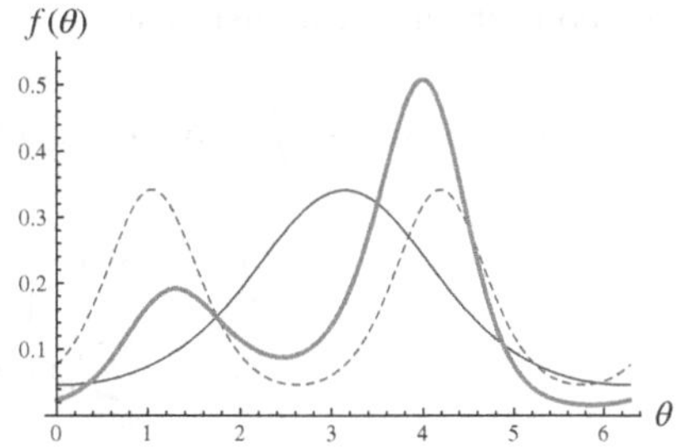

Figure 5.1 Linear plots of symmetric and unimodal von Mises ($\mu_1 = \pi$, $\kappa_1 = 1$; thin), symmetric and bimodal von Mises ($\mu_2 = \pi/3$, $\kappa_2 = 1$; dashed), and asymmetric and bimodal generalized von Mises ($\mu_1 = \pi$, $\mu_2 = \pi/3$, $\kappa_1 = \kappa_2 = 1$; thick) distributions.

Table 5.1 Modes and antimodes for special values of μ_2.

μ_2	Relation between κ_1 and κ_2	Modality	Modes	Antimodes
0 or π	$4\kappa_2 > \kappa_1$	Bimodal	$0, \pi$	$\pi - \theta_0, \pi + \theta_0$
0 or π	$4\kappa_2 < \kappa_1$	Unimodal	0	π
$\pi/2$ or $3\pi/2$	$4\kappa_2 > \kappa_1$	Bimodal	$\theta_0, 2\pi - \theta_0$	$0, \pi$
$\pi/2$ or $3\pi/2$	$4\kappa_2 < \kappa_1$	Unimodal	0	π
$\pi/4$ or $5\pi/4$	$2\kappa_2 > \kappa_1$	Bimodal	θ_1, θ_3	θ_2, θ_4
$\pi/4$ or $5\pi/4$	$2\kappa_2 < \kappa_1$	Unimodal	θ_1	θ_2
$3\pi/4$ or $7\pi/4$	$2\kappa_2 > \kappa_1$	Bimodal	θ_2^*, θ_4^*	θ_1^*, θ_3^*
$3\pi/4$ or $7\pi/4$	$2\kappa_2 < \kappa_1$	Unimodal	θ_2^*	θ_1^*

Without loss of generality, we assume that $\mu_1 = 0$. Table 5.1 (Mardia and Sutton, 1975; Yfantis and Borgman, 1982) shows a summary of the behavior of (5.2) for some values of μ_2. The notation in the table is:

$$\theta_0 = \cos^{-1}\left(\frac{\kappa_1}{4\kappa_2}\right),\ \theta_1 = \sin^{-1}\left(\frac{-\kappa_1 + \sqrt{\kappa_1^2 + 32\kappa_2^2}}{8\kappa_2}\right),\ \theta_2 = \pi - \theta_1,$$

$$\theta_3^* = \sin^{-1}\left(\frac{\kappa_1 + \sqrt{\kappa_1^2 + 32\kappa_2^2}}{8\kappa_2}\right),\ \theta_3 = \pi + \theta_3^*,\ \theta_4 = 2\pi - \theta_3^*,\ \theta_1^* = \pi + \theta_1,$$

$$\theta_2^* = 2\pi - \theta_1,\ \theta_4^* = \pi - \theta_3^*.$$

5.3.1.3 Trigonometric Moments

Let Θ denote a random variable whose distribution is $\text{GvM}(\mu_1, \mu_2, \kappa_1, \kappa_2)$ with pdf given by (5.2). The rth cosine moment of Θ is expressed by using the Fourier series expansion as

$$
\begin{aligned}
E\{\cos(r\Theta)\} &= \int_0^{2\pi} \cos(r\theta) f(\theta) d\theta \\
&= C \int_0^{2\pi} \cos(r\theta) \left[I_0(\kappa_1) I_0(\kappa_2) + 2I_0(\kappa_2) \sum_{n\geq 1} I_n(\kappa_1) \cos\{n(\theta - \mu_1)\} \right. \\
&\quad + 2I_0(\kappa_1) \sum_{n\geq 1} I_n(\kappa_2) \cos\{2n(\theta - \mu_2)\} \\
&\quad \left. + 4 \sum_{m,n\geq 1} I_m(\kappa_1) I_n(\kappa_2) \cos\{m(\theta - \mu_1)\} \cos\{2n(\theta - \mu_2)\} \right] d\theta \\
&= 2\pi C \left[I_r(\kappa_1) I_0(\kappa_2) \cos(r\mu_1) \right. \\
&\quad + \sum_{2n\geq r} I_{2n-r}(\kappa_1) I_n(\kappa_2) \cos\{2n(\mu_2 - \mu_1) + r\mu_1\} \\
&\quad + \sum_{n\geq 1} I_{2n+r}(\kappa_1) I_n(\kappa_2) \cos\{2n(\mu_2 - \mu_1) - r\mu_1\} \\
&\quad \left. + \sum_{n+2m=r} I_n(\kappa_1) I_m(\kappa_2) \cos(n\mu_1 + 2m\mu_2) \right].
\end{aligned}
$$

Here $\sum_{2n\geq r}$ denotes the sum for all possibilities of positive integers n which satisfy $2n \geq r$ when r is given, and $\sum_{n+2m=r}$ implies that the sum is taken for all possibilities of positive integers n and m which satisfy $n + 2m = r$ when r is given. Similarly,

$$
\begin{aligned}
E\{\sin(r\Theta)\} &= 2\pi C \left[I_r(\kappa_1) I_0(\kappa_2) \sin(r\mu_1) \right. \\
&\quad + \sum_{2n\geq r} I_{2n-r}(\kappa_1) I_n(\kappa_2) \sin\{2n(\mu_2 - \mu_1) + r\mu_1\} \\
&\quad - \sum_{n\geq 1} I_{2n+r}(\kappa_1) I_n(\kappa_2) \sin\{2n(\mu_2 - \mu_1) - r\mu_1\} \\
&\quad \left. + \sum_{n+2m=r} I_n(\kappa_1) I_m(\kappa_2) \sin(n\mu_1 + 2m\mu_2) \right].
\end{aligned}
$$

As a special case, the first cosine moment is, from (B.28),

$$E(\cos\Theta) = C(a\cos\mu_1 - b\sin\mu_1)\,(= \alpha,\ \text{say})$$

and the first sine moment is

$$E(\sin\Theta) = C(a\sin\mu_1 + b\cos\mu_1)\,(= \beta,\ \text{say}),$$

where

$$a = 2\pi \left[I_1(\kappa_1) I_0(\kappa_2) + 2\sum_{n=1}^{\infty} \left\{ I_{2n-1}(\kappa_1) - \frac{2n}{\kappa_1} I_{2n}(\kappa_1) \right\} I_n(\kappa_2) \cos\{2n(\mu_2 - \mu_1)\} \right],$$

$$b = \frac{8\pi}{\kappa_1} \sum_{n=1}^{\infty} n I_{2n}(\kappa_1) I_n(\kappa_2) \sin\{2n(\mu_2 - \mu_1)\}.$$

The mean direction μ is obtained from the equation $\alpha + i\beta = \rho e^{i\mu}$. Here $\rho = \sqrt{\alpha^2 + \beta^2} = C\sqrt{a^2 + b^2}$ is the mean resultant length, which is a function of $\mu_2 - \mu_1$.

5.3.1.4 Maximum Entropy

The generalized von Mises distribution is obtained by maximizing the entropy (3.1) under the restrictions on the first and second trigonometric moments

$$\int_0^{2\pi} \cos\theta f(\theta)\,d\theta = \lambda_1, \qquad \int_0^{2\pi} \sin\theta f(\theta)\,d\theta = \lambda_2,$$
$$\int_0^{2\pi} \cos(2\theta) f(\theta)\,d\theta = \lambda_3, \qquad \int_0^{2\pi} \sin(2\theta) f(\theta)\,d\theta = \lambda_4 \tag{5.3}$$

for specified constants $\lambda_1, \lambda_2, \lambda_3$, and λ_4. From the argument in Section 3.2.1, the maximizing entropy distribution has pdf

$$f(\theta) = \exp\{a_0 + a_1 \cos\theta + a_2 \sin\theta + a_3 \cos(2\theta) + a_4 \sin(2\theta)\}$$

for constants a_0, a_1, a_2, a_3, and a_4 fixed, such that $f(\theta)$ satisfies the moment conditions (5.3). After reparameterization we have the pdf (5.2).

5.3.2 Derivation by Conditioning Method

Similar to the fact that the von Mises distribution is provided by the conditioning method (Section 4.3.3.6), the generalized von Mises distribution is derived by a bivariate normal distribution (Gatto and Jammalamadaka, 2007). Let a random vector $(X_1, X_2)'$ follow $N_2(\boldsymbol{\eta}, \Sigma)$ with mean vector $\boldsymbol{\eta} = (\eta_1, \eta_2)'$ and variance–covariance matrix

$$\Sigma = \begin{pmatrix} \sigma_1^2 & \sigma_1\sigma_2\rho \\ \sigma_1\sigma_2\rho & \sigma_2^2 \end{pmatrix}, \quad \sigma_1, \sigma_2 > 0, -1 < \rho < 1.$$

Making the polar transformation $X_1 = R\cos\Theta$, $X_2 = R\sin\Theta$ with $\eta_1 = \xi\cos\tau$, $\eta_2 = \xi\sin\tau$, we have the pdf of (R, Θ)

$$g(r,\theta) = \frac{r}{2\pi\sigma_1\sigma_2\sqrt{1-\rho^2}} \exp\left[-\frac{1}{2(1-\rho^2)}\left\{\left(\frac{r\cos\theta - \xi\cos\tau}{\sigma_1}\right)^2\right.\right.$$
$$\left.\left. -2\rho\left(\frac{r\cos\theta - \xi\cos\tau}{\sigma_1}\right)\left(\frac{r\sin\theta - \xi\sin\tau}{\sigma_2}\right) + \left(\frac{r\sin\theta - \xi\sin\tau}{\sigma_2}\right)^2\right\}\right].$$

From $\cos^2\theta = \{1+\cos(2\theta)\}/2$, $\sin^2\theta = \{1-\cos(2\theta)\}/2$, and $\cos\theta\sin\theta = \{\sin(2\theta)\}/2$, the conditional distribution of Θ given $R = r$ and evaluated at $r = 1$ has pdf (5.2) after reparameterization.

5.3.3 Possibly Asymmetric 3-Parameter Bimodal Generalized von Mises

As an extension of the von Mises distribution and a subclass of the generalized von Mises distribution in Section 5.3.1, Kim and SenGupta (2013) studied a three-parameter generalized von Mises distribution, GvM$_3$ (the suffix corresponds to the number of parameters), whose pdf is given by

$$f(\theta) = \frac{1}{2\pi G_0(\pi/4,\kappa_1,\kappa_2)} \exp\left[\kappa_1\cos(\theta-\mu) + \kappa_2\sin\{2(\theta-\mu)\}\right], \quad 0 \le \theta < 2\pi,$$

with a location parameter $0 \le \mu < 2\pi$, concentration parameter $\kappa_1 > 0$, and skewness parameter $-1 \le \kappa_2 \le 1$. The distribution is GvM$(\mu, \mu+\pi/4, \kappa_1, \kappa_2)$ by using the notation in Section 5.3.1. Thus, the parametric function $G_0(\pi/4,\kappa_1,\kappa_2)$ in the normalizing constant is expressed by

$$G_0(\pi/4,\kappa_1,\kappa_2) = I_0(\kappa_1)I_0(\kappa_2) + 2\sum_{n=1}^{\infty} I_{2n}(\kappa_1)I_n(\kappa_2)\cos\left(\frac{n\pi}{2}\right).$$

The modes and antimodes are obtained from the result in Section 5.3.1 as the solutions of the equation $\kappa_1\sin(\theta-\mu) = 2\kappa_2\cos\{2(\theta-\mu)\}$. In the general case, the equation could be solved numerically. For the case when $\mu = 0$, the solutions are given by $\sin^{-1}\{(-\kappa_1 \pm \sqrt{\kappa_1^2 + 32\kappa_2^2})/(8\kappa_2)\}$ provided that $\kappa_2 \neq 0$. Hence, the distribution is unimodal if $\kappa_1 \ge 2|\kappa_2|$, and bimodal, otherwise. The densities have the property of "mirror images" of each other for κ_2 and $-\kappa_2$, i.e. $f(\theta|\kappa_1,-\kappa_2) = f(-\theta|\kappa_1,\kappa_2)$.

GvM$_3$ generalizes GvM in several ways: it replaces the $\cos(2\theta)$ term by $\sin(2\theta)$ to enhance asymmetry and reduces one parameter to provide a location parameter in the model. It is stressed in Kim and SenGupta (2013) that the distribution is useful in practice due to easy interpretation of its parameters.

5.3.4 Possibly Asymmetric Multimodal Generalized von Mises

Kim and SenGupta (2021) proposed two multimodal circular distributions which are suitable for modeling circular datasets with two or more modes. Both distributions belong to the regular exponential family of distributions and are considered as extensions of the von Mises distribution. Hence, they possess highly desirable properties, such as the existence of nontrivial sufficient statistics and optimal inferences for their parameters.

It is possible to generalize the von Mises density $f(\theta) = \exp\{\kappa\cos(\theta - \mu)\}/\{2\pi I_0(\kappa)\}$ in different forms. Since $\kappa\cos(\theta - \mu) = \kappa\cos\mu\cos\theta + \kappa\sin\mu\sin\theta$, a possible generalization (Maksimov, 1967) is to use the form

$$f(\theta) \propto \exp\left[\sum_{j=1}^{n}\{a_j\cos(j\theta) + b_j\sin(j\theta)\}\right]. \tag{5.4}$$

The main term of the density may be viewed as the exponential to the Fourier series or trigonometric sums of fixed order. The truncated Fourier series in (5.1) has the restriction to be a density. However, the distribution with density (5.4) is a maximizing entropy distribution under the restrictions on the trigonometric moments

$$\int_0^{2\pi}\cos(j\theta)f(\theta)d\theta = \lambda_j, \quad \int_0^{2\pi}\sin(j\theta)f(\theta)d\theta = \eta_j$$

for specified constants λ_j and η_j for $j = 1, \ldots, n$. The distribution is possibly asymmetric and multimodal. It is easy to see that the distribution contains the von Mises distribution ($n = 1$) and the generalized von Mises distribution ($n = 2$) with density (5.2) given by $f(\theta) \propto \exp\left[\kappa_1\cos(\theta - \mu_1) + \kappa_2\cos\left\{2(\theta - \mu_2)\right\}\right]$.

Relating the frequency of the periodic functions $\cos(p\theta)$ and $\sin(q\theta)$ having the degrees of the trigonometric polynomials as p and q, respectively, with their number of turn points, one can conveniently achieve the desired number of modes in the distribution. This fact was already utilized by Kim and SenGupta (2013) for primarily achieving bimodal asymmetric distributions GvM_3. It was further exploited by Kim and SenGupta (2021) to present the $q\,(> 2)$-modal, possibly asymmetric, distribution GvM_4. This distribution has only four parameters, including a location parameter, in striking contrast to the large number needed for the corresponding q-modal mixture density with even, e.g. von Mises components. The pdf of GvM_4 is given by

$$f(\theta) = G_4(\kappa_1, \kappa_2, q)\exp\left[\kappa_1\cos(\theta - \mu) + \kappa_2\sin\{q(\theta - \mu)\}\right], \quad 0 \le \theta < 2\pi,$$

where $G_4(\cdot)$ is the normalizing constant. The modes and antimodes (corresponding to the turning points of the associated trigonometric polynomial) are obtained as the solutions of

$$\kappa_1\sin(\theta - \mu) = q\kappa_2\cos\{q(\theta - \mu)\}.$$

A variety of shapes of this pdf exhibiting the multimodality together with varied degrees of skewness and kurtosis is given in this paper.

Another multimodal asymmetric generalization of von Mises distribution which is a member of the exponential family, named as the multimodal generalized von Mises (MGvM), was given in Kim and SenGupta (2021) as

$$f(\theta) = G_M \exp\left[\sum_{s=1}^{S} \kappa_s \cos(s\theta - \mu_s)\right], \quad -\pi \le \theta < \pi,$$

where G_M is the normalizing constant, and κ_s $(s = 1, \dots, S)$ are shape parameters. The modes and antimodes are obtained as the solutions of

$$\sum_{s=1}^{S} s\kappa_s \sin\{s(\theta - \mu_s)\} = 0,$$

which shows that none of μ_s $(s = 1, \dots, S)$ are either mode or antimode in general. The special case of the trimodal $(S = 3)$ distribution in this family has been considered in this paper in details, requiring applications of Descartes' rule of sign changes and solving for the roots of a sixth degree polynomial.

5.4 Power Batschelet Distributions

A flexible symmetric family, Power Batschelet family, of distributions was introduced by Mulder et al. (2020) which covers very low to very high peaks. This family overcomes the undesirable hurdle of not possessing an analytical representation faced by the Inverse Batschelet distributions. Multimodality can be achieved by taking, as usual, mixtures of these distributions. In this section, we discuss its genesis and trace its development.

5.4.1 Asymmetric Papakonstantinou and Batschelet Distributions

Recall the family of Papakonstantinou and Batschelet distributions discussed in Chapter 4. If cos–sin in (4.14) is replaced with cos–cos or sin–sin, then a possibly asymmetric or skew symmetric distribution with pdf

$$f(\theta) = \frac{1}{2\pi}\,[1 + \kappa \cos\{(\theta - \mu) + \nu \cos(\theta - \mu)\}]$$

is obtained. Note that due to the periodicity of $\cos\alpha$ and the linearity of the kernel in it, the normalizing constant remains unaffected. The arguments about distributional properties proceed in similar ways as given in Chapter 4 for the symmetric case.

5.4.2 Inverse Batschelet Distribution

A general approach for constructing skew distributions by Jones and Pewsey (2012) is as follows. Let $f(\theta), -\pi \le \theta < \pi$, be the pdf of a symmetric unimodal circular distribution with mode and center of symmetry at 0 and antimode at $-\pi$, which is necessarily periodic. Define $t_\nu(\theta) = t_{1,\nu}^{-1}(\theta)$, where $t_{1,\nu}(\theta) = \theta - \nu - \nu\cos\theta$, $|\nu| \le 1$. Then the functions $t_\nu(\theta)$ and $t_{1,\nu}(\theta)$ are bijections from $[-\pi, \pi)$ to $[-\pi, \pi)$ since $t_{1,\nu}(\pm\pi) = \pm\pi$, $t'_{1,\nu}(\theta) = dt_{1,\nu}(\theta)/d\theta = 1 + \nu\sin\theta \ge 0$ for $-\pi \le \theta < \pi$, and $t'_{1,\nu}(\theta) = 0$ if and only if $\nu = 1$ and $\theta = -\pi/2$ or $\nu = -1$ and $\theta = \pi/2$. We have

$$\int_{-\pi}^{\pi} f\{t_\nu(\theta)\}d\theta = \int_{-\pi}^{\pi} (1 + \nu\sin\omega)f(\omega)d\omega = 1$$

by the symmetry of f. Periodicity of $g_\nu(\theta) = f\{t_\nu(\theta)\}$ can be shown as follows. Since, for any integer k, $t_{1,\nu}(\omega + 2k\pi) = t_{1,\nu}(\omega) + 2k\pi$ and thus $\omega + 2k\pi = t_\nu\{t_{1,\nu}(\omega) + 2k\pi\}$, which equals $t_\nu\{t_{1,\nu}(\omega)\} + 2k\pi$ by the definition of t_ν and $t_{1,\nu}$, we have

$$g_\nu(\theta + 2k\pi) = f\{t_\nu(\theta + 2k\pi)\} = f\{t_\nu(\theta) + 2k\pi\} = f\{t_\nu(\theta)\} = g_\nu(\theta)$$

using the periodicity of f. As a result, the function $g_\nu(\theta)$ gives a circular pdf. The approach to construction of distributions is based on transformation of scale. More generally (Jones and Pewsey, 2012), it is possible to consider pdfs of the form $g_\tau(\theta) \propto f\{\tau(\theta)\}$ with a monotone skewing function $\tau(\theta)$, depending on ν, instead of $t_\tau(\theta)$.

Unimodality of $g_\nu(\theta) = f\{t_\nu(\theta)\}$ follows from that of $f(\theta)$ because $t_{1,\nu}(\theta)$ is a monotonically increasing function and $g'_\nu(\theta) = t'_\nu(\theta)f'\{t_\nu(\theta)\}$. We have the mode of $g_\nu(\theta)$ at $t_{1,\nu}(0) = -2\nu$, while the antimode remains at $-\pi$. Also note that $t_{-\nu}(\theta) = -t_\nu(-\theta)$ since $t_{1,-\nu}(\theta) = \theta + \nu + \nu\cos\theta = -t_{1,\nu}(-\theta)$, and hence $g_{-\nu}(\theta) = f\{t_{-\nu}(\theta)\} = f\{-t_\nu(-\theta)\} = f\{t_\nu(-\theta)\} = g_\nu(-\theta)$ using the symmetry of f, which means that the pdf associated with $-\nu$ is the reflection in zero of the pdf associated with ν.

The von Mises-based inverse Batschelet distribution (Jones and Pewsey, 2012) is defined by the pdf

$$g(\theta) = \frac{1}{2\pi I_0(\kappa)} e^{\kappa\cos\{t_\nu(\theta)\}}, \quad -\pi \le \theta < \pi$$

as a generally skew distribution. The distribution is closely related to the sine-skewed von Mises distribution (Section 5.11.3.1). Density plots of the skew von Mises distribution with different values of parameters are given in the paper by Jones and Pewsey (2012). The function $t_\nu(\theta)$ does not have an explicit analytical formula. However, the normalizing constant of the pdf $g(\theta)$ is not changed from that of the von Mises distribution. This feature is retained in the linear (in $\cos\theta$) version but not in the exponentiated von Mises version of the skew or asymmetric Batschelet distribution (Batschelet, 1981) with pdf $f_B(\theta) = \{c(\kappa, \nu)\}^{-1}e^{\kappa\cos(\theta+\nu\cos\theta)}$, whose normalizing constant is given by $c(\kappa, \nu) = \int_{-\pi}^{\pi} e^{\kappa\cos(\theta+\nu\cos\theta)}d\theta$. The above distributions are in general skewed but unimodal.

5.4.3 A Wider Family of Symmetric Distributions

To construct a wider family of symmetric distributions (Jones and Pewsey, 2012), define a transformation by

$$t_\lambda(\theta) = \frac{1-\lambda}{1+\lambda}\theta + \frac{2\lambda}{1+\lambda}s_\lambda^{-1}(\theta), \quad -1 < \lambda \le 1, \tag{5.5}$$

where

$$s_\lambda(\omega) = \omega - \frac{1}{2}(1+\lambda)\sin\omega. \tag{5.6}$$

Since

$$t_\lambda\{s_\lambda(\theta)\} = \frac{1-\lambda}{1+\lambda}s_\lambda(\theta) + \frac{2\lambda}{1+\lambda}\theta = s_{-\lambda}(\theta), \quad -1 < \lambda \le 1,$$

$t_{-1}(\theta)$ is well defined by $t_{-1}(\theta) = \lim_{\lambda\to -1} t_\lambda(\theta) = s_1(\theta) = \theta - \sin\theta$. Thus, $t_\lambda\{s_\lambda(\theta)\} = s_{-\lambda}(\theta)$ for $-1 \le \lambda \le 1$. The transformation $t_\lambda(\theta)$ is one-to-one increasing from $[-\pi, \pi)$ to itself, $t_\lambda(0) = 0$, $t_\lambda(\pm\pi) = \pm\pi$, $t_\lambda(\theta + 2k\pi) = t_\lambda(\theta) + 2k\pi$ for any integer k, and $t_\lambda^{-1}(\theta) = t_{-\lambda}(\theta)$ for $-1 \le \lambda \le 1$ because $t_{-\lambda}\{s_{-\lambda}(\theta)\} = s_\lambda(\theta) = t_\lambda^{-1}\{s_{-\lambda}(\theta)\}$.

Suppose that f involves a concentration parameter $\kappa > 0$, whose typical example is the von Mises pdf. Then the pdf

$$g_\lambda(\theta) = K_{\kappa,\lambda}^{-1} f\{t_\lambda(\theta)\}$$

provides a family of symmetric distributions on the circle, unimodal with mode at 0 and antimode at $-\pi$. Here $K_{\kappa,\lambda}$ denotes the normalizing constant

$$\begin{aligned}
K_{\kappa,\lambda} &= \int_{-\pi}^{\pi} f\left\{\frac{1-\lambda}{1+\lambda}\omega + \frac{2\lambda}{1+\lambda}s_\lambda^{-1}(\omega)\right\} d\omega \\
&= \int_{-\pi}^{\pi} \left\{1 - \frac{1}{2}(1+\lambda)\cos\theta\right\} f\left\{\frac{1-\lambda}{1+\lambda}s_\lambda(\theta) + \frac{2\lambda}{1+\lambda}\theta\right\} d\theta \\
&= \int_{-\pi}^{\pi} \left\{1 - \frac{1}{2}(1+\lambda)\cos\theta\right\} f\left\{\theta - \frac{1}{2}(1-\lambda)\sin\theta\right\} d\theta \\
&= \frac{1+\lambda}{1-\lambda} - \frac{2\lambda}{1-\lambda}\int_{-\pi}^{\pi} f\left\{\theta - \frac{1}{2}(1-\lambda)\sin\theta\right\} d\theta
\end{aligned} \tag{5.7}$$

for $-1 \le \lambda < 1$. The last equality follows from the fact that

$$\int_{-\pi}^{\pi} \left\{1 - \frac{1}{2}(1-\lambda)\cos\theta\right\} f\left\{\theta - \frac{1}{2}(1-\lambda)\sin\theta\right\} d\theta = 1.$$

For $\lambda = 1$, we have $K_{\kappa,1} = 1 - E_f(\cos\Theta)$, where Θ has pdf f. It is shown (Jones and Pewsey, 2012) that the special case where f is the von Mises pdf gives wide ranges of shapes ranging from the very flat-topped to the very sharply peaked.

5.4.4 A Family of Possibly Skewed Distributions

From the general approach for constructing skew distributions in this section, the form

$$g_{\nu,\lambda}(\theta) = K_{\kappa,\lambda}^{-1} f[t_\lambda\{t_\nu(\theta)\}]$$

provides skew and symmetric unimodal pdfs, with $t_\nu(\theta) = t_{1,\nu}^{-1}(\theta)$, where $t_{1,\nu}(\theta) = \theta - \nu - \nu\cos\theta$, $|\nu| \leq 1$, as before, $t_\lambda(\theta)$ by (5.5), and $K_{\kappa,\lambda}$ by (5.7). If f is the von Mises pdf, then we have

$$g_{\mathrm{VM};\nu,\lambda}(\theta) = \frac{1}{2\pi I_0(\kappa) K_{\kappa,\lambda}} \exp\left(\kappa\cos\left[\frac{1-\lambda}{1+\lambda} t_\nu(\theta) + \frac{2\lambda}{1+\lambda} s_\lambda^{-1}\{t_\nu(\theta)\}\right]\right), \quad \kappa > 0, \tag{5.8}$$

where s_λ is given by (5.6) and

$$K_{\kappa,\lambda} = \frac{1+\lambda}{1-\lambda} - \frac{2\lambda}{(1-\lambda)2\pi I_0(\kappa)} \int_{-\pi}^{\pi} \exp\left[\kappa\cos\left\{\theta - \frac{1}{2}(1-\lambda)\sin\theta\right\}\right] d\theta$$

for $-1 \leq \lambda < 1$, with $K_{\kappa,1} = 1 - \{I_1(\kappa)/I_0(\kappa)\}$, where $I_p(\cdot)$ denotes the modified Bessel function of the first kind and order p. Observe that, $t_\nu(\theta) = \theta$ in (5.8) yields the von Mises-based symmetric inverse Batschelet family. If $\nu = \lambda = 0$, then $g_{\mathrm{VM};\nu,\lambda}(\theta)$ clearly reduces to the von Mises pdf as a special case. Density plots of symmetric and skew members for the pdf $g_{\mathrm{VM};\nu,\lambda}(\theta)$ are given in Jones and Pewsey (2012). The figures show wide range of symmetric and skew, flat-topped and sharply peaked, pdfs. However, these families suffer from several drawbacks. The foremost hurdle is that their pdfs do not have any analytical representations (since t_λ does not) – evaluation of the pdf needs both numerical inversion and numerical integration. Their Fourier series representations are not available and neither are their trigonometric moments in any closed forms. They do not possess any nontrivial sufficient statistic. Further, they do not possess the desirable (for optimal inference) property of being a member of the regular exponential family, even though, e.g. though the generating von Mises distribution does. Also, the distributions in this family are unimodal, thus restricting high flexibility. Also, for this wider family, the normalizing constant does not remain the same as that of the base distribution, but becomes more complicated to compute. The first and the resulting last hurdles primarily become even more discouraging for implementing the Inverse Batschelet family of distributions for real-life complex models, e.g. mixture models, which are often advocated for multi-modal data. These facts led Mulder et al. (2020) to propose the symmetric Power Batschelet family of distributions discussed next.

5.4.5 Flexible Power Batschelet Family

Mulder et al. (2020) obtained an explicit function $t_\lambda^*(\theta)$ to replace $t_\lambda(\theta)$, which is virtually similar in shape to it and can lead to the evaluation of the resulting Power Batschelet distribution "more than several hundred times faster" than the corresponding Inverse Batschelet distribution.

The Power Batschelet distribution was defined by Mulder et al. (2020) as

$$f_{\text{PB}}(\theta;\mu,\kappa,\lambda) = K_{\kappa,\lambda}^{*-1} \exp[\kappa \cos\{t_\lambda^*(\theta-\mu)\}],$$

where

$$t_\lambda^*(\theta) = \text{sgn}(\theta)\,\pi \left(\frac{|\theta|}{\pi}\right)^{\frac{1-0.40522284\lambda}{1+0.40522284\lambda}}$$

with $\text{sgn}(\theta) = 1\,(\theta > 0)$, $0\,(\theta = 0)$ and $-1\,(\theta < 0)$, and $K_{\kappa,\lambda}^*$ is the normalizing constant, which can be easily computed numerically. Several appealing properties of this family have been presented by Mulder et al. (2020). This distribution has found popularity in very recently emerging applications, see e.g. virtual reality studies with vision aids by García et al. (2024).

Such an useful approach as above is not available for the skew-symmetric case.

5.5 Generalized Power Cardioid Family

The power cardioid family in Section 4.6.3 is derived by the reparameterization of a t-distribution on the circle. We use this method for constructing a generalized power cardioid family, which includes the generalized von Mises distribution as a special case.

5.5.1 Definition and Properties

A family of distributions with pdf

$$f(\theta) = C_{\rho_1,\rho_2,r,\mu,\nu}^{-1}\left[1+\rho_1\cos(\theta-\mu)+\rho_2\cos\{2(\theta-\nu)\}\right]^r, \quad 0 \le \theta < 2\pi \tag{5.9}$$

with $|\rho_1| + |\rho_2| < 1$, $0 \le \mu, \nu < 2\pi$, $-\infty < r < \infty$ and the normalizing constant

$$C_{\rho_1,\rho_2,r,\mu,\nu} = \int_0^{2\pi} \left[1+\rho_1\cos(\theta-\mu)+\rho_2\cos\{2(\theta-\nu)\}\right]^r d\theta$$

is called the generalized power cardioid family in this book. The distribution appears as the generalized t-distribution in Siew et al. (2008) under different parameterization. If $\rho_2 = 0$ in (5.9), then the generalized power cardioid distribution reduces to the power cardioid distribution with $C_{\rho_1,\rho_2,r,\mu,\nu} = 2\pi\,{}_2F_1(-r/2,(1-r)/2;1;\rho_1^2)$. The generalized von Mises distribution is obtained if $\rho_1 = \kappa_1/r$, $\rho_2 = \kappa_2/r$ and $r \to \infty$. The case $r = -1$ corresponds to a generalized wrapped Cauchy distribution.

For $r < -1$, it is proved below that the normalizing constant is expressed as

$$C_{\rho_1,\rho_2,r,\mu,\nu} = 2\pi \left\{ F_4\left(-\frac{r}{2}, \frac{1-r}{2}; 1, 1; \rho_1^2, \rho_2^2\right) + 2\sum_{j=1}^{\infty} \frac{(\rho_1/2)^{2j}(-\rho_2/2)^j(-r)_{3j}}{j!(2j)!} \cos\{2j(\mu-\nu)\} \times F_4\left(-\frac{r}{2} + \frac{3j+1}{2}, \frac{1-r}{2} + \frac{3j}{2}; j+1, 2j+1; \rho_1^2, \rho_2^2\right)\right\} \tag{5.10}$$

in terms of Appell's double hypergeometric function (B.40). However, the expression (5.10) looks valid for all r. For example, it is obvious that $C_{\rho_1,\rho_2,r,\mu,\nu} = 2\pi$ for $r = 1$, whose case corresponds to a generalized cardioid distribution.

Similar to the modality of a generalized von Mises distribution, the modes and antimodes of (5.9) are provided by the solution θ^* of an equation $\rho_1 \sin(\theta - \mu) + 2\rho_2 \sin\{2(\theta - \nu)\} = 0$. A mode or antimode depends on the sign of $h(\theta^*) = r[\rho_1 \cos(\theta^* - \mu) + 4\rho_2 \cos\{2(\theta^* - \nu)\}]$.

5.5.2 Derivation by Conditioning Method

Let $\boldsymbol{W}|(U = u)$ be a random variable given $U = u$ having a bivariate normal distribution $N_2(\boldsymbol{\eta}, \Sigma/u)$ with mean vector $\boldsymbol{\eta} \in \mathbb{R}^2$ and variance–covariance matrix Σ/u, where $\Sigma = \begin{pmatrix} \sigma_1^2 & \rho\sigma_1\sigma_2 \\ \rho\sigma_1\sigma_2 & \sigma_2^2 \end{pmatrix}$, $\sigma_1 > 0$, $\sigma_2 > 0$, $|\rho| < 1$, and $u > 0$. Suppose that U is a gamma variable with shape parameter α (>0) and scale parameter $1/\beta$ ($\beta > 0$) whose pdf is given by $g(u) = \beta^\alpha u^{\alpha-1} e^{-\beta u}/\Gamma(\alpha)$. Then the unconditional distribution of $\boldsymbol{W}$ is the bivariate Pearson type VII distribution with pdf

$$\begin{aligned} f(\boldsymbol{w}) &= \int_0^\infty f(\boldsymbol{w}|u)g(u)du \\ &= \frac{\beta^\alpha}{2\pi|\Sigma|^{1/2}\Gamma(\alpha)} \int_0^\infty u^\alpha \exp\left\{-\frac{u}{2}\left(2\beta + (\boldsymbol{w}-\boldsymbol{\eta})'\Sigma^{-1}(\boldsymbol{w}-\boldsymbol{\eta})\right)\right\} du \\ &= \frac{\alpha}{2\pi\beta|\Sigma|^{1/2}} \left[1 + \frac{(\boldsymbol{w}-\boldsymbol{\eta})'\Sigma^{-1}(\boldsymbol{w}-\boldsymbol{\eta})}{2\beta}\right]^{-\alpha-1}. \end{aligned}$$

Now make polar transformation from $\boldsymbol{W} = (W_1, W_2)' \in \mathbb{R}^2$ to a random vector $(R, \Theta)'$, i.e. $\boldsymbol{W} = \begin{pmatrix} W_1 \\ W_2 \end{pmatrix} = R \begin{pmatrix} \cos\Theta \\ \sin\Theta \end{pmatrix}$, where $R = ||\boldsymbol{W}||$ is the length of $\boldsymbol{W}$ and $(\cos\Theta, \sin\Theta)'$ is the direction cosine of $\boldsymbol{W}$, whose domain is $\{(\xi, \theta)' : \xi \geq 0, \theta \in [0, 2\pi)\}$. Similarly, we transform $\boldsymbol{\eta} = \zeta(\cos\tau, \sin\tau)', \zeta > 0$, $\tau \in [0, 2\pi)$. Then the conditional distribution of a circular variable Θ given $R = \xi$ has pdf (5.9) after appropriate reparameterization with $r = -\alpha - 1$. The asymmetric t-distribution with n degrees of freedom on the unit circle is its special case when $\alpha = \beta = n/2$.

5.6 Wrapping-Based Distributions and Families

The method of wrapping as discussed for the symmetric case in Chapter 4 is a powerful one for the asymmetric case also. To derive asymmetric circular distributions one can adopt two approaches: (i) wrap asymmetric linear and (ii) asymmetrize (skew) wrapped circular distributions and families. Distributions resulting from both these approaches are presented below – for (ii), see the Remarks at the end of the discussions on Families of Sine-Skewed Distributions (Section 5.11.2).

5.6.1 Wrapped Kumaraswamy Distribution

The pdf of Kumaraswamy's (1980) distribution on the interval $(0, 1)$ as its support is represented by

$$f_K(x) = \alpha\beta x^{\alpha-1}(1 - x^\alpha)^{\beta-1}$$

for $\alpha, \beta > 0$. The distribution function is obviously given by $F_K(x) = 1 - (1 - x^\alpha)^\beta$ for $0 < x < 1$. If $\beta = 1$, the Kumaraswamy distribution is a power function distribution and a beta distribution. If $\alpha = 1$, it is a beta distribution. If $\alpha = \beta = 1$, it reduces to the uniform distribution. It is considered that the Kumaraswamy distribution has an advantage as a beta-type distribution because the distribution function has a simple expression. See Jones (2009) for basic properties of the distribution.

A wrapped Kumaraswamy distribution is defined by the density function

$$\begin{aligned} f(\theta) &= \sum_{k=0}^{m} f_T(\theta + 2k\pi) \\ &= \frac{\alpha\beta}{2(m+1)\pi} \sum_{k=0}^{m} \left\{ \frac{\theta + 2k\pi}{2(m+1)\pi} \right\}^{\alpha-1} \left[1 - \left\{ \frac{\theta + 2k\pi}{2(m+1)\pi} \right\}^{\alpha} \right]^{\beta-1} \end{aligned}$$

on $(0, 2\pi)$ for a fixed (but otherwise arbitrarily chosen) $m (= 0, 1, 2, \dots)$, using $f_T(y)dy = f_K(x)dx$ with the transformation $2(m+1)\pi x = y$ for $0 < x < 1$ $(0 < y < 2(m+1)\pi)$. The distribution function has an expression

$$F(\theta) = \sum_{k=0}^{m} \left(\left\{ 1 - \left(\frac{k}{m+1} \right)^{\alpha} \right\}^{\beta} - \left[1 - \left\{ \frac{\theta + 2k\pi}{2(m+1)\pi} \right\}^{\alpha} \right]^{\beta} \right).$$

If $m = 0$, the density function reduces to $f(\theta) = (2\pi)^{-1} f_K(\theta/(2\pi))$ and the distribution function $F(\theta) = F_K(\theta/(2\pi)) = 1 - \left[1 - \{\theta/(2\pi)\}^\alpha\right]^\beta$ on $(0, 2\pi)$. It looks difficult to obtain trigonometric moments in a simple form.

5.6.2 Wrapped Skew-Normal Distribution

A random variable X is said to have a skew normal distribution $\mathrm{SN}(\lambda)$ (Azzalini, 1985) with skewness parameter $\lambda\,(-\infty < \lambda < \infty)$ if its pdf is given by

$$f(x) = 2\phi(x)\Phi(\lambda x), \quad -\infty < x < \infty, \tag{5.11}$$

where ϕ denotes the standard normal pdf and Φ the standard normal distribution function. The distribution belongs to the general skew or asymmetric class of Azzalini and Capitanio (2003) and Azzalini (2005) shown in Section 3.12: $f(z) = 2f_0(z)G(w(z)), -\infty < z < \infty$ for a pdf f_0 symmetric about zero, a distribution function G such that its derivative G' exists and is a pdf symmetric about zero, and an odd function w. The proof of the result is given in Section 3.12. The fact that (5.11) provides a pdf can be shown by using the identity

$$E\left[\Phi(Y)\right] = E\left[P(U \le Y|Y)\right] = P(U - Y \le 0) = \Phi\left(\frac{\mu}{\sqrt{1+\sigma^2}}\right) \tag{5.12}$$

when Y and U are independently distributed as a normal distribution with mean μ and variance σ^2 and as a standard normal distribution respectively. Equation (5.12) is also derived from differentiation of

$$I(b) = \int_{-\infty}^{\infty} \Phi(ax+b)\phi(x)dx$$

with respect to b. The moment generating function of $\mathrm{SN}(\lambda)$ is, from (5.12),

$$E(e^{tX}) = 2\,e^{t^2/2}\,\Phi\left(\frac{\lambda t}{\sqrt{1+\lambda^2}}\right).$$

Pewsey (2000) studied two parameterizations of the wrapped skew-normal distribution on the circle induced by $\mathrm{SN}(\lambda)$ on the real line. A wrapped skew-normal distribution $\mathrm{WSNC}_D(\xi, \eta, \lambda)$ with location parameter $\xi\,(-\infty < \xi < \infty)$ and scale parameter $\eta\,(>0)$ obtained from $\Theta_D = Y_D\,(\mathrm{mod}\,2\pi)$, $Y_D = \xi + \eta X$, $X \sim \mathrm{SN}(\lambda)$, has pdf

$$f(\theta) = \frac{2}{\eta}\sum_{k=-\infty}^{\infty} \phi\left(\frac{\theta + 2\pi k - \xi}{\eta}\right)\Phi\left[\lambda\left(\frac{\theta+2\pi k - \xi}{\eta}\right)\right], \quad 0 \le \theta < 2\pi.$$

It is pointed out (Pewsey, 2000) that the above direct parameterization has inconvenience when one would like to get numerical solutions based on the maximum-likelihood estimation. To solve this problem, the centered parameterization

$$Y_C = \mu + \sigma\left(\frac{X - E(X)}{\mathrm{Var}(X)}\right), \quad -\infty < \mu < \infty, \sigma > 0$$

was introduced. A wrapped skew-normal distribution $\mathrm{WSNC}_C(\mu, \sigma, \gamma_1)$ obtained from $\Theta_C = Y_C \,(\mathrm{mod}\, 2\pi)$ has pdf

$$f(\theta) = \frac{2}{\sigma\sqrt{1+c^2\gamma_1^{2/3}}} \sum_{k=-\infty}^{\infty} \phi\left[\frac{1}{\sqrt{1+c^2\gamma_1^{2/3}}}\left(\frac{\theta+2\pi k-\mu}{\sigma}+c\gamma_1^{1/3}\right)\right]$$
$$\times \Phi\left[\frac{c\gamma_1^{1/3}}{\sqrt{\left\{b^2+c^2(b^2-1)\gamma_1^{2/3}\right\}(1+c^2\gamma_1^{2/3})}}\left(\frac{\theta+2\pi k-\mu}{\sigma}+c\gamma_1^{1/3}\right)\right],$$
$$0 \le \theta < 2\pi.$$

The parameters of two parameterizations are related by

$$\xi = \mu - c\gamma_1^{1/3}\sigma, \; \eta = \sigma\sqrt{1+c^2\gamma_1^{2/3}}, \; \lambda = \frac{c\gamma_1^{1/3}}{\sqrt{b^2+c^2(b^2-1)\gamma_1^{2/3}}},$$

where $b = \sqrt{2/\pi}$, $c = \{2/(4-\pi)\}^{1/3}$ and γ_1 represents the skewness of Y_C $(-0.99527 < \gamma < 0.99527)$. Particular cases of $\mathrm{WSNC}_C(\mu, \sigma, \gamma_1)$ include

a) circular uniform distribution if $\eta \to \infty \,(\sigma \to \infty)$,
b) symmetric wrapped normal distribution if $\lambda = 0 \,(\gamma_1 = 0)$,
c) wrapped half-normal distribution if $\lambda \to \pm\infty \,(\gamma_1 \to \pm 0.99527)$.

5.7 Characteristic Function-Based Method

Even if the pdf does not exist, the method of wrapping may be invoked on the characteristic function (c.f.) to produce asymmetric, just as for the symmetric, circular distributions. This fact will be enhanced in Sections 5.7.1 and 5.7.2 dealing with the stable and Linnik families of linear densities for which in general no density exists, save for a very few special cases.

5.7.1 Wrapped Asymmetric α-Stable Family

The wrapped symmetric α-stable family of distributions was discussed in Section 4.5.1. The present section presents its generalization to the wrapped asymmetric α-stable family of distributions. This family appeared in Wintner (1947) and Mardia and Jupp (2000, p. 52), Jammalamadaka and SenGupta (2001, pp. 46–48), SenGupta (2004), Pewsey (2008), and several other references therein.

A stable distribution with parameters $\alpha\,(0 < \alpha \leq 2)$, $\beta\,(-1 \leq \beta \leq 1)$, $\tau\,(\geq 0)$ and $\mu\,(-\infty < \mu < \infty)$ on the real line (cf. Samorodnitsky and Taqqu, 1994) is defined by the distribution of a random variable X whose c.f. has the form

$$\varphi_S(t; \alpha, \beta, \tau, \mu) = E(e^{itX}) = \begin{cases} \exp\left[-\tau^\alpha |t|^\alpha \left\{1 - i\beta\, \mathrm{sgn}(t) \tan\left(\frac{\pi\alpha}{2}\right)\right\} + i\mu t\right], & \alpha \in (0,1) \cup (1,2], \\ \exp\left[-\tau|t| \left\{1 + i\beta \frac{2}{\pi}\, \mathrm{sgn}(t) \log|t|\right\} + i\mu t\right], & \alpha = 1, \end{cases} \tag{5.13}$$

where sgn(·) is the sign function used in Section 5.4.5. The random variable X is called α-stable and we write $X \sim S(\alpha, \beta, \tau, \mu)$. The parameters α, β, τ, and μ play the role of the index of stability, asymmetry or skewness, scale, and location or shift, respectively. The distribution $S(\alpha, \beta, \tau, 0)$ is said to be skewed to the right (left) if $\beta > 0\,(\beta < 0)$ and totally skewed to the right (left) if $\beta = 1\,(\beta = -1)$. For $1 < \alpha \leq 2$, μ equals the mean of X. The distribution is symmetric about μ if $\beta = 0$.

As a special case, $S(2, 0, \tau, \mu)$ is the normal distribution $N(\mu, 2\tau^2)$ with variance $2\tau^2$. Note that τ controls the scale, but is not equal to the standard deviation. The choice $\beta = 0$ is conventional because β is not specified from the equation $\beta \tan \pi = 0$ when $\alpha = 2$. Because of this constraint, α has the full support (0, 2] for the symmetric, while it is constrained to (0, 2) for the asymmetric cases.

Other special cases are $S(1, 0, \tau, \mu)$, the Cauchy distribution, and $S(1/2, 1, \tau, \mu)$, the Lévy distribution with pdf

$$f(x) = \left(\frac{\tau}{2\pi}\right)^{1/2} (x - \mu)^{-3/2} \exp\left\{-\frac{\tau}{2(x - \mu)}\right\}, \quad \mu < x < \infty.$$

The α-stable distribution links the Cauchy and normal distributions in a different sense from that done by the t-distribution.

The c.f. (5.13) can be expressed by

$$E(e^{itX}) = \exp\left[\tau^\alpha\{-|t|^\alpha + it\omega(t, \alpha, \beta)\} + i\mu_1 t\right], \tag{5.14}$$

due to Zolotarev's (1986) remark (cf. Samorodnitsky and Taqqu, 1994), where

$$\mu_1 = \begin{cases} \mu + \beta\tau^\alpha \tan\left(\frac{\pi\alpha}{2}\right), & \alpha \neq 1, \\ \mu, & \alpha = 1, \end{cases}$$

and

$$\omega(t, \alpha, \beta) = \begin{cases} \beta(|t|^{\alpha-1} - 1) \tan\left(\frac{\pi\alpha}{2}\right), & \alpha \neq 1, \\ -\beta \frac{2}{\pi} \log|t|, & \alpha = 1, \end{cases}$$

which is continuous in both α and β since

$$\lim_{\alpha\to 1}(|t|^{\alpha-1}-1)\tan\left(\frac{\pi\alpha}{2}\right) = -\frac{2}{\pi}\log|t|.$$

Some properties of $S(\alpha,\beta,\tau,\mu)$ are given by using the c.f. (5.13) as follows.

The α-stable distribution has the reproductive property. To see this, let X_1 and X_2 be independent random variables, and X_j be distributed as $S(\alpha,\beta_j,\tau_j,\mu_j)$ for $j=1,2$ with the common index of stability, α. Then the sum X_1+X_2 has $S(\alpha,\beta,\tau,\mu)$ with

$$\beta = \frac{\beta_1\tau_1^\alpha+\beta_2\tau_2^\alpha}{\tau_1^\alpha+\tau_2^\alpha},\ \tau = (\tau_1^2+\tau_2^2)^{1/\alpha},\ \mu=\mu_1+\mu_2$$

because the c.f. of X_1+X_2 is, for $\alpha\neq 1$, given by

$$\begin{aligned}&E\{e^{it(X_1+X_2)}\}\\&= \exp\left[-(\tau_1^\alpha+\tau_2^\alpha)|t|^\alpha\left\{1-i\frac{\beta_1\tau_1^\alpha+\beta_2\tau_2^\alpha}{\tau_1^\alpha+\tau_2^\alpha}\,\mathrm{sgn}(t)\tan\left(\frac{\pi\alpha}{2}\right)\right\}+i(\mu_1+\mu_2)t\right].\end{aligned}$$

The case of $\alpha=1$ is similarly obtained.

It is easy to see that if $X\sim S(\alpha,\beta,\tau,\mu)$ and a is a real constant, then $X+a\sim S(\alpha,\beta,\tau,\mu+a)$. If $X\sim S(\alpha,\beta,\tau,\mu)$ and a is a nonzero real constant, then

$$aX \sim \begin{cases} S(\alpha,\mathrm{sgn}(a)\,\beta,|a|\tau,a\mu), & \alpha\neq 1,\\ S(1,\mathrm{sgn}(a)\,\beta,|a|\tau,a\mu-\dfrac{2}{\pi}a(\log|a|)\tau\beta), & \alpha=1.\end{cases}$$

For any $0<\alpha<2$, if $X\sim S(\alpha,\beta,\tau,0)$, then $-X\sim S(\alpha,-\beta,\tau,0)$, vice versa.

As mentioned earlier, the α-stable distribution $S(\alpha,\beta,\tau,\mu)$ is symmetric about μ if $\beta=0$. We assume here that $\mu=0$. Then the c.f. of X distributed as $S(\alpha,0,\tau,0)$ has the simple form

$$E(e^{itX}) = e^{-\tau^\alpha|t|^\alpha},\quad 0<\alpha\le 2,\ \tau\ge 0.$$

If X is distributed as $S(2,0,\tau,0)\equiv N(0,2\tau^2)$ and a positive random variable A independent of X has the Laplace transform

$$E(e^{-\gamma A}) = \exp\left(-\gamma^{\alpha/2}\right),\quad \gamma>0,\ 0<\alpha<2,$$

i.e. $A\sim S(\alpha/2,1,\{\cos(\pi\alpha/4)\}^{2/\alpha},0)$, then $Z=A^{1/2}X\sim S(\alpha,0,\tau,0)$ since

$$E(e^{itZ}) = E\{E(e^{itA^{1/2}X}|A)\} = E[\exp\{-\tau^2(tA^{1/2})^2\}] = e^{-\tau^\alpha|t|^\alpha}.$$

This means that $S(\alpha,0,\tau,0)$ can be interpreted as a variance mixture of normal distribution.

For $p=0,\pm1,\pm2,\dots$, the c.f. or trigonometric moment, $\varphi(p)$, of the corresponding wrapped α-stable random variable Θ, whose distribution is denoted by

$\text{WS}(\alpha, \beta, \tau, \mu)$, is given by $E(e^{ip\Theta})$ in (5.13) or (5.14), where t is replaced by p and X by Θ. For $\alpha \in (0,1) \cup (1,2]$, the pdf of $\text{WS}(\alpha, \beta, \tau, \mu)$ is obtained by

$$\begin{aligned} f(\theta) &= \frac{1}{2\pi} \sum_{p=-\infty}^{\infty} \varphi(p) \exp(-ip\theta) \\ &= \frac{1}{2\pi} \left[1 + 2 \sum_{p=1}^{\infty} \exp(-\tau^\alpha p^\alpha) \cos\left\{ p(\theta - \mu) - \tau^\alpha p^\alpha \beta \tan\left(\frac{\pi\alpha}{2}\right) \right\} \right] \end{aligned} \tag{5.15}$$

for $0 \le \theta < 2\pi$, where μ is redefined by $\mu \,(\text{mod}\, 2\pi) \in [0, 2\pi)$. Similarly, for $\alpha = 1$, the pdf is

$$f(\theta) = \frac{1}{2\pi} \left[1 + 2 \sum_{p=1}^{\infty} \exp(-\tau p) \cos\left\{ p(\theta - \mu) + \tau p \beta \frac{2}{\pi} \log p \right\} \right]. \tag{5.16}$$

The distribution with pdfs (5.15) and (5.16) has the reproductive property because the α-stable distribution on the real line has this property.

For $p = 1, 2, 3, \ldots$, the pth mean resultant length is

$$\rho_p = \exp\{-(\tau p)^\alpha\} = \rho^{p^\alpha}, \quad 0 < \alpha \le 2,$$

where $\rho = \rho_1 = \exp(-\tau^\alpha)$, and the pth mean direction is

$$\mu_p = \begin{cases} \mu p + (\tau p)^\alpha \beta \tan\left(\frac{\pi\alpha}{2}\right), & \alpha \ne 1, \\ \mu p - \tau p \beta \frac{2}{\pi} \log p, & \alpha = 1, \end{cases}$$

or

$$\mu_p = \mu_1 p + \tau^\alpha p\, \omega(p, \alpha, \beta)$$

with

$$\mu_1 = \begin{cases} \mu + \beta \tau^\alpha \tan\left(\frac{\pi\alpha}{2}\right), & \alpha \ne 1, \\ \mu, & \alpha = 1, \end{cases}$$

and

$$\omega(p, \alpha, \beta) = \begin{cases} \beta(p^{\alpha-1} - 1) \tan\left(\frac{\pi\alpha}{2}\right), & \alpha \ne 1, \\ -\beta \left(\frac{2}{\pi}\right) \log p, & \alpha = 1. \end{cases}$$

The circular variance is $\nu = 1 - \rho = 1 - \exp(-\tau^\alpha)$ and the circular standard deviation is $\sigma = \{-2\log(1-\nu)\}^{1/2} = \{-2\log\rho\}^{1/2} = \sqrt{2}\tau^{\alpha/2}$.

The pth central cosine and sine moments, $\overline{\alpha}_p = E\{\cos p(\Theta - \mu)\}$ and $\overline{\beta}_p = E\{\sin(\Theta - \mu)\}$ about the mean direction μ_1, are, for $p = 1, 2, 3, \ldots$,

$$\overline{\alpha}_p = \begin{cases} \rho_p \cos\left\{ \beta\tau^\alpha (p^\alpha - p) \tan\left(\frac{\pi\alpha}{2}\right) \right\}, & \alpha \ne 1, \\ \rho_p \cos\left(\beta\tau p \frac{2}{\pi} \log p \right), & \alpha = 1, \end{cases}$$

and

$$\overline{\beta}_p = \begin{cases} \rho_p \sin\left\{\beta\tau^{\alpha}(p^{\alpha} - p)\tan\left(\frac{\pi\alpha}{2}\right)\right\}, & \alpha \neq 1, \\ -\rho_p \sin\left(\beta\tau p \frac{2}{\pi} \log p\right), & \alpha = 1, \end{cases}$$

respectively. The circular skewness and kurtosis, $s = \overline{\beta}_2/(1-\rho)^{3/2}$ and $k = (\overline{\alpha}_2 - \rho^4)/(1-\rho)^2$, are obtained from the equations above.

Often τ is reparametrized to a concentration parameter $\rho = \exp(-\tau^{\alpha})$. This family is skewed or asymmetric in general, but is symmetric about μ if $\beta = 0$. Wrapped asymmetric Cauchy and Lévy distributions, for $\alpha = 1$ and 0.5 respectively, are also included in this family. However, to a wrapped normal distribution if $\alpha = 2$ whatever be β. Thus interestingly, since 2 is excluded from the parameter space of α in the asymmetric case, a wrapped asymmetric normal distribution is not available from this family.

5.7.2 Wrapped Asymmetric Linnik Family

The Linnik family of distributions, with some attractive properties and relation to the stable family has been enhanced in Chapter 4 as a highly flexible family which can also serve as a complementary one of the stable family. Further detailed discussions on asymmetric Linnik families are given in SenGupta and Roy (2024).

Based on property (ii) in Section 4.5.2 and using the same notations as used for the stable family of distributions above, the c.f. of a Linnik rv in general is given by

$$\varphi_L(t; \alpha, \beta, \tau, \mu) = [1 - \log \varphi_S(t; \alpha, \beta, \tau, \mu)]^{-1}. \tag{5.17}$$

The parameters in (5.17) have the same interpretations but do not of course lead to the same distributions – in fact the two families have distinct members.

The wrapped asymmetric Linnik family (WAL) of distributions corresponding to the above c.f. $\varphi_L(t; \alpha, \beta, \tau, \mu)$ follows by the method of wrapping. Denote the c.f. of the wrapped asymmetric Linnik family as $\varphi_{WAL}(p) = \alpha_p + i\beta_p$. Then the pdf of WAL is given by SenGupta and Roy (2024) as

$$f(\theta) = \frac{1}{2\pi}\left[1 + 2\sum_{p=1}^{\infty}\{\alpha_p \cos(p\theta) + \beta_p \sin(p\theta)\}\right],$$

where

$$\alpha_p = \gamma_p/(\gamma_p^2 + \eta_p^2),\ \beta_p = \eta_p/(\gamma_p^2 + \eta_p^2),$$

$$\gamma_p = 1 + (p\tau)^{\alpha},\ \eta_p = (p\tau)^{\alpha}\beta\tan\left(\frac{\pi\alpha}{2}\right) + p\mu,\ \text{for } \alpha \in (0,2)/\{1\};$$

$$\gamma_p = 1 + p\tau,\ \eta_p = p\mu - p\tau\beta\log p,\ \text{for } \alpha = 1.$$

Since (α_p, β_p), $p = 1, 2, 3, \ldots$, give the pth trigonometric moments of WAL explicitly, properties of WAL can be derived from these conveniently (SenGupta and Roy, 2024). Wrapped asymmetric Mittag–Leffler distribution is obtained from this family with its usual supports for the parameters, i.e. $\beta = 1$, $\alpha < 1$, $0 < \mu < 1$. However, because of the restriction on α, just as the wrapped symmetric but no wrapped asymmetric normal is included in the wrapped stable family, for the wrapped Linnik family also the wrapped symmetric but no wrapped asymmetric Laplace distribution is included in it. We give below different approaches to the construction of wrapped asymmetric Laplace distribution(s).

5.8 Wrapped Exponential and Laplace Distributions

This section deals with properties of wrapped exponential and several asymmetric double-exponential (Laplace) distributions (Jammalamadaka and Kozubowski, 2003, 2004; SenGupta and Roy, 2024) as well as a brief discussion of wrapped generalized Laplace distributions (Arnold and SenGupta, 2025a).

5.8.1 Wrapped Exponential Distribution

Let a random variable X be distributed as an exponential distribution Exp(λ) with parameter λ (>0) on the real line with pdf

$$g(x) = \lambda e^{-\lambda x}, \quad x \geq 0,$$

and c.f.

$$E(e^{itX}) = \frac{1}{1 - it/\lambda}, \quad -\infty < t < \infty.$$

Then the pdf of the corresponding wrapped exponential distribution WE(λ) is, for $0 \leq \theta < 2\pi$,

$$f(\theta) = \sum_{k=0}^{\infty} g(\theta + 2k\pi) = \sum_{k=0}^{\infty} \lambda e^{-\lambda(\theta+2k\pi)} = \frac{\lambda e^{-\lambda\theta}}{1 - e^{-2\lambda\pi}}, \tag{5.18}$$

which is the same as the pdf of the truncated exponential distribution on the interval $[0, 2\pi)$ or the pdf of $(X - 2k\pi \mid 2k\pi \leq X < 2(k+1)\pi)$. Since a circular pdf is a periodic function, (5.18) is extended to the whole real line:

$$f(\theta + 2k\pi) = f(\theta), \quad k = 0, \pm 1, \pm 2, \ldots.$$

If $\Theta \sim$ WE(λ), then the distribution function of Θ is

$$F(\theta) = \frac{1 - e^{-\lambda\theta}}{1 - e^{-2\lambda\pi}}, \quad 0 \leq \theta < 2\pi, \tag{5.19}$$

and the c.f. is, for $p = 0, \pm 1, \pm 2, \dots,$

$$\phi_p = E(e^{ip\Theta}) = \frac{1}{1 - ip/\lambda} = \frac{\lambda^2 + ip\lambda}{p^2 + \lambda^2}. \tag{5.20}$$

Thus, the pdf of WE(λ) is represented by a Fourier series as

$$f(\theta) = \frac{1}{2\pi}\left[1 + 2\sum_{p=1}^{\infty}\left\{\frac{\lambda^2}{p^2 + \lambda^2}\cos(p\theta) + \frac{p\lambda}{p^2 + \lambda^2}\sin(p\theta)\right\}\right],$$

which is a periodic function on the whole real line. The wrapped exponential distribution WE(λ) is asymmetric and the value $f(0)$ is different from $\lim_{\theta \to 2\pi} f(\theta)$.

For $\lambda = 0$ in (5.18), we have the circular uniform distribution by continuity, i.e.

$$\lim_{\lambda \to 0} f(\theta) = \frac{1}{2\pi}, \quad 0 \le \theta < 2\pi.$$

For $\lambda < 0$, the wrapped negative exponential distribution is obtained from the negative exponential distribution on the real line with pdf

$$g(x) = |\lambda| e^{|\lambda| x}, \quad x < 0.$$

Thus, we can consider a more general wrapped exponential distribution WE(λ) for $\lambda \in \mathbb{R}$, with the same form of the pdf, distribution function, and c.f. as (5.18), (5.19), and (5.20), respectively. The pdf of WE(λ) is strictly decreasing (increasing) on $[0, 2\pi)$ if $\lambda > 0\,(\lambda < 0)$. We have the relation: if $\Theta \sim$ WE(λ), then $2\pi - \Theta \sim$ WE($-\lambda$), and vice versa. Since the exponential distribution has the property of infinite divisibility, WE(λ) also has this property: for any integer $n \ge 1$, there exist independent and identically distributed random variables $\Theta_1, \dots, \Theta_n$ such that $\Theta \stackrel{d}{=} \Theta_1 + \cdots + \Theta_n \,(\mathrm{mod}\, 2\pi)$, where $\stackrel{d}{=}$ means that the distributions of both sides are the same. In fact, for $\lambda = 0$, Θ_j have circular uniform distributions, and for $\lambda \neq 0$, wrapped gamma distributions with the c.f. $(1 - ip/\lambda)^{-1/n}$. The pth cosine and sine moments, pth mean direction and mean resultant length, circular variance and standard deviation, pth central cosine and sine moments, circular skewness and kurtosis, and median are obtained by using (5.18) and (5.20) and summarized as Table 3 in Jammalamadaka and Kozubowski (2003, 2004).

The argument about maximizing entropy is given in Section 3.2.1. The condition corresponding to (3.2)

$$\int_0^{2\pi} \theta f(\theta) d\theta = m$$

for a given constant $m\,(0 < m < 2\pi)$ leads to the wrapped exponential distribution WE(λ) which attains the maximum entropy. Its value is

$$H(f) = -\int_0^{2\pi} f(\theta)\log f(\theta) d\theta = \log\left(\frac{1 - e^{-2\lambda\pi}}{\lambda}\right) + 1 - \frac{2\lambda\pi e^{-2\lambda\pi}}{1 - e^{-2\lambda\pi}},$$

and we have the relation between m and λ

$$\frac{m}{2\pi} = \xi - \frac{1}{e^{1/\xi} - 1},$$

where $\lambda = (2\xi\pi)^{-1}$, since

$$m = \frac{1}{\lambda} - \frac{2\pi e^{-2\lambda\pi}}{1 - e^{-2\lambda\pi}}.$$

For $0 < m < \pi$, WE(λ) corresponds to the case $\lambda > 0$, $\lambda < 0$ for $\pi < m < 2\pi$, and $\lambda = 0$ for $m = \pi$, since the function $g(\xi) = \xi - 1/(e^{1/\xi} - 1)$ has the properties $g(-\infty) = 1/2$, $g(-0) = 1$, $g(+0) = 0$, $g(\infty) = 1/2$ and $g(\xi)$ is monotonically increasing.

5.8.2 Wrapped Asymmetric Laplace Distributions

It is possible to obtain several asymmetric generalizations of the symmetric wrapped Laplace distribution. We discuss these below. The natural approach is that of deriving the wrapped distribution corresponding to the c.f. of a skewed Laplace distribution, to be termed Wrapped Asymmetric Laplace (WAL) distribution (SenGupta and Roy, 2024). Another approach is to wrap various skewed versions of the symmetric Laplace pdf.

5.8.2.1 CF-Based WAL Distribution

Recall that the c.f. of Laplace distribution is obtained from that of the symmetric Linnik family with $\alpha = 2$. However, a wrapped asymmetric Laplace (WAL) distribution corresponding to $\alpha = 2$ is not available in the same vein.

5.8.2.2 PDF-Based WAL Distributions

Let us write the density of the generalized asymmetric Laplace (Arnold and SenGupta, 2025b) as

$$f^{(\mathrm{GAL1})}(x) = p\lambda_1 e^{-\lambda_1 x} I(x > 0) + (1 - p)\lambda_2 e^{\lambda_2 x} I(x < 0),$$

where p denotes the mixing parameter. The corresponding wrapped generalized asymmetric Laplace (WGAL1) is the same mixture of the wrapped versions of its components, i.e.

$$f^{(\mathrm{WGAL1})}(\theta) = p\frac{\lambda_1 e^{-\lambda_1 \theta}}{1 - e^{-\lambda_1 (2\pi)}} + (1 - p)\frac{\lambda_2 e^{-\lambda_2 (2\pi - \theta)}}{1 - e^{-\lambda_2 (2\pi)}}. \tag{5.21}$$

The three parameters in (5.21), λ_1, λ_2, and p do not constrain each other and thus lend flexibility to this model. However, the above density is discontinuous at $\theta = 0$. This lacuna may be removed, as shown below, however at the cost of losing the flexibility introduced through the parameter p, i.e. letting p be determined completely by the other two parameters λ_1 and λ_2.

A random variable X is said to have an asymmetric Laplace distribution with parameters $\lambda\,(>0)$ and $\kappa\,(>0)$, denoted by $L(\lambda, \kappa)$, if its pdf is defined by

$$g(x) = \frac{\lambda}{1/\kappa + \kappa} \begin{cases} e^{-\lambda\kappa x}, & x \geq 0, \\ e^{(\lambda/\kappa)x}, & x < 0. \end{cases} \tag{5.22}$$

Equation (5.22) is represented by

$$g(x) = pg_1(x) + (1-p)g_2(x), \qquad -\infty < x < \infty$$

as a mixture of positive and negative exponential distributions, where $p = 1/(\kappa^2 + 1)$ and

$$\begin{aligned} g_1(x) &= \lambda_1 e^{-\lambda_1 x}, \quad && x \geq 0, \\ g_2(x) &= \lambda_2 e^{\lambda_2 x}, \quad && x \leq 0, \end{aligned}$$

with $\lambda_1 = \lambda\kappa$ and $\lambda_2 = \lambda/\kappa$.

The c.f. of X is given by

$$\begin{aligned} \phi(t) &= E(e^{itX}) \\ &= \left(\frac{1}{\kappa^2+1}\right)\frac{1}{1 - it/(\lambda\kappa)} + \left(\frac{\kappa^2}{\kappa^2+1}\right)\frac{1}{1 + it/(\lambda/\kappa)} \\ &= \left\{\frac{1}{1 - it/(\lambda\kappa)}\right\}\left\{\frac{1}{1 + it/(\lambda/\kappa)}\right\} \\ &= \frac{\lambda^2\kappa^2}{(t^2 + \lambda^2\kappa^2)(t^2\kappa^2 + \lambda^2)}\left\{t^2 + \lambda^2 + i\frac{t\lambda}{\kappa}(1-\kappa^2)\right\}, \qquad -\infty < t < \infty. \end{aligned}$$

As a special case, $L(\lambda, \kappa)$ reduces to a symmetric Laplace distribution if $\kappa = 1$.

The wrapped Laplace distribution $\mathrm{WL}(\lambda, \kappa)$ (Jammalamadaka and Kozubowski, 2003, 2004) is defined through $X \sim L(\lambda, \kappa)$ as $X_w = X\,(\mathrm{mod}\,2\pi)$, with pdf

$$f_w(\theta) = \sum_{j=-\infty}^{\infty} g(\theta + 2j\pi) = \frac{\lambda\kappa}{1+\kappa^2}\left\{\frac{e^{-\lambda\kappa\theta}}{1 - e^{-2\lambda\kappa\pi}} + \frac{e^{(\lambda/\kappa)\theta}}{e^{2(\lambda/\kappa)\pi} - 1}\right\}, \qquad 0 \leq \theta < 2\pi. \tag{5.23}$$

The distribution is unimodal and generally asymmetric. If $\kappa = 1$, $\mathrm{WL}(\lambda, \kappa)$ reduces to a symmetric wrapped Laplace distribution with pdf

$$f_w(\theta) = \frac{\lambda}{2}\left(\frac{e^{-\lambda\theta}}{1 - e^{-2\lambda\pi}} + \frac{e^{\lambda\theta}}{e^{2\lambda\pi} - 1}\right), \qquad 0 \leq \theta < 2\pi.$$

The c.f. or pth trigonometric moment of $\mathrm{WL}(\lambda, \kappa)$, $\phi_p = E(e^{ipX_w})$, $p = 0, \pm 1, \pm 2, \ldots$, has the same form as $\phi(t)$, i.e. $\phi_p = \phi(p)$. Thus, the pth cosine and sine moments are

$$E\{\cos(pX_w)\} = \frac{\lambda^2\kappa^2(p^2 + \lambda^2)}{(p^2 + \lambda^2\kappa^2)(p^2\kappa^2 + \lambda^2)}, \quad E\{\sin(pX_w)\} = \frac{p\lambda^3\kappa(1-\kappa^2)}{(p^2 + \lambda^2\kappa^2)(p^2\kappa^2 + \lambda^2)},$$

respectively, and the Fourier series representation is

$$f_w(\theta) = \frac{1}{2\pi}\left[1 + 2\sum_{p=1}^{\infty}\frac{\lambda\kappa\{\lambda\kappa(p^2+\lambda^2)\cos(p\theta) + p\lambda^2(1-\kappa^2)\sin(p\theta)\}}{(p^2+\lambda^2\kappa^2)(p^2\kappa^2+\lambda^2)}\right],$$

which gives a periodic function on the whole real line. The distribution function of $\mathrm{WL}(\lambda,\kappa)$ is given by

$$F_w(\theta) = \left(\frac{1}{1+\kappa^2}\right)\frac{1-e^{-\lambda\kappa\theta}}{1-e^{-2\lambda\kappa\pi}} + \left(\frac{\kappa^2}{1+\kappa^2}\right)\frac{e^{(\lambda/\kappa)\theta}-1}{e^{2(\lambda/\kappa)\pi}-1}, \quad 0 \le \theta < 2\pi.$$

From the c.f., a random variable Θ of $\mathrm{WL}(\lambda,\kappa)$ has the following representations: (a) $\Theta \stackrel{d}{=} I\Theta_1 + (1-I)\Theta_2$ and (b) $\Theta \stackrel{d}{=} \Theta_1 + \Theta_2$, where Θ_1 and Θ_2 are independent and distributed as $\mathrm{WE}(\lambda\kappa)$ and $\mathrm{WE}(-\lambda/\kappa)$, respectively, and I is an indicator random variable, independent of Θ_1 and Θ_2, which takes the values 1 and 0 with probabilities $1/(1+\kappa^2)$ and $\kappa^2/(1+\kappa^2)$, respectively. As the wrapped exponential distribution has the property of infinite divisibility, the wrapped Laplace distribution also has this property.

It is remarked that the range of the parameter λ can be extended to $\lambda \in \mathbb{R}$ in (5.23) and the pdf (5.23) is extended to the whole real line through $f_w(\theta + 2j\pi) = f_w(\theta)$ for $j = 0, \pm 1, \pm 2, \ldots$. For $\lambda = 0$ by continuity, or $\kappa \to 0$, or $\kappa \to \infty$, the distribution reduces to the circular uniform distribution. Since $L(\lambda,\kappa)$ is viewed as a mixture of positive and negative exponential distributions, $\mathrm{WL}(\lambda,\kappa)$ can be viewed as a mixture of positive and negative wrapped exponential distributions $\mathrm{WE}(\lambda\kappa)$ and $\mathrm{WE}(-\lambda/\kappa)$. It holds that $\mathrm{WL}(-\lambda,\kappa) = \mathrm{WL}(\lambda, 1/\kappa)$ and $\Theta \sim \mathrm{WL}(\lambda,\kappa)$ if and only if $2\pi - \Theta \sim \mathrm{WL}(\lambda, 1/\kappa)$. This means that if the pdf of $\mathrm{WL}(\lambda,\kappa)$ is denoted by $f(\theta;\lambda,\kappa)$, then $f(\theta;-\lambda,\kappa) = f(\theta;\lambda,1/\kappa) = f(2\pi-\theta;\lambda,\kappa)$. From Jammalamadaka and Kozubowski (2003, 2004), the following results are obtained for $\lambda > 0$ from the graphs:

a) The pdf $f(\theta;\lambda,\kappa)$ is strictly decreasing on $(0,\theta^*)$ and strictly increasing on $(\theta^*, 2\pi)$, where

$$\theta^* = \frac{1}{\lambda(\kappa + 1/\kappa)}\log\left(\kappa^2\frac{e^{2(\lambda/\kappa)\pi}-1}{1-e^{-2\lambda\kappa\pi}}\right).$$

Moreover, $\theta^* > \pi$ for $\kappa < 1$, $\theta^* = \pi$ for $\kappa = 1$ and $\theta^* < \pi$ for $\kappa > 1$.

b) The maximum and the minimum values of $f(\theta;\lambda,\kappa)$ are

$$f(0;\lambda,\kappa) = \lim_{\theta\to 2\pi} f(\theta;\lambda,\kappa) = \frac{\lambda}{\kappa+1/\kappa}\left\{\frac{1}{e^{2(\lambda/\kappa)\pi}-1} + \frac{1}{1-e^{-2\lambda\kappa\pi}}\right\}$$

and

$$f(\theta^*;\lambda,\kappa) = \frac{1}{2\pi}e^{2\lambda\kappa\pi/(1+\kappa^2)}\left\{\frac{e^{2(\lambda/\kappa)\pi}-1}{2(\lambda/\kappa)\pi}\right\}^{-\kappa^2/(1+\kappa^2)}\left(\frac{e^{2\lambda\kappa\pi}-1}{2\lambda\kappa\pi}\right)^{-1/(1+\kappa^2)}.$$

The pth mean direction and mean resultant length, circular variance and standard deviation, pth central cosine and sine moments, circular skewness, and kurtosis are obtained from the c.f. and summarized in Table 4 also in the same papers. The median direction ξ_0 of WL(λ, κ), which satisfies

$$\int_{\xi_0}^{\xi_0+\pi} f(\theta)d\theta = \int_{\xi_0+\pi}^{\xi_0+2\pi} f(\theta)d\theta = \frac{1}{2},$$

where $f(\xi_0) > f(\xi_0 + \pi)$, is given by

$$\xi_0 = \begin{cases} \xi^*, & \text{for } \lambda > 0,\ 0 < \kappa < 1, \\ \xi^* + \pi, & \text{for } \lambda > 0,\ \kappa > 1, \end{cases}$$

where $\xi^* \in [0, \pi]$ is the unique solution of the equation

$$\frac{1}{1+\kappa^2}\left\{\frac{e^{-\lambda\kappa\xi}}{1+e^{-\lambda\kappa\pi}} + \kappa^2\,\frac{e^{(\lambda/\kappa)\xi}}{1+e^{(\lambda/\kappa)\pi}}\right\} = \frac{1}{2}.$$

We briefly describe a wrapped generalized symmetric Laplace distribution, which can be constructed from the generalized symmetric Laplace distribution on the real line (cf. Kotz et al., 2001, pp. 189–190). Its pdf is expressed as

$$f(\theta) = \frac{2\beta^{\alpha/2+1/4}}{\sqrt{\pi}\Gamma(\alpha)} \sum_{k=-\infty}^{\infty} |\theta - \mu + 2\pi k|^{\alpha-1/2} K_{1/2-\alpha}(2\sqrt{\beta}|\theta - \mu + 2\pi k|), \quad 0 \le \theta < 2\pi, \tag{5.24}$$

where $\alpha > 0, \beta > 0, 0 \le \mu < 2\pi$, and $K_\nu(z)$ denotes the modified Bessel function of the third kind and order ν, defined by (B.30). With a simpler form (B.33), if $\alpha = 1$ and $\mu = 0$, (5.24) reduces to

$$\begin{aligned} f(\theta) &= \frac{2\beta^{3/4}}{\sqrt{\pi}} \sum_{k=-\infty}^{\infty} |\theta + 2\pi k|^{1/2} \frac{\sqrt{\pi}\exp(-2\sqrt{\beta}|\theta + 2\pi k|)}{2\beta^{1/4}|\theta + 2\pi k|^{1/2}} \\ &= \sqrt{\beta}\left[\frac{\exp\{2\sqrt{\beta}(2\pi - \theta)\} + \exp\{2\sqrt{\beta}\theta\}}{\exp\{2\sqrt{\beta}(2\pi)\} - 1}\right], \end{aligned}$$

which is the wrapped symmetric Laplace distribution WL$(\lambda, 1)$ with $\lambda = 2\sqrt{\beta}$.

It is possible to construct the wrapped generalized asymmetric Laplace distribution (WGAL2) based on, to be termed here as the generalized asymmetric Laplace distribution type 2 (GAL2), the real line (cf. Kotz et al., 2001, p. 180) with c.f.

$$\Psi(t) = \frac{e^{i\delta t}}{(1 + \sigma^2 t^2/2 - i\mu t)^\tau}, \qquad -\infty < t < \infty,$$

where $\delta, \mu \in \mathbb{R}$ and $\sigma, \tau \ge 0$. For $\sigma > 0$, the GAL2 c.f. is factored as

$$\Psi(t) = e^{i\delta t}\left\{\frac{1}{1 + i(\sqrt{2}/2)\sigma\kappa t}\right\}^\tau \left\{\frac{1}{1 - i\{\sqrt{2}/(2\kappa)\}\sigma t}\right\}^\tau$$

with the additional parameter $\kappa > 0$, which satisfies $\mu = \sigma(1/\kappa - \kappa)/\sqrt{2}$. It is clear that the distribution reduces to the asymmetric Laplace distribution $L(\lambda, 1/\kappa)$ if $\delta = 0$, $\tau = 1$ and $(\sqrt{2}/2)\sigma = 1/\lambda$.

5.9 Wrapped Gamma Distribution

As a direct extension of the wrapped exponential distribution in Section 5.8, Coelho (2007) studied the wrapped gamma distribution. We briefly introduce the definition of the distribution with mixture representations and trigonometric moments. Their details and other properties of the distribution can be seen in Coelho (2007).

5.9.1 Definition and Properties

Let X be a random variable which follows a gamma distribution $\Gamma(r, \lambda)$ with pdf

$$g(x) = \frac{\lambda^r}{\Gamma(r)} e^{-\lambda x} x^{r-1}, \quad x > 0$$

with shape parameter $r\,(>0)$ and rate parameter $\lambda\,(>0)$. Then the distribution function of $\Gamma(r, \lambda)$ is given by

$$F(x) = \frac{\gamma(r; \lambda x)}{\Gamma(r)}, \quad x > 0$$

with the incomplete gamma function (B.4). The corresponding wrapped gamma distribution $W\Gamma(r, \lambda)$ has pdf

$$f(\theta) = \sum_{k=0}^{\infty} g(\theta + 2k\pi) = \frac{\lambda^r}{\Gamma(r)} \sum_{k=0}^{\infty} e^{-\lambda(\theta+2k\pi)}(\theta + 2k\pi)^{r-1}, \quad 0 \le \theta < 2\pi.$$

If we use the Lerch function $\Phi(z, s, a)$ given by (B.42), the wrapped gamma pdf is written as

$$f(\theta) = \frac{\lambda^r}{\Gamma(r)} e^{-\lambda\theta}(2\pi)^{r-1}\Phi\left(e^{-2\lambda\pi}, 1 - r, \frac{\theta}{2\pi}\right).$$

If $\lambda \to 0$ or $r \to \infty$, $W\Gamma(r, \lambda)$ reduces to the circular uniform distribution, and if $r = 1$, the wrapped exponential distribution in Section 5.8.

The mode and antimode of $W\Gamma(r, \lambda)$ are at the values of θ in the interval $[0, 2\pi)$ for which $df(\theta)/d\theta$ does not exist or $df(\theta)/d\theta = 0$, equivalently,

$$\Phi\left(e^{-2\lambda\pi}, 1 - r, \frac{\theta}{2\pi}\right) = \frac{r-1}{2\lambda\pi}\,\Phi\left(e^{-2\lambda\pi}, 2 - r, \frac{\theta}{2\pi}\right).$$

For values of $r \le 1$, the mode is at the point $\theta = 0$, where $df(\theta)/d\theta$ is not defined.

Like the wrapped negative exponential distribution in Section 5.8, we can extend $W\Gamma(r, \lambda)$ to the wrapped negative gamma distribution for positive integer r. Since

$$g(x; r, \lambda) = \frac{\lambda^r}{\Gamma(r)} e^{-\lambda x} x^{r-1} = -g(-x; r, -\lambda), \quad x > 0,$$

it holds that, for $\theta \in [0, 2\pi)$,

$$\begin{aligned} f(2\pi - \theta; r, -\lambda) &= \sum_{k=-\infty}^{-1} \{-g(2\pi - \theta + 2k\pi; r, -\lambda)\} = \sum_{k=0}^{\infty} \{-g(-\theta - 2k\pi; r, -\lambda)\} \\ &= f(\theta; r, \lambda). \end{aligned}$$

We use hereafter the following notation in this section:

$$\gamma^*(r; x) = \frac{\gamma(r; x)}{\Gamma(r)}$$

and $\Gamma(r, \lambda, [0, b))$ for a gamma distribution with shape parameter $r\,(>0)$ and rate parameter $\lambda\,(>0)$, truncated at $b\,(>0)$ with pdf

$$g(x) = \frac{\lambda^r e^{-\lambda x} x^{r-1} / \Gamma(r)}{\gamma^*(r; \lambda b)} = \frac{\lambda^r}{\gamma(r; \lambda b)} e^{-\lambda x} x^{r-1}, \quad x > b.$$

5.9.2 Mixtures

Let n be either finite positive integer or infinite. If $f_j(\theta)\,(j = 0, 1, \dots, n)$ are pdfs of wrapped distributions and if $\sum_{j=0}^{n} p_j = 1$ with $0 < p_j < 1$, then $f(\theta) = \sum_{j=0}^{n} p_j f_j(\theta)$ is also the pdf of a wrapped distribution. And if $\Phi_j(p)\,(j = 0, 1, \dots, n)$ are the c.f.s of the wrapped distributions with pdfs $f_j(\theta)$, then $\Phi(p) = \sum_{j=0}^{n} p_j \Phi_j(p)(\theta)$ is the c.f. of the wrapped mixture distribution with pdf $f(\theta)$.

Such mixtures as mentioned above have been constructed for von Mises components (SenGupta and Bhattacharya, 2015). Mixtures with peaked power Batschelet components have also been studied by Mulder et al. (2020). One disadvantage with these type (additive) of mixture distributions is that the number of parameters, $\sum_{j=0}^{n} n_j + n$, where $n_j\,(j = 0, 1, \dots, n)$ are the number of parameters in $f_j(\theta)\,(j = 0, 1, \dots, n)$, increases rapidly with $n + 1$, the number of modes desired. This poses both computational and inferential problems. Another problem is that often one will not have a priori knowledge of n, and it may be needed to choose it sequentially. A solution to the last problem, however, was given by Bhattacharya and SenGupta (2009a) in the Bayesian framework using the Dirichlet process prior. Another situation arises when the underlying experiment or data may indicate that the components of the mixture need not be of the same form, e.g. when isotropic behavior can be suspected in behavorial studies for a part of the population. It trivially follows that wrapping and such (additive) mixing commute. Exploiting this fact, SenGupta and Pal (2001) proposed the symmetric wrapped stable – circular uniform mixture family for such scenarios.

For a non-integer shape parameter r, $\mathrm{W}\Gamma(r,\lambda)$ is represented as an infinite mixture with weights

$$p_j = \gamma^*(j+1; 2\lambda\pi)(-1)^j \frac{\lambda^{r-j-1}\Gamma(1-r+j)}{\Gamma(r)\Gamma(1-r)} \,\mathrm{PLog}(e^{-2\lambda\pi}, j+1-r), \quad j = 0,1,2,\dots,$$

of infinitely many $\Gamma(j+1,\lambda;[0,2\pi))$ $(j=0,1,2,\dots)$ and with weight $p^* = \gamma^*(r;2\lambda\pi)$ of an additional $\Gamma(r,\lambda;[0,2\pi))$. Thus, the pdf of the wrapped gamma distribution $\mathrm{W}\Gamma(r,\lambda)$ has the form

$$\begin{aligned} f(\theta) = {} & \frac{\lambda^r}{\gamma(r;2\lambda\pi)} e^{-\lambda\theta}\theta^{r-1}\gamma^*(r;2\lambda\pi) + \sum_{j=0}^{\infty} \frac{\lambda^{j+1}}{\gamma(j+1;2\lambda\pi)} e^{-\lambda\theta}\theta^j \\ & \times \gamma^*(j+1;2\lambda\pi)(-1)^j \frac{\lambda^{r-j-1}\Gamma(1-r+j)}{\Gamma(r)\Gamma(1-r)} \,\mathrm{PLog}(e^{-2\lambda\pi}, j+1-r). \end{aligned}$$

Here the polylogarithm function $\mathrm{PLog}(z,s)$ is defined by (B.43). Obviously we have $p^* + \sum_{j=0}^{\infty} p_j = 1$ since the integral value of $f(\theta)$ over the interval $[0,2\pi)$ is unity.

For a positive-integer shape parameter r, $\mathrm{W}\Gamma(r,\lambda)$ is represented as a finite mixture with weights

$$p_h = \gamma^*(h+1;2\lambda\pi)\frac{(2\pi\lambda)^{r-h-1}}{(r-h-1)!}\Phi^*(e^{-2\lambda\pi}, r-h-1), \quad h = 0,\dots,r-1,$$

of r distributions $\Gamma(h+1,\lambda;[0,2\pi))$ $(h=0,\dots,r-1)$. Here

$$\Phi^*(z,n) = \sum_{k=0}^{\infty} z^k k^n = \frac{t_{n,0} + \sum_{j=1}^{n} t_{n,j} z^{n+1-j}}{(1-z)^{n+1}}$$

for a positive integer n and a real number z with $t_{n,j}$ defined by

$$t_{n,j} = (n-j+1)t_{n-1,j-1} + j\,t_{n-1,j}$$

for $1 < j < n$ with $t_{n,n} = t_{n,1} = 1$, $t_{n,j} = 0$ for $j > n$ or $j < 0$, $t_{n,0} = 1\,(n=0)$, $0\,(n>0)$, and $t_{n,j} = 0$ for $n < 0$.

From the mixture representation of $\mathrm{W}\Gamma(r,\lambda)$, the distribution function, both for integer and non-integer shape r, is expressed by using the incomplete gamma functions $\gamma(r;x)$ and $\gamma^*(r;x)$, i.e. for non-integer shape parameter r,

$$F(\theta) = p^*\frac{\gamma(r;\lambda\theta)}{\gamma(r;2\pi\lambda)} + \sum_{j=0}^{\infty} p_j \frac{\gamma(j+1;\lambda\theta)}{\gamma(j+1;2\pi\lambda)}$$

and for integer shape parameter r,

$$F(\theta) = \sum_{h=0}^{r-1} p_h \frac{\gamma(h+1;\lambda\theta)}{\gamma(h+1;2\pi\lambda)}.$$

5.9.3 Trigonometric Moments

The c.f. of $W\Gamma(r, \lambda)$ is, for $p = 0, \pm 1, \pm 2, \dots,$

$$\phi_p = \frac{\lambda^r}{(\lambda - ip)^r}, \qquad i = \sqrt{-1}.$$

For $\lambda > 0$, the pth trigonometric moment is represented by

$$\phi_p = \lambda^r(\lambda^2 + p^2)^{-r/2} e^{ir \arctan(p/\lambda)} = \rho_p e^{i\mu_p}$$

with the pth mean resultant length $\rho_p = \lambda^r(\lambda^2 + p^2)^{-r/2}$ and the pth mean direction $\mu_p = r \arctan(p/\lambda)$. The characteristics such as pth cosine and sine moments, circular variance, circular standard deviation, skewness, and kurtosis are easily obtained from the pth trigonometric moment.

For $\lambda < 0$, the pth trigonometric moment is

$$\phi_p = |\lambda|^r(\lambda^2 + p^2)^{-r/2} e^{ir\pi} e^{ir\{\pi + \arctan(p/\lambda)\}} = \rho_p e^{i\mu_p}$$

with $\rho_p = |\lambda|^r(\lambda^2 + p^2)^{-r/2}$ (>0) and $\mu_p = r\{2\pi + \arctan(p/\lambda)\}$.

5.10 Other Wrapped Distributions

As in Section 3.3, any linear random variable X on the real line leads to the corresponding circular wrapped distribution with random variable $\Theta = X \pmod{2\pi}$. Sections 4.4–4.6 and 5.6.1–5.9 deal with wrapped Cauchy, normal, t, stable, Linnik, Kumaraswamy, skew-normal, α-stable and Linnik, exponential and Laplace, and gamma distributions, respectively. See Dattatreya Rao et al. (2007) for wrapped lognormal, logistic and Weibull distributions as other wrapped distributions.

5.11 Sine-Skewed Distribution

5.11.1 Definition and Properties

As in Section 3.12, Umbach and Jammalamadaka (2009) have given a general class of skew distributions:

$$f_\mu(\theta) = 2f(\theta - \mu)G(w(\theta - \mu)), \quad -\pi \le \theta < \pi, -\pi \le \mu < \pi.$$

Here f is a circular pdf on $[-\pi, \pi)$ which is symmetric about zero, $G(\theta) = \int_{-\pi}^{\theta} g(t)dt$ is the distribution function of a circular pdf g on $[-\pi, \pi)$ which is symmetric about zero, and $w(\theta)$ on $[-\pi, \pi)$ is an odd function and periodic.

Umbach and Jammalamadaka (2009) study the case when G is the distribution function of the circular uniform distribution, $G(\theta) = (\pi + \theta)/(2\pi)$, and

$w(\theta) = \lambda\pi \sin(k\theta)$ for some integer k and $-1 \leq \lambda \leq 1$. If $f(\theta - \mu)$ follows the wrapped Cauchy distribution and $w(\theta) = \lambda\pi \sin(2\theta)$, we have a skew wrapped Cauchy distribution with pdf

$$f(\theta; \mu, \rho, \lambda) = \frac{1}{2\pi} \frac{1-\rho^2}{1+\rho^2 - 2\rho\cos(\theta-\mu)} [1 + \lambda \sin\{2(\theta - \mu)\}]$$

for $-\pi \leq \mu < \pi$ and $0 < \rho < 1$; If $f(\theta - \mu)$ follows the von Mises distribution and $w(\theta) = \lambda\pi \sin\theta$, we have a skew von Mises distribution with pdf

$$f(\theta; \mu, \kappa, \lambda) = \frac{1}{2\pi I_0(\kappa)} e^{\kappa\cos(\theta-\mu)}\{1 + \lambda\sin(\theta - \mu)\}$$

for $-\pi \leq \mu < \pi$ and $\kappa \geq 0$. Clearly, if $\lambda = 0$, then the skew wrapped Cauchy and skew von Mises distributions respectively reduce to the wrapped Cauchy and von Mises distributions.

Abe and Pewsey (2011a) study properties of a sine-skewed distribution with pdf of the form

$$f(\theta) = f_0(\theta - \eta)\{1 + \lambda\sin(\theta - \eta)\}.$$

Here f_0 is a base pdf symmetric about η. Assume that $\eta = 0$ without loss of generality. Then the distribution function of Θ with pdf $f(\theta)$ is expressed as

$$F(\theta) = \int_{-\pi}^{\theta} f_0(\phi)(1 + \lambda\sin\phi)d\phi = F_0(\theta) + \lambda\int_{-\pi}^{\theta} f_0(\phi)\sin\phi\, d\phi,$$

where $F_0(\theta)$ is the distribution function of the base pdf $f_0(\theta)$ symmetric about zero.

The trigonometric moments $\alpha_p = E\{\cos(p\Theta)\}$ and $\beta_p = E\{\sin(p\Theta)\}$ can be expressed in terms of the cosine moments $\alpha_{0,p} = \int_{-\pi}^{\pi}\cos(p\theta)f_0(\theta)d\theta$ of the base pdf $f_0(\theta)$. In fact, for $p = \pm1, \pm2, \pm3, \ldots,$

$$\alpha_p = \int_{-\pi}^{\pi}\cos(p\theta)f_0(\theta)(1 + \lambda\sin\theta)d\theta = \alpha_{0,p}$$

since $f_0(\theta)$ is symmetric about zero, and

$$\beta_p = \lambda\int_{-\pi}^{\pi} f_0(\theta)\frac{1}{2}[\cos\{(p-1)\theta\} - \cos\{(p+1)\theta]d\theta = \frac{\lambda}{2}(\alpha_{0,p-1} - \alpha_{0,p+1}).$$

Therefore, the pth mean resultant length is $\rho_p = \sqrt{\alpha_{0,p}^2 + \lambda^2(\alpha_{0,p-1} - \alpha_{0,p+1})^2/4}$ and the pth mean direction is $\mu_p = \arg\{\alpha_{0,p} + i\lambda(\alpha_{0,p-1} - \alpha_{0,p+1})/2\}$. Assuming that $\lambda(1 - \alpha_{0,2}) \neq 0$, we have $\rho \equiv \rho_1 \neq 0$ and $\mu \equiv \mu_1 \neq 0$, and the circular variance and circular standard deviation are respectively given by $\nu = 1 - \rho = 1 - \sqrt{\alpha_{0,1}{}^2 + \lambda^2(1-\alpha_{0,2})^2/4}$ and $\sigma = \{-2\log(1-\nu)\}^{1/2} = [-\log\{\alpha_{0,1}^2 + \lambda^2(1-\alpha_{0,2})^2/4\}]^{1/2}$. From these equations, the second cosine and sine moments about the mean direction μ, $\overline{\alpha}_2 = E[\cos\{2(\Theta - \mu)\}]$ and $\overline{\beta}_2 = E[\sin\{2(\Theta - \mu)\}]$, and the measures of circular skewness and kurtosis, $s = \overline{\beta}_2/\nu^{3/2}$ and $k = (\overline{\alpha}_2 - \rho^4)/\nu^2$, are obtained.

Concerning a finite mixture of two sine-skewed pdfs $f_1(\theta) = f_{10}(\theta)(1 + \lambda \sin\theta)$ and $f_2(\theta) = f_{20}(\theta)(1 + \lambda \sin\theta)$ with common λ, we have

$$pf_1(\theta) + (1-p)f_2(\theta) = (1 + \lambda \sin\theta)\{pf_{10}(\theta) + (1-p)f_{20}(\theta)\}, \quad 0 \le p \le 1,$$

i.e. the finite mixture distribution is also the sine-skewed distribution with the corresponding mixture of the two base symmetric pdfs.

5.11.2 Families of Sine-Skewed Distributions

As in Section 4.6.3.1, the family of symmetric unimodal power-cardioid distributions includes von Mises, wrapped Cauchy, and cardioid distributions. The sine-skewed version of this family has pdf

$$f(\theta) = C_{\rho,r}^{-1}(1 + \rho\cos\theta)^r(1 + \lambda\sin\theta), \tag{5.25}$$

and if $r \neq -1$ and $\rho \neq 0$, the distribution function of (5.25) is given by

$$F(\theta) = F_0(\theta) + \frac{\lambda C_{\rho,r}^{-1}}{\rho(r+1)}\left\{(1-\rho)^{r+1} - (1+\rho\cos\theta)^{r+1}\right\},$$

where $F_0(\theta)$ is the distribution function of the power cardioid distribution.

The modality of the sine-skewed distribution has been discussed by Abe and Pewsey (2011a) with $\rho = \tanh(\kappa\psi)$ and $r = 1/\psi$.

Remark 5.1 Other symmetric unimodal circular families, e.g. the wrapped symmetric stable and Linnik families, can be sine-skewed trivially. These also yield the corresponding sine-skewed forms for the pdfs of their popular members, e.g. wrapped normal, wrapped Cauchy, and wrapped Linnik.

5.11.3 Some Special Distributions and Their Properties

5.11.3.1 Sine-Skewed von Mises Distribution

The sine-skewed von Mises distribution has pdf

$$f(\theta) = \frac{e^{\kappa\cos\theta}}{2\pi I_0(\kappa)}(1 + \lambda\sin\theta), \quad -\pi \le \theta < \pi, \tag{5.26}$$

where $\kappa \ge 0$ and $I_r(\kappa)$ is the modified Bessel function of the first kind of order r, and the distribution function of equation (5.26) is given by

$$F(\theta) = F_0(\theta) + \frac{\lambda}{2\pi\kappa I_0(\kappa)}(e^{-\kappa} - e^{\kappa\cos\theta}),$$

where $F_0(\theta)$ is the distribution function of the base von Mises distribution.

The pth cosine and sine moments are, respectively, $\alpha_p = I_p(\kappa)/I_0(\kappa)$ and

$$\beta_p = \frac{\lambda\{I_{p-1}(\kappa) - I_{p+1}(\kappa)\}}{2I_0(\kappa)} = \frac{p\lambda}{\kappa}\frac{I_p(\kappa)}{I_0(\kappa)} = \frac{p\lambda}{\kappa}\alpha_p,$$

which are obtained from the pth cosine moment of the von Mises distribution given in Section 4.3.3.1 and the formula (B.28). Other circular measures are easily obtained from α_p and β_p. In particular, the mean resultant length $\rho \equiv \rho_1 = I_1(\kappa)\sqrt{\kappa^2+\lambda^2}/\{\kappa I_0(\kappa)\}$ has range $0 \le \rho \le 1$ and thus the ranges of the circular variance $\nu = 1-\rho$ and circular standard deviation $\sigma = \{-2\log(1-\nu)\}^{1/2}$ are $[0,1]$ and $[0,\infty)$, respectively. The ranges of other circular measures are studied in Abe and Pewsey (2011a).

5.11.3.2 Sine-Skewed Cardioid Distribution

The pdf of the sine-skewed cardioid distribution is given by

$$f(\theta|\kappa,\lambda) = \frac{1}{2\pi}(1+k\cos\theta)(1+\lambda\sin\theta), \quad -\pi \le \theta < \pi, \tag{5.27}$$

where $0 \le k \le 1$ and $0 \le \lambda \le 1$. The restriction against λ is due to the fact that $f(\theta|k,-\lambda) = f(\theta+\pi/2|\lambda,k)$. Note that (5.27) has an alternative representation

$$f(\theta) = \frac{1}{2\pi}\left[1+\sqrt{\lambda^2+k^2}\cos(\theta-\xi)+\frac{k\lambda}{2}\cos\left\{2\left(\theta-\frac{\pi}{4}\right)\right\}\right],$$

where $\cos\xi = k/\sqrt{\lambda^2+k^2}$ and $\sin\xi = \lambda/\sqrt{\lambda^2+k^2}$. This representation is related to (5.9).

The trigonometric moments for the sine-skewed cardioid distribution are given by $\alpha_1 = k/2$, $\alpha_p = 0\,(p \ge 2)$ and

$$\beta_1 = \frac{\lambda}{2}, \quad \beta_2 = \frac{k\lambda}{4}, \quad \beta_p = 0\,(p \ge 3).$$

The mean resultant length is $\rho = \sqrt{k^2+\lambda^2}/2$ and has range $0 \le \rho \le 1/\sqrt{2}$. Thus, the circular variance and circular standard deviation have range $[1-1/\sqrt{2},1]$ and $[\sqrt{\log 2},\infty)$, respectively. The ranges of the second cosine and sine moments about the mean direction and the measures of circular skewness and kurtosis are $0 \le \overline{\alpha}_2 \le 1/4$, $-0.075 \le \overline{\beta}_2 \le 0.075$, $-0.2627 \le s \le 0.2627$, and $-1/4 \le k \le 0.3151$.

5.11.3.3 Sine-Skewed Wrapped Cauchy Distribution

The sine-skewed wrapped Cauchy distribution has pdf

$$f(\theta) = \frac{1-\rho_0^2}{2\pi(1+\rho_0^2-2\rho_0\cos\theta)}(1+\lambda\sin\theta), \quad -\pi \le \theta < \pi, \tag{5.28}$$

where $0 \le \rho_0 < 1$ and $-1 \le \lambda \le 1$. The distribution function of equation (5.28) is expressed as

$$F(\theta) = \frac{1}{2}+\frac{1}{\pi}\arctan\left\{\frac{1+\rho_0}{1-\rho_0}\tan\left(\frac{\theta}{2}\right)\right\}+\frac{\lambda(1-\rho_0^2)}{4\pi\rho_0}\log\left\{\frac{1+\rho_0^2-2\rho_0\cos\theta}{(1+\rho_0)^2}\right\}.$$

Since differentiation of the pdf is

$$f'(\theta) = \frac{\sqrt{1-a^2}\sqrt{\lambda^2+a^2}}{2\pi(1-a\cos\theta)^2}\{\cos(\theta+\xi)-\zeta\},$$

where $a = 2\rho_0/(1+\rho_0^2)$ $(\in [0,1))$, $\zeta = a\lambda/\sqrt{\lambda^2+a^2}$ $(\in [-1/\sqrt{2}, 1/\sqrt{2}])$, and $\xi = \arg(\lambda + ia)$, the distribution is always unimodal. Let $\xi \in [0,\pi]$. If $0 < \lambda \leq 1$, then $\xi \in [0, \pi/2)$ and the mode and antimode occur at $\theta^* = -\xi \pm \cos^{-1}\zeta$. If $-1 \leq \lambda < 0$, then $\xi \in [\pi/2, \pi]$. The mode and antimode occur at $\theta^* = -\xi + \cos^{-1}\zeta$ and $\theta^* = -\xi - \cos^{-1}\zeta + 2\pi$, respectively. The density values at the mode and antimode are the same as those when $0 < \lambda \leq 1$:

$$f(\theta^*) = \frac{(1-\rho_0^2)\left\{\sqrt{\lambda^2+a^2-(a\lambda)^2} \pm \lambda^2\right\}}{2\pi(1+\rho_0^2)\left\{\sqrt{\lambda^2+a^2-(a\lambda)^2} \mp a^2\right\}}.$$

The pth cosine and sine moments of the sine-skewed wrapped Cauchy distribution are given by $\alpha_p = \rho_0^{|p|}$ and $\beta_p = \lambda(\rho_0^{|p-1|} - \rho_0^{|p+1|})/2$. The mean resultant length has range $0 \leq \rho \leq 1$. The ranges of the second cosine and sine moments about the mean direction and the measures of circular skewness and kurtosis are $0 \leq \overline{\alpha}_2 \leq 1$, $-\sqrt{3}/9 \leq \overline{\beta}_2 \leq \sqrt{3}/9$, $-\infty < s < \infty$ and $-1/4 \leq k < \infty$.

The sine-skewed wrapped Cauchy distribution has a remarkable convolution property. Let Θ_1 be a sine-skewed wrapped Cauchy rv with parameter (ρ_1, λ_1) and c.f. $\phi_p = \rho_1^{|p|} + i\lambda_1(\rho_1^{|p-1|} - \rho_1^{|p+1|})/2$, and Θ_2 be independent of Θ_1, a wrapped Cauchy rv with parameter $(\rho_1, 0)$ and c.f. $\psi_p = \rho_2^{|p|}$, Then

$$\phi_p\psi_p = (\rho_1\rho_2)^{|p|} + i\rho_2^{|p|}\lambda_1(\rho_1^{|p-1|} - \rho_1^{|p+1|})/2.$$

This is the c.f. of a sine-skewed wrapped Cauchy distribution with parameter (ρ_3, λ_3), where $\rho_3 = \rho_1\rho_2$ and $\lambda_3 = \rho_2^{|p|}\lambda_1(\rho_1^{|p-1|} - \rho_1^{|p+1|})/\{(\rho_1\rho_2)^{|p-1|} - (\rho_1\rho_2)^{|p+1|}\}$, if $\lambda_3 \in [-1, 1]$.

5.12 A Family of Cosine-Perturbed Distributions

There are some bipolar data, the data from a circular distribution that has two diametrically opposite modes. A way to generate a model for such data is to use doubling the argument of the base pdf. However, Abe and Pewsey (2011b) give a cosine perturbed family of symmetric circular distributions with pdf

$$f^*(\theta|\lambda) = (1+\lambda\cos\theta)f(2\theta), \quad -\pi \leq \theta < \pi, \tag{5.29}$$

where $f(\theta)$ is unimodal on $[-\pi, \pi)$ and symmetric about $\theta = 0$ and $-1 \leq \lambda \leq 1$. The resulting distributions are generally skew. The range of λ can be constrained $0 \leq \lambda \leq 1$ because $f^*(\theta|-\lambda) = (1-\lambda\cos\theta)f(2\theta) = \{1+\lambda\cos(\theta+\pi)\}f(2(\theta+\pi)) = f^*(\theta+\pi|\lambda)$ if $-\pi \leq \theta \leq 0$ and $f^*(\theta|-\lambda) = f^*(\theta-\pi|\lambda)$ if $0 \leq \theta < \pi$. Clearly, the pdf reduces to $f^*(\theta|\lambda) = f(2\theta)$ if $\lambda = 0$, which is antipodally symmetric and bimodal. A general construction via duplication is given in the paper by Abe and Pewsey (2011b).

Basic properties of the cosine perturbed family with pdf (5.29) and a circular random variable Θ^* are as follows. The distribution function of Θ^* is given by

$$F^*(\theta) = \int_{-\pi}^{\theta} f^*(t|\lambda)dt = \int_{-\pi}^{\theta} f(2t)dt + \lambda \int_{-\pi}^{\theta} \cos t f(2t)dt$$

in terms of the base pdf $f(\theta)$. The pth sine and cosine moments are $\beta_p^* = E\{\sin(p\Theta^*)\} = 0, \alpha_{2p+1}^* = E[\cos\{(2p+1)\Theta^*\}] = \lambda(\alpha_p + \alpha_{p+1})/2$ and $\alpha_{2p}^* = E\{\cos(2p\Theta^*)\} = \alpha_p$, where α_p is the pth cosine moment of $f(\theta)$.

The cosine perturbed power cardioid family has pdf

$$f^*(\theta|\rho, r, \lambda) = C_{\rho,r}^{-1}(1 + \lambda\cos\theta)\{1 + \rho\cos(2\theta)\}^r.$$

Special cases include the cosine perturbed von Mises, cardioid, and wrapped Cauchy distributions.

5.13 Transforming Circular Distributions via Möbius Transformation

5.13.1 Transforming von Mises Distribution

The Möbius circle transformation (Section 3.6) from $\tilde{\theta}$ to θ on the unit circle in the complex plane is given by

$$e^{i\theta} = e^{i\mu}\frac{e^{i\tilde{\theta}} + \rho e^{i\nu}}{1 + \rho e^{i(\tilde{\theta}-\nu)}} \tag{5.30}$$

and the angular representation is expressed as

$$\theta = M(\tilde{\theta}) = \mu + \nu + 2\arctan\left[w_\rho \tan\left\{\frac{1}{2}(\tilde{\theta} - \nu)\right\}\right],$$

where $0 \le \mu, \nu < 2\pi$, $0 \le \rho < 1$ are the three parameters of the transformation (μ controlling location) with $w_\rho = (1-\rho)/(1+\rho)$. If the transformation is applied to a circular uniform random variable, then the resulting distribution is a wrapped Cauchy distribution because the Möbius inverse transformation is

$$M^{-1}(\theta) = \nu + 2\arctan\left[w_\rho^{-1}\tan\left\{\frac{1}{2}(\theta - \gamma)\right\}\right]$$

with $\gamma = \mu + \nu$ and

$$\frac{d}{d\theta}M^{-1}(\theta) = \frac{1-\rho^2}{1+\rho^2 - 2\rho\cos(\theta - \gamma)},$$

while wrapped Cauchy distributions are closed under the transformation.

Kato and Jones (2010) consider the family derived by transforming the von Mises distribution VM$(0, \kappa)$ with pdf

$$f_{\mathrm{VM}}(\theta) = \frac{1}{2\pi I_0(\kappa)} \exp(\kappa \cos\theta), \quad 0 \le \theta < 2\pi,$$

via the Möbius transformation, where $\kappa \ge 0$ is the concentration parameter and $I_0(\kappa)$ is the modified Bessel function of the first kind and order 0. Since the von Mises distribution contains the uniform distribution as a special case, the family of transforming the von Mises distribution via the Möbius transformation contains the wrapped Cauchy distribution and gives a wide family of distributions, generally asymmetry. Let $\tilde{\Theta}$ be distributed as VM$(0, \kappa)$. Then the distribution of $M(\tilde{\Theta})$ has pdf given by

$$\begin{aligned} f(\theta) = {} & \frac{1}{2\pi I_0(\kappa)} \exp\left[\frac{\kappa\{\xi\cos(\theta-\eta) - 2\rho\cos\nu\}}{1+\rho^2-2\rho\cos(\theta-\gamma)}\right] \\ & \times \frac{1-\rho^2}{1+\rho^2-2\rho\cos(\theta-\gamma)}, \quad 0 \le \theta < 2\pi, \end{aligned} \tag{5.31}$$

where $\gamma = \mu + \nu$, $\xi = \sqrt{\rho^4 + 2\rho^2\cos(2\nu) + 1}$ and $\eta = \mu + \arg\{\rho^2\cos(2\nu) + 1 + i\rho^2\sin(2\nu)\}$ with $i = \sqrt{-1}$. This family of distributions has four parameters μ, ν, ρ, and κ. The distribution with pdf (5.31) clearly reduces to the von Mises ($\rho = 0$), wrapped Cauchy with parameters γ and ρ ($\kappa = 0$), and uniform ($\rho = \kappa = 0$) distributions, and goes to a degenerate distribution at $\theta = \gamma$ as $\kappa \to \infty$ or $\rho \to 1$.

The distribution with pdf (5.31) is generally asymmetrically distributed. Conditions for symmetry and unimodality are investigated in Kato and Jones (2010): The pdf (5.31) is symmetric if and only if $\rho = 0$ (von Mises), $\nu = 0, \pi$, or $\kappa = 0$ (wrapped Cauchy), and areas of bimodality/unimodality for two pairs of parameters (ν, ρ) and (ν, κ) are displayed in their paper.

In the special symmetric case corresponding to $\nu = 0$ or π, the pdf (5.31) reduces to

$$\begin{aligned} f(\theta) = {} & \frac{1-\rho^2}{2\pi I_0(\kappa)} \exp\left[\frac{\kappa\{(1+\rho^2)\cos(\theta-\mu) - 2\rho\}}{1+\rho^2-2\rho\cos(\theta-\mu)}\right] \\ & \times \frac{1}{1+\rho^2-2\rho\cos(\theta-\mu)}, \quad 0 \le \theta < 2\pi, \end{aligned} \tag{5.32}$$

where $0 \le \mu < 2\pi$, $\kappa \ge 0$, and $-1 < \rho < 1$. Note that the domain of the parameter ρ is extended so that it includes negative values. This model also contains the von Mises ($\rho = 0$), wrapped Cauchy ($\kappa = 0$), and uniform ($\rho = \kappa = 0$) distributions. The condition for unimodality is that $\kappa > -2\rho/(1+\rho)^2$ or $\kappa < -2\rho/(1-\rho)^2$. The moments of $\cos\Theta$ when Θ has pdf (5.32) with $\mu = 0$ can be expressed in terms of the beta function and confluent hypergeometric series of two variables Φ_1 given by (B.41) as

$$E(\cos^n\Theta) = \frac{e^{-\kappa}}{\pi I_0(\kappa)} \sum_{j=0}^{n} \binom{n}{j} (-1)^j \left\{ \frac{2(r^2+1)}{(1-r)^2} \right\}^{n-j}$$
$$\times B\left(\frac{1}{2}+n-j, \frac{1}{2}\right) \Phi_1 \left(\frac{1}{2}+n-j, n, 1+n-j, d_r, 2\kappa\right),$$

where $d_\rho = -4\rho/(1-\rho)^2$.

A three-parameter special case when $\nu = \pm\pi/2$ has pdf

$$f(\theta;\mu,\rho) = \frac{1-\rho^2}{2\pi I_0(\kappa)} \exp\left\{ \frac{\kappa(1-\rho^2)\cos(\theta-\mu)}{1+\rho^2-2\rho\sin(\theta-\mu)} \right\}$$
$$\times \frac{1}{1+\rho^2-2\rho\sin(\theta-\mu)}, \quad 0 \le \theta < 2\pi, \tag{5.33}$$

where again $0 \le \mu < 2\pi$, $\kappa \ge 0$, and $-1 < \rho < 1$. It is clear that $f(\mu+\theta;\mu,\rho) = f(\mu-\theta;\mu,-\rho)$ for any θ. This model with pdf (5.33) is asymmetric except for the special cases where $r = 0$ (von Mises) and $\kappa = 0$ (wrapped Cauchy).

A circular–circular regression model with (5.31) as angular error is studied by Kato and Jones (2010). The model contains circular–circular regression models by Downs and Mardia (2002) with von Mises error and Kato et al. (2008) with wrapped Cauchy error because (5.31) contains von Mises and wrapped Cauchy pdfs as special cases.

5.13.2 Transforming Cardioid Distribution

Wang and Shimizu (2012) dealt with the Möbius transformation of a random variable distributed as a cardioid distribution. The von Mises distribution VM(μ,κ) with small κ approximates the cardioid distribution $C(\mu,\rho)$ with the mean direction μ and the concentration $\rho = \kappa/2$ by an approximation $e^x \approx 1+x$. Thus, the pdf of the Möbius transformation of the von Mises random variable by Kato and Jones (2010) contains the Möbius transformed pdf of the cardioid distribution. However, the distribution function of the Möbius transformed cardioid distribution is obtainable in an explicit form.

The cardioid distribution is closed under convolution, but generally the Möbius transformed distribution does not have this property. Moments and other characteristics of the Möbius transformed distribution, and conditions for unimodality and symmetry are studied in Wang and Shimizu (2012).

5.13.3 A More General Distribution

If the Möbius transformation (5.30) is applied to a random variable $\tilde{\Theta}$ which is distributed as a member of the power cardioid family (4.37) of symmetric distributions, then the resulting distribution has a more general, possibly asymmetric and bimodal, pdf than (5.31) because the family includes the von Mises distribution as a special case. With $\mu = 0$ and $\rho = \xi$ in (4.37), as

$$\cos\tilde{\theta} = \mathrm{Re}\left[\frac{e^{i(\theta-\mu)} - \rho e^{i\nu}}{1 - \rho e^{i(\tilde{\theta}-\gamma)}}\right]$$
$$= \frac{\cos(\theta-\mu) + \rho^2\cos\{\theta-(\nu+\gamma)\} - 2\rho\cos\nu}{1+\rho^2-2\rho\cos(\theta-\gamma)} \quad (\equiv h(\theta),\ \text{say})$$

using the inverse Möbius transformation of (5.30) and the notation Re[z] which denotes the real part of a complex number z, the resulting pdf is

$$f(\theta) = \frac{1}{2\pi\ {}_2F_1(-r/2,(1-r)/2;1;\xi^2)}\{1+\xi h(\theta)\}^r \times \frac{1-\rho^2}{1+\rho^2-2\rho\cos(\theta-\gamma)}$$

or

$$f(\theta) \propto \frac{\{1+\alpha\cos(\theta-\beta)\}^r}{\{1+\rho^2-2\rho\cos(\theta-\gamma)\}^{r+1}}$$

with appropriate constants α and β. If $\nu = 0$, then

$$f(\theta) = \frac{(1-\rho_2^2)^{r+1/2}}{2\pi(1+\rho_1\rho_2)^r\ {}_2F_1(-r/2,(1-r)/2;1;\xi^2)} \times \frac{\{1+\rho_1\cos(\theta-\mu)\}^r}{\{1+\rho_2\cos(\theta-\mu)\}^{r+1}} \tag{5.34}$$

with $\rho_1 = \{\xi(1+\rho^2) - 2\rho\}/(1+\rho^2-2\rho\xi)$ and $\rho_2 = 2\rho/(1+\rho^2)$, and hence $\xi = (\rho_1+\rho_2)/(1+\rho_1\rho_2)$. A symmetric distribution with the pdf (5.34) is related to (4.49) as the Möbius circle transformation may be viewed as a combination of the stereographic projection $x = \tan(\theta/2)$, a scale change $y = \{(1-\rho)/(1+\rho)\}x$, and the inverse stereographic projection.

Kato and Jones (2010) discussed the Möbius transformation of a power cardioid random variable in a different parameterization with $r = 1/\psi$ and $\rho = \tan(\kappa\psi)$ in (4.37). The normalizing constant is expressed using the associated Legendre function $P_{1/\psi}(\cdot)$. The relation between $P_{1/\psi}(\cdot)$ and the Gauss hypergeometric function can be seen in Section 4.6.3.1.

5.14 Projected Lognormal Distribution

We call a projected distribution on the quadrant $(0, \pi/2)$ constructed from a bivariate lognormal distribution $\mathrm{LN}_2(\boldsymbol{\eta}, \Sigma)$ with parameters $\boldsymbol{\eta}$ and Σ the projected lognormal distribution and denote by $\mathrm{PLN}(\boldsymbol{\eta}, \Sigma)$, where

$$\boldsymbol{\eta} = (\eta_1, \eta_2)', -\infty < \eta_1, \eta_2 < \infty,$$
$$\Sigma = \begin{pmatrix} \sigma_1^2 & \sigma_1\sigma_2\rho \\ \sigma_1\sigma_2\rho & \sigma_2^2 \end{pmatrix}, \quad 0 < \sigma_1, \sigma_2 < \infty, -1 < \rho < 1.$$

The joint pdf of $\mathrm{LN}_2(\boldsymbol{\eta}, \Sigma)$ is given by

$$f(x_1, x_2) = \frac{1}{2\pi\sigma_1\sigma_2\sqrt{1-\rho^2}x_1x_2} \exp\left\{-\frac{1}{2(1-\rho^2)}Q\right\}, \quad 0 < x_1, x_2 < \infty$$

$$Q = \frac{(\log x_1 - \eta_1)^2}{\sigma_1^2} - \frac{2\rho(\log x_1 - \eta_1)(\log x_2 - \eta_2)}{\sigma_1\sigma_2} + \frac{(\log x_2 - \eta_2)^2}{\sigma_2^2}.$$

After making a polar transformation $x_1 = r\cos\theta$, $x_2 = r\sin\theta\,(0 < r < \infty, 0 < \theta < \pi/2)$, we have $\mathrm{PLN}(\boldsymbol{\eta}, \Sigma)$ as the marginal distribution of Θ with pdf

$$f(\theta) = \frac{\sqrt{2}}{\sqrt{\pi}(\boldsymbol{t}'\Sigma\boldsymbol{t})^{1/2}\sin(2\theta)} \exp\left\{-\frac{(\log\tan\theta + \eta_1 - \eta_2)^2}{2(\boldsymbol{t}'\Sigma\boldsymbol{t})}\right\},$$

$$\boldsymbol{t} = (1, -1)', \quad 0 < \theta < \frac{\pi}{2} \tag{5.35}$$

with $\boldsymbol{t}'\Sigma\boldsymbol{t} = \sigma_1^2 - 2\sigma_1\sigma_2\rho + \sigma_2^2$.

An alternative way to get (5.35) is as follows. If a random vector $(X_1, X_2)'$ has a bivariate lognormal distribution $\mathrm{LN}_2(\boldsymbol{\eta}, \Sigma)$, then $Y = X_2/X_1$ has a univariate lognormal distribution $\mathrm{LN}(\eta_2 - \eta_1, \sigma_1^2 - 2\sigma_1\sigma_2\rho + \sigma_2^2)$ with pdf

$$g(y) = \frac{1}{\sqrt{2\pi}\sigma y} \exp\left\{-\frac{(\log y - \mu)^2}{2\sigma^2}\right\}, \quad y > 0,$$

where $\mu = \eta_2 - \eta_1$ and $\sigma^2 = \sigma_1^2 - 2\sigma_1\sigma_2\rho + \sigma_2^2$. Transforming $\Theta = \arctan^{-1} Y$ $(y \in (0, \infty) \longleftrightarrow \theta \in (0, \pi/2))$, we have (5.35).

5.15 Other Distributions

Kato and Jones (2013) have given a distribution with pdf

$$f(\theta) = C\frac{1 + \rho_1^2 - 2\rho_1\cos(\theta - \mu_1)}{1 + \rho_2^2 - 2\rho_2\cos(\theta - \mu_2)}, \quad -\pi \le \theta < \pi$$

derived from a Brownian motion on $\mathbb{R}^2$, where $-\pi \le \mu_1, \mu_2 < \pi$, $0 \le \rho_1, \rho_2 < 1$, and the normalizing constant is expressed as

$$C = \frac{(1 - \rho_1^2)(1 - \rho_2^2)\{1 + \rho_1^2\rho_2^2 - 2\rho_1\rho_2\cos(\mu_1 - \mu_2)\}}{2\pi(1 - \rho_1^2\rho_2^2)}.$$

The distribution reduces to a wrapped Cauchy distribution if $\rho_1 = 0$.

In Kato and Jones (2015), a four-parameter wrapped Cauchy distribution with

$$f(\theta) = \frac{1}{2\pi}\left\{1 + 2\gamma\frac{\cos(\theta - \mu) - \rho\cos\lambda}{1 + \rho^2 - 2\rho\cos(\theta - \mu - \lambda)}\right\}, \quad -\pi \le \theta < \pi$$

is presented, where $-\pi \le \mu, \lambda < \pi$, $0 \le \gamma < 1$, and $0 \le \rho < 1$ with $(\rho\cos\lambda - \gamma)^2 + (\rho\sin\lambda)^2 \le (1-\gamma)^2$. The pth trigonometric moment is expressed as

$\phi(p) = \gamma(\rho e^{i\lambda})^{-1}\{\rho e^{i(\mu+\lambda)}\}^p$, $p = 1, 2, 3, \ldots$. If $\lambda = 0$ and $\gamma = \rho$, then the distribution reduces to a wrapped Cauchy distribution.

5.16 Statistical Inference

To mitigate the general hurdle of normalizing constants for circular distributions not generally available in closed analytical or "unknown" forms, Bhattacharya and SenGupta (2009b) enhanced novel Bayesian inference approaches.

Pewsey (2006) carried out a simulation study in a wrapped skew-normal distribution $\mathrm{WSNC}_C(\mu, \sigma, \gamma_1)$ to show that maximum-likelihood estimation is superior to the method of moments except the case when the underlying distribution is close to symmetric and relatively dispersed. Pewsey (2008) studied parameter estimation based on the method of moments and the method of maximum likelihood in the wrapped α-stable distribution and presented an illustrative analysis of data on the headings of migrating birds.

As noted in Section 4.12, Ley and Verdebout (2014) proposed locally and asymptotically optimal (in the maximin sense) tests for reflective symmetry about a fixed central direction against k-sine-skewed alternatives and reported simulation study and applied the tests to the ant data by Jander (1957) (cf. Fisher, 1993, Appendix B.7). The data were also analyzed by Umbach and Jammalamadaka (2009) and Abe and Pewsey (2011a).

5.17 Applications

SenGupta and Pal (2001) proposed the symmetric wrapped stable-circular uniform mixture family and provided a good fit to the interesting ant dataset from Jander, for which the usual von Mises distribution did not give an acceptable fit. Pewsey (2006) analyzed the mallard dataset of vanishing angles and showed a good fit of a mixture of the circular uniform and a wrapped skew-normal $\mathrm{WSNC}_D(\xi, \eta, \lambda)$ distributions

$$f(\theta) = p\frac{1}{2\pi} + (1-p)\frac{2}{\eta}\sum_{k=-\infty}^{\infty}\phi\left(\frac{\theta + 2\pi k - \xi}{\eta}\right)\Phi\left[\lambda\left(\frac{\theta + 2\pi k - \xi}{\eta}\right)\right], \quad 0 < p < 1$$

to the dataset.

Two examples of cosine perturbed distributions given by Abe and Pewsey (2011b) are: the directions in which the 76 female turtles moved after laying their eggs on a beach (cf. Fisher, 1993, Appendix B.3, Mardia and Jupp, 2000, Table 1.5) were reanalyzed. The histogram of the data shows two modes separated by approximately π radians. The cosine perturbed von Mises, wrapped Cauchy, and power cardioid distributions give adequate fits to the data. The second example is about the directions of flight of 33 homing pigeons (cf. Batschelet, 1981, Figure 1.6.6). In this case, the cosine perturbed cardioid distribution gives an optimal model among the power cardioid full model and its three submodels in the sense of minimum AIC.

Kim and SenGupta (2013) dealt with an example of the spawning times of a particular fish (Lund, 1999) in a marine biology study. It is shown that the fit of the three-parameter generalized von Mises distribution is quite well. Kim and SenGupta (2021) were motivated by incidences of multiple air pollution (PM2.5 level) peak times in a day in big cities and showed that their simple yet asymmetric and multimodal generalized von Mises distribution fitted a representative data set quite well.

References

Abe, T. and Pewsey, A. (2011a). Sine-skewed circular distributions, *Statistical Papers*, 52, 683–707.

Abe, T. and Pewsey, A. (2011b). Symmetric circular models through duplication and cosine perturbation, *Computational Statistics & Data Analysis*, 55, 3271–3282.

Arnold, B. C. and SenGupta, A. (2025a). Symmetric and asymmetric axial and directional models and bivariate extensions thereof, *Quantitative Frontiers: Applications of Mathematics and Statistics*, Editors: A. Adhikary, E. Trell and A. SenGupta, Springer. Chapter 13.

Arnold, B. C. and SenGupta, A. (2025b). Wrapping generalized Laplace and generalized discrete Laplace distributions, *Axioms*, 14, 529, 1–17.

Azzalini, A. (1985). A class of distributions which includes the normal ones, *Scandinavian Journal of Statistics*, 12, 171–178.

Azzalini, A. (2005). The skew-normal distribution and related multivariate families, *Scandinavian Journal of Statistics*, 32, 159–188.

Azzalini, A. and Capitanio, A. (2003). Distributions generated by perturbation of symmetry with emphasis on a multivariate skew t-distribution, *Journal of the Royal Statistical Society*, B65, 367–389.

Azzalini, A. and Capitanio, A. (2014). *The Skew-normal and Related Families*, Cambridge University Press, Cambridge.

Batschelet, E. (1981). *Circular Statistics in Biology*, Academic Press, London.

Beran, R. (1979). Exponential models for directional data, *The Annals of Statistics*, 7, 1162–1178.

Bhattacharya, S. and SenGupta, A. (2009a). Bayesian analysis of semiparametric linear-circular models, *Journal of Agricultural, Biological and Environmental Statistics*, 14, 33–65.

Bhattacharya, S. and SenGupta, A. (2009b). Bayesian inference for circular distributions with unknown normalizing constants, *Journal of Statistical Planning and Inference*, 139, 4179–4192.

Coelho, C. A. (2007). The wrapped Gamma distribution and wrapped sums and linear combinations of independent Gamma and Laplace distributions, *Journal of Statistical Theory and Practice*, 1, 1–29.

Dattatreya Rao, A. V., Ramabhadra Sarma, I. and Girija, S. V. S. (2007). On wrapped version of some life testing models, *Communications in Statistics - Theory and Methods*, 36, 2027–2035.

Downs, T. D. and Mardia, K. V. (2002). Circular regression, *Biometrika*, 89, 683–697.

Fejér, L. (1916). Über trigonometrische Polynome, *Journal für die reine und angewandte Mathematik*, 146, 53–82.

Fernández-Durán, J. J. (2004). Circular distributions based on nonnegative trigonometric sums, *Biometrics*, 60, 499–503.

Fisher, N. I. (1993). *Statistical Analysis of Circular Data*, Cambridge University Press, Cambridge.

García, M. G., Sauer, Y., Watson, T. and Wahl, S. (2024). Virtual reality (VR) as a testing bench for consumer optical solutions: a machine learning approach (GBR) to visual comfort under simulated progressive addition lenses (PALs) distortions, *Virtual Reality*, 28:36.

Gatto, R. and Jammalamadaka, S. R. (2007). The generalized von Mises distribution, *Statistical Methodology*, 4, 341–353.

Jammalamadaka, S. R. and Kozubowski, T. J. (2003). A new family of circular models: The wrapped Laplace distributions, *Advances and Applications in Statistics*, 3, 77–103.

Jammalamadaka, S. R. and Kozubowski, T. J. (2004). New families of wrapped distributions for modeling skew circular data, *Communications in Statistics - Theory and Methods*, 33, 2059–2074.

Jammalamadaka, S. R. and SenGupta, A. (2001). *Topics in Circular Statistics*, World Scientific, Singapore.

Jander, R. (1957). Die optische Richtungsorientierung der Roten Waldameise (*Formica Ruea* L.), *Zeitschrift für vergleichende Physiologie*, 40, 162–238.

Jones, M. C. (2009). Kumaraswamy's distribution: A beta-type distribution with some tractability advantages, *Statistical Methodology*, 6, 70–81.

Jones, M. C. and Pewsey, A. (2012). Inverse Batschelet distributions for circular data, *Biometrics*, 68, 183–193.

Kato, S. and Jones, M. C. (2010). A family of distributions on the circle with links to, and applications arising from, Möbius transformation, *Journal of the American Statistical Association*, 105, 249–262.

Kato, S. and Jones, M. C. (2013). An extended family of circular distributions related to wrapped Cauchy distributions via Brownian motion, *Bernoulli*, 19, 154–171.

Kato, S. and Jones, M. C. (2015). A tractable and interpretable four-parameter family of unimodal distributions on the circle, *Biometrika*, 102, 181–190.

Kato, S., Shimizu, K. and Shieh, G. S. (2008). A circular-circular regression model, *Statistica Sinica*, 18, 633–645.

Kim, S. and SenGupta, A. (2013). A three-parameter generalized von Mises distribution, *Statistical Papers*, 54, 685–693.

Kim, S. and SenGupta, A. (2021). Multimodal exponential families of circular distributions with application to daily peak hours of PM2.5 level in a large city, *Journal of Applied Statistics*, 48, 3193–3207.

Kotz, S., Kozubowski, T. J. and Podgórski, K. (2001). *The Laplace Distribution and Generalizations: A Revisit with Applications to Communications, Economics, Engineering, and Finance*, Birkhäuser, Boston, MA.

Kumaraswamy, P. (1980). A generalized probability density function for double-bounded random processes, *Journal of Hydrology*, 46, 79–88.

Ley, C. and Verdebout, T. (2014). Simple optimal tests for circular reflective symmetry about a specified median direction, *Statistica Sinica*, 24, 1319–1339.

Lund, U. (1999). Least circular distance regression for directional data, *Journal of Applied Statistics*, 26, 723–733.

Maksimov, V. M. (1967). Necessary and sufficient statistics for the family of shifts of probability distributions on continuous bicompact groups, *Teoriya Verojatnosteĭ i ee Primeneniya*, 12, 307–321 (in Russian). English translation: *Theory of Probability and Its Applications*, 12, 267–280.

Mardia, K. V. (1972). *Statistics of Directional Data*, Academic Press, London.

Mardia, K. V. (1975). Statistics of directional data (with discussion), *Journal of the Royal Statistical Society*, B37, 349–393.

Mardia, K. V. and Jupp, P. E. (2000). *Directional Statistics*, Wiley, Chichester.

Mardia, K. V. and Sutton, T. W. (1975). On the modes of a mixture of two von Mises distributions, *Biometrika*, 62, 699–701.

Mitsuyasu, H., Tasai, F., Suhara, T., Mizuno, S., Ohkusu, M., Honda, T. and Rikiishi, K. (1975). Observations of the directional spectrum of ocean waves using a cloverleaf buoy, *Journal of Physical Oceanography*, 5, 750–760.

Mulder, K., Klugkist, I., van Renswoude, D. and Visser, I. (2020). Mixtures of peaked power Batschelet distributions for circular data with application to saccade directions, *Journal of Mathematical Psychology*, 95, 1–12.

Pewsey, A. (2000). The wrapped skew-normal distribution on the circle, *Communications in Statistics - Theory and Methods*, 29, 2459–2472.

Pewsey, A. (2006). Modelling asymmetrically distributed circular data using the wrapped skew-normal distribution, *Environmental and Ecological Statistics*, 13, 257–269.

Pewsey, A. (2008). The wrapped stable family of distributions as a flexible model for circular data, *Computational Statistics & Data Analysis*, 52, 1516–1523.

Rattihalli, R. N. and SenGupta, A. (2002). Classes of asymmetric circular distributions, *Technical Report*, 267, Department of Statistics, University of California, Riverside.

Rukhin, A. L. (1972). Some statistical decisions about distributions on a circle for large samples, *Sankhyā: The Indian Journal of Statistics, Series A*, 34, 243–250.

Samorodnitsky, G. and Taqqu, M. S. (1994). *Stable Non-Gaussian Random Processes: Stochastic Models with Infinite Variance*, Chapman and Hall, New York.

SenGupta, A. (2004). On the constructions of probability distributions for directional data, *Bulletin of the Calcutta Mathematical Society*, 96, 139–154.

SenGupta, A. and Bhattacharya, S. (2015). Finite mixture-based Bayesian analysis of linear-circular models, Special Issue on Statistical Models for Environmental Issues, Guest Editor: A. SenGupta, *Environmental and Ecological Statistics*, 22, 667–679.

SenGupta, A. and Pal, C. (2001). On optimal tests for isotropy against the symmetric wrapped stable-circular uniform mixture family, *Journal of Applied Statistics*, 28, 129–143.

SenGupta, A. and Roy, M. (2024). A characteristic function based circular distribution family and its goodness of fit: The flexible wrapped Linnik family, *Journal of Applied Statistics*, 51, 2364–2381.

Siew, H. Y., Kato, S. and Shimizu, K. (2008). The generalized t-distribution on the circle, *Japanese Journal of Applied Statistics*, 37, 1–16.

Umbach, D. and Jammalamadaka, S. R. (2009). Building asymmetry into circular distributions, *Statistics & Probability Letters*, 79, 659–663.

Wang, M. Z. and Shimizu, K. (2012). On applying Möbius transformation to cardioid random variables, *Statistical Methodology*, 9, 604–614.

Wintner, A. (1947). On the shape of the angular case of Cauchy's distribution curves, *The Annals of Mathematical Statistics*, 18, 589–593.

Yfantis, E. A. and Borgman, L. E. (1982). An extension of the von Mises distribution, *Communications in Statistics - Theory and Methods*, 11, 1695–1706.

Zolotarev, V. M. (1986). *One-Dimensional Stable Distributions*, Translations of Mathematical Monographs, Volume 65, American Mathematical Society, Providence, RI, Translation from the original 1983 Russian edition.

6

Miscellaneous Distributions on the Circle

A variety of circular distributions is considered in this chapter. First, we deal with discrete distributions, which though have not been studied as much as the ones for the linear case, have yet been recently emerging in the literature. A typical example of such a data set as discrete circular arises in the study of wind directions where often the data are represented summarily in terms of the 8 (or 16, with further subcategories) directions: N, NE, E, SE, S, SW, W, and NW, with N representing the North at 0° of the (incoming) wind direction and the angles in $[-\pi, \pi)$ being measured in the clockwise direction. Next, we discuss compound distributions, and establish their useful property of infinite divisibility which plays a very important role in many areas including actuarial and biological sciences. Mixed distributions and distributions arising out of Hidden Truncation models are needed to analyze yet some other types of circular data. These are taken up toward the end of this chapter.

6.1 Preliminary Notions

Circular data cover a wide spectrum and a variety of probability distributions are necessary to model those. While some data sets have simple structures and can be modeled by the usual distributions as derived by the general methods in the earlier chapters, there remain many others which require special distributions to be derived by methods specific for those cases. In this chapter we present a collection of several such distributions as under the miscellaneous group.

6.1.1 Distributions Under Specifications

Real-life data will often dictate specifications on the form of the probability distribution suitable for modeling it. These specifications could be in terms of the support, e.g. discrete, both continuous and discrete, etc. as well as in terms of the shape of the distribution to accommodate, e.g. inflation or deflation in certain regions of the support, possible contamination, etc. In this chapter we present several such distributions.

6.1.2 Definitions and Properties

A few definitions and properties are given below which will be useful for developments in the later parts of this chapter.

A discrete circular random variable Θ is defined to assume values on a finite number of points $0 = \theta_0 < \theta_1 < \theta_2 < \cdots < \theta_{k-1} < 2\pi$ $(k \geq 2)$ on the circle. Often these points are equally spaced, i.e. $\theta_j = 2\pi j/k$ $(j = 0, 1, 2, \ldots, k-1)$. However, one may envisage unequally spaced points and also $k = \infty$, but such distributions have not been enhanced yet.

A finite mixture of given c.d.f.s $F_1, \ldots, F_k$, $F_j \in \mathcal{F}$, a family of c.d.f.s, is defined as any convex sum $G(\cdot) = \sum_{j=1}^{k} \pi_j F_j(\cdot)$, $\pi_j \in (0, 1)$, $\sum_{j=1}^{k} \pi_j = 1$. The (suppressed) rv here could be defined on a $\mathbb{R}$, circle, torus, cylinder, sphere, disc, or their higher-dimensional analogues.

A mixed circular random variable Θ is defined to have probability masses on a single or finite number of points (also termed atoms), $0 \leq \theta_0 < \theta_1 < \theta_2 < \cdots < \theta_{k-1} < 2\pi$ $(k \geq 1)$, as well as assume values continuously in the arcs, (θ_j, θ_{j+1}), $j = 0, 1, 2, \ldots, k-1$, $\theta_k \equiv \theta_0$, on the circle. When one (or more) of the distribution(s) on the arc(s) is (are) mixture distribution(s), then Θ will be said to have a mixed mixture distribution.

A rv S (and its distribution and characteristic functions) is defined to be infinitely divisible or equivalently said to possess the infinite divisibility (ID) property if $\forall\, n \in \mathbb{Z}^+$, the set of all positive integers, there exist i.i.d. rvs $X_{1,n}, \ldots, X_{n,n}$, such that $S \stackrel{d}{=} X_{1,n} + \cdots + X_{n,n}$. The summation defining S is termed geometric summation.

A rv S is said to follow a compound N-X distribution with primary N and secondary X distribution when it is represented as $S \equiv \sum_{n=1}^{N} X_n$, with X_n, $n = 1, 2, \ldots$, being i.i.d. rvs each following the X distribution. A specialization of the ID property to the case when n is a rv N. In this case, each summand X_n can have a distribution with parameters related to that of N. Letting N follow, say geometric (negative binomial, Poisson, etc.) distribution, leads to geometrically (negative binomial, Poisson, etc.) ID (GID) distribution. Thus all compound distributions with geometric primary distribution by virtue of their definition are GID. Further, Y will be said to have the corresponding closure property if its distribution is in the same family as that of the $X_{j,n}$s. Also, it is known that a GID family of distributions is a proper subset of the ID family of distributions. GID property is considered below in the context of compound circular distributions. Such properties play useful roles in applied, e.g. actuarial and biological sciences.

Let $\mathcal{G}$ be the class of finite mixtures of $\mathcal{F}$ defined above. Then, $\mathcal{G}$ is said to be identifiable iff for any $G_1, G_2 \in \mathcal{G}$, $G_j = \sum_{\ell=1}^{k_j} \pi_{\ell j} F_{\ell j}$, $j = 1, 2$, $G_1 = G_2$ holds iff $k_1 = k_2$ and $(\pi_{\ell 1}, F_{\ell 1})$, $\ell = 1, \ldots, k_1$, are a permutation of $(\pi_{\ell 2}, F_{\ell 2})$, $\ell = 1, \ldots, k_2$ (see, e.g. Atienza et al., 2006).

Another approach to defining identifiability of finite mixtures of a family of densities is as follows (see, e.g. Holzmann et al., 2004). Let $\{f(\theta; \zeta) : \zeta \equiv (\mu,\ \rho)' \in \Omega_m \equiv (\mu \in (\mathbb{S}^1)^s,\ \rho \in (\mathbb{R})^r),\ m = s + r\}$ be a family of circular densities w.r.t. Lebesgue or counting measure. Then, finite mixtures of the family of densities $\{f(\theta; \zeta)\}$ are said to be identifiable iff for any distinct set of parameters ζ_j, $j = 1, \ldots, k$, $k \geqslant 1$, the functions $f(\theta; \zeta_j)$ are linearly independent. The property of identifiability of a finite mixture distribution is crucial for not only the development of optimal statistical inference for the parameters but even for the meaningful interpretation of the distribution itself.

6.2 Discrete Distributions

A discrete circular distribution with support on k equally spaced points $2\pi j/k$, $j = 0, \ldots, k-1$, on the unit circle, is given by its pmf $P(\Theta = 2\pi j/k) = q_j$ say, where q_j satisfies,

$$\text{(i) } 0 \le q_j \le 1,\ \text{(ii) } \sum_{j=0}^{k-1} q_j = 1 \ \text{ and } \ \text{(iii) } q_{j+kr} = q_j,\ r = 0, \pm 1, \pm 2, \ldots, \pm\infty.$$

(iii) needs to hold due to the periodicity of Θ.

Let $\phi_t(X)$, $t \in \mathbb{R}$ and $\phi_p(\Theta)$, $p \in \mathbb{Z}$ denote the c.f.s of the generating linear discrete and the corresponding circular wrapped discrete random variables X and Θ with supports on $\mathbb{Z}$, the set of all integers, and on $2\pi j/k$, $j = 0, 1, \ldots, k-1$, respectively. Then proceeding in an analogous way to that in Section 3.3, we have,

Result 6.1 $\phi_p(\Theta) = \phi_{t=2\pi p/k}(X)$.

Proof: The characteristic function of Θ is given by,

$$\phi_p(\Theta) = E(e^{ip\Theta}) = \sum_{j=0}^{k-1} e^{2\pi i p j/k} \sum_{r=-\infty}^{\infty} g(j + rk)$$

$$= \sum_{j=-\infty}^{\infty} e^{i(2\pi p/k)j} g(j) = \phi_{t=2\pi p/k}(X).$$

6.2.1 Uniform Distribution

As a special case, the pmf of the discrete circular uniform distribution on the equally spaced finite number of points $0 = \theta_0 < \theta_1 < \theta_2 < \cdots < \theta_{k-1} < 2\pi$ $(k \geq 2)$ on the circle has a constant value, i.e.

$$f(\theta_j) = 1/k, \quad \theta_j = \frac{2\pi j}{k} \quad (j = 0, 1, 2, \ldots, k-1),$$

since $\sum_{j=0}^{k-1} f(\theta_j) = 1$.

The distribution function is

$$F(\theta) = \frac{j+1}{k}, \quad \theta_j \leq \theta < \theta_{j+1} \quad (j = 0, 1, 2, \ldots, k-1)$$

with $\theta_k = 2\pi$, and the pth trigonometric moment is

$$\phi_p = \frac{1}{k}\sum_{j=0}^{k-1} e^{ip\theta_j} = \frac{1}{k}\sum_{j=0}^{k-1} \omega_k^{jp} = \begin{cases} 1, & p = 0, \\ 0, & p \neq 0, \end{cases} \quad (\text{mod } k)$$

where $\omega_k = e^{2\pi i/k}$, a complex kth root of 1. No mean direction exists for the discrete uniform distribution.

From the characteristic function of the discrete uniform distribution, and using **Result 6.1**, if Θ_1 is a uniform distribution on the support $\theta_j = 2\pi j/k$ $(j = 0, 1, 2, \ldots, k-1)$ and Θ_2 which is independent of Θ_1 is a discrete distribution on the same support, then the sum $\Theta_1 + \Theta_2$ is uniformly distributed. Note that this result is (i) parallel and (ii) its proof is similar to that for the continuous case.

6.2.2 General Methods of Constructing Discrete Circular Distributions

The beginning of Section 6.2 introduced the lattice distribution as a discrete distribution on the circle. In this section we study more discrete distributions and discretization methods. These methods can be broadly classified into those applied to (A) linear discrete and (B) circular or linear continuous distributions.

6.2.2.1 Linear Discrete to Circular Discrete Distributions

A convenient and simple method is to wrap a discrete linear distribution analogous to the approach we had discussed for deriving continuous circular distributions. We illustrate this method by the examples given below.

(A) Wrapping Linear Discrete Distributions

Let the linear discrete rv X be wrapped as $M = X$ (mod k) and correspondingly define $\Theta = 2\pi j/k$ be the circular wrapped discrete rv. Then,

$$P(\Theta = 2\pi j/k) \equiv P(M = j) = \sum_{r=0}^{\infty} g(j + rk), \quad j = 0, 1, 2, \ldots, k-1,$$

defines the pmf of Θ. Choice of various distributions for X leads to the corresponding variety of circular wrapped discrete distributions. We illustrate this approach below by several examples.

(A.1) Wrapped Poisson Distribution

The wrapped Poisson distribution (see, e.g. Mardia and Jupp, 2000, p. 49) WP(λ) with random variable $M = X \pmod{k}$ is obtained by wrapping the Poisson distribution with pmf

$$g(x) = P(X = x) = \frac{\lambda^x e^{-\lambda}}{x!}, \quad x = 0, 1, 2, \ldots, \ \lambda > 0.$$

The pmf of WP(λ) can be written as

$$\begin{aligned} P(M = j) &= \frac{1}{k}\, e^{-\lambda} \sum_{r=0}^{k-1} \omega_k^{-jr} e^{\omega_k^r \lambda} \\ &= \frac{1}{k}\, e^{-\lambda} \sum_{r=0}^{k-1} e^{\lambda\cos(2\pi r/k)} \cos\left\{ \frac{2\pi jr}{k} - \lambda \sin\left(\frac{2\pi r}{k}\right)\right\} \end{aligned}$$

in a finite form (Ball and Blackwell, 1992), since it holds that

$$\sum_{r=0}^{\infty} \frac{\lambda^{j+rk}}{(j+rk)!} = \frac{1}{k} \sum_{r=0}^{k-1} \omega_k^{-jr} e^{\omega_k^r \lambda}$$

with $\omega_k = e^{2\pi i/k}$, a complex kth root of 1.

On recalling the c.f. of Poisson distribution and using **Result 6.1**, the c.f. of Θ is given by,

$$\phi_p = E(e^{ip\Theta}) = \exp\{\lambda(\omega_k^p - 1)\}.$$

The wrapped Poisson distribution has the reproductive property, since the Poisson distribution has the reproductive property, i.e. if $\Theta_1 \sim \mathrm{WP}(\lambda_1)$, $\Theta_2 \sim \mathrm{WP}(\lambda_2)$, and Θ_1 is independent of Θ_2, then $\Theta_1 + \Theta_2 \sim \mathrm{WP}(\lambda_1 + \lambda_2)$. The wrapped Poisson distribution is infinitely divisible since the Poisson distribution is infinitely divisible.

(A.2) Wrapped Negative Binomial Distribution

If X follows a negative binomial distribution with pmf

$$g(x) = P(X = x) = \binom{-m}{x} p^m (-q)^x, \quad x = 0, 1, 2, \ldots$$

for $m > 0$ and $0 < p = 1 - q < 1$, then the c.f. of X is given by

$$E(e^{itX}) = \left(\frac{p}{1 - qe^{it}}\right)^m.$$

Thus, the wrapped negative binomial random variable Θ on the lattice $\{2\pi r/k, r = 0, 1, 2, \ldots, k\}$ corresponding to X has pmf

$$P(\Theta = 2\pi j/k) = \sum_{r=0}^{\infty} g(j + rk), \quad j = 0, 1, 2, \ldots, k - 1$$

and c.f.

$$E(e^{ip'\Theta}) = \left(\frac{p}{1 - qe^{2\pi ip'/k}}\right)^m, \quad p' \in \mathbb{Z}.$$

In particular, the wrapped geometric distribution (Jacob and Jayakumar, 2013) in the case $m = 1$ of the wrapped negative binomial distribution has an expression of its pmf

$$P(\Theta = 2\pi j/k) = \sum_{r=0}^{\infty} q^{j+rk} p = \frac{pq^j}{1 - q^k}, \quad j = 0, 1, 2, \ldots, k - 1$$

and the c.f. is expressed by

$$\phi_{p'} = \frac{p}{1 - qe^{2\pi ip'/k}} = \frac{p}{x_{p'} - iy_{p'}}, \quad p' \in \mathbb{Z},$$

where $x_{p'} = 1 - q\cos(2\pi p'/k)$ and $y_{p'} = q\sin(2\pi p'/k)$. For the wrapped geometric distribution, the mean direction is $\mu = \tan^{-1}\{q\sin(2\pi/k)\}/\{1 - q\cos(2\pi/k)\}$ and the mean resultant length is $\rho = p/\sqrt{p^2 + 2q(1 - \cos(2\pi/k))}$. It is obvious that if independent and identically distributed random variables $\Theta_1, \ldots, \Theta_m$ follow the wrapped geometric distribution with parameter p, then the sum $\Theta_1 + \cdots + \Theta_m \pmod{2\pi}$ is distributed as a wrapped negative binomial distribution with parameters p and m. The wrapped geometric distribution is infinitely divisible since the geometric distribution is infinitely divisible. The wrapped discrete Mittag–Leffler distribution as a generalization of the wrapped geometric distribution has been considered in Jacob and Jayakumar (2013).

(A.3) Wrapped Discrete Skew Laplace Distribution

In Chapters 4 and 5, we had derived continuous circular distributions, e.g. wrapped exponential, Laplace, and several generalized symmetric and asymmetric distributions by wrapping corresponding continuous distributions on the real line. In this section we derive discrete circular distributions by wrapping discrete (after discretization of continuous) distributions on integers.

For the pdf (5.22) of the asymmetric (skew) Laplace distribution, we define a discrete skew Laplace distribution on integers with pmf

$$P(Y = j) = \frac{g(j)}{\sum_{r=-\infty}^{\infty} g(r)}, \quad j = 0, \pm 1, \pm 2, \ldots,$$

which is represented by

$$P(Y=j)=\frac{(1-p^*)(1-q^*)}{1-p^*q^*}\begin{cases} p^{*j}, & j=0,1,2,\dots \\ q^{*|-j|}, & j=-1,-2,-3,\dots \end{cases}$$

in terms of the parameters $p^*=e^{-\lambda\kappa}\in(0,1)$ and $q^*=e^{-\lambda/\kappa}\in(0,1)$. The c.f. of Y is

$$E(e^{itY})=\frac{(1-p^*)(1-q^*)}{1-p^*q^*}\left(\sum_{j=0}^{\infty}e^{itj}p^{*j}+\sum_{j=1}^{\infty}e^{-itj}q^{*j}\right)=\frac{(1-p^*)(1-q^*)}{(1-p^*e^{it})(1-q^*e^{-it})}.$$

The corresponding wrapped discrete skew Laplace distribution is given by

$$\begin{aligned} P(\Theta=2\pi j/k) &= \sum_{r=-\infty}^{\infty}P(Y=j+rk) \\ &= \frac{(1-p^*)(1-q^*)}{1-p^*q^*}\left\{\sum_{r=1}^{\infty}p^{*(j+rk)}+p^{*j}+\sum_{r=1}^{\infty}q^{*(-j+rk)}\right\} \\ &= \frac{(1-p^*)(1-q^*)}{1-p^*q^*}\times\frac{\{p^{*j}(1-q^{*k})+q^{*(k-j)}(1-p^{*k})\}}{(1-p^{*k})(1-q^{*k})} \end{aligned}$$

for $j=0,1,2,\dots,k$. As a special case, when q^* goes to zero, the discrete skew Laplace distribution tends to a wrapped geometric distribution with pmf $(1-p^*)p^{*j}/(1-p^{*k})$. The wrapped discrete skew Laplace distribution is infinitely divisible (Jayakumar and Jacob, 2012).

(A.4) Wrapped Generalized Discrete Asymmetric Laplace

The presentation in this subsection follows Arnold and SenGupta (2025). The generalized discrete asymmetric Laplace (GDAL) distribution has the mixture representation given by

$$X=IV_1+(1-I)(-V_2),$$

where $V_j\sim geo(p_j)$, $j=1,2$, are independent and I is an independent Bernoulli random variable with $P(I=1)=p_0$.

If $X^{(1)}\sim geo(p^{(1)})$ then it readily follows that its wrapped version, $\Theta^{(1)}=X^{(1)}$ (mod 2π) has a truncated geometric distribution, i.e.

$$P(\Theta^{(1)}=2\pi j/k)\equiv P(Y^{(1)}=j)=f_{Y^{(1)}}(j)=\frac{p^{(1)}(1-p^{(1)})^j}{1-(1-p^{(1)})^k},\quad j=0,1,2,\dots,k-1.$$

Let $X^{(2)}$ be a negative geometric variable, i.e. has the pmf

$$f_{X^{(2)}}(x)=p^{(2)}(1-p^{(2)})^{-y},\quad y=0,-1,-2,-3,\dots.$$

Then, $\Theta^{(2)} = X^{(2)}$ (mod 2π), its wrapped version, analogous to that for $\Theta^{(1)}$ has the form

$$P(\Theta^{(2)} = 2\pi j/k) \equiv P(Y^{(2)} = j) = f_{Y^{(2)}}(j) = \frac{p^{(2)}(1-p^{(2)})^{(k-1-j)}}{1-(1-p^{(2)})^j}, \quad j = 0, 1, 2, \dots, k-1.$$

Then the pmf of the wrapped GDAL (WGDAL) distribution easily follows as

$$f_Y(y) = p_0 \frac{p^{(1)}(1-p^{(1)})^y}{1-(1-p^{(1)})^k} + (1-p_0)\frac{p^{(2)}(1-p^{(2)})^{(k-1-y)}}{1-(1-p^{(2)})^k}, \quad y = 0, 1, 2, \dots, k-1. \tag{6.1}$$

As in the continuous case, this circular wrapped pmf is typically bimodal.

(B) Discretization of Continuous Linear and Circular Distributions

We first consider methods for discretizing continuous circular distributions.

(B.1) Normalized Point Masses

Imoto et al. (2020) (see also Mardia and Sriram, 2022) have studied a method of discretization. Define a pmf $p_k(j)$ from a circular pdf $f(\theta)$, with the corresponding random variable Θ, as

$$p_k(j) = C_k^{-1} f(2\pi j/k), \quad j = 0, 1, 2, \dots, k-1,$$

on k pre-fixed equidistant points, where C_k is the normalizing constant obtained by $C_k = \sum_{j=0}^{k-1} f(2\pi j/k)$. Denote the corresponding discrete random variable by Θ_k.

Suppose that the circular pdf $f(\theta)$ is represented by a Fourier series

$$f(\theta) = \frac{1}{2\pi}\left[1 + 2\sum_{p=1}^{\infty}\{\alpha_p \cos(p\theta) + \beta_p \sin(p\theta)\}\right].$$

Here α_p and β_p are, respectively, the pth cosine moment and pth sine moment of Θ. Then, for $m = 0, 1, 2, \dots, k-1$,

$$\sum_{j=0}^{m} f(2\pi j/k) = \frac{1}{2\pi}\left[(m+1)\left(1 + 2\sum_{p=1}^{\infty}\alpha_{pk}\right)\right.$$
$$\left. + 2 \sum_{\substack{p = 1, 2, 3, \dots, \\ p \neq k, 2k, 3k, \dots}} \frac{\sin\{\pi p(m+1)/k\}}{\sin(\pi p/k)} \left\{\alpha_p \cos\left(\frac{\pi p m}{k}\right) + \beta_p \sin\left(\frac{\pi p m}{k}\right)\right\}\right].$$

The equation is useful for expressing the distribution function of the discrete distribution. For $m = k-1$ in the above equation as a special case, we have

$$C_k = \sum_{j=0}^{k-1} f(2\pi j/k) = \frac{k}{2\pi}\left(1 + 2\sum_{p=1}^{\infty}\alpha_{pk}\right).$$

Note that the normalizing constant is expressed in terms of the cosine moments α_{pk} of the corresponding circular distribution. Intuitively the discrete distribution approaches the base continuous distribution as k tends to infinity. More precisely it holds that, assuming that $f \in C^2(0, 2\pi)$,

$$\frac{2\pi}{k}\sum_{j=0}^{[(k-1)a]} f(2\pi j/k) = \int_0^{2\pi a} f(\theta)d\theta + \mathrm{O}(1/k^2)$$

for any $0 < a < 1$, where $[x]$ denotes the greatest integer less than or equal to x. The qth cosine and sine moments of the discrete distribution are, respectively, expressed as

$$\alpha_{k,q} = \frac{\alpha_q + \sum_{p=1}^{\infty}\left(\alpha_{pk+q} + \alpha_{pk-q}\right)}{1 + 2\sum_{p=1}^{\infty}\alpha_{pk}}, \quad \beta_{k,q} = \frac{\beta_q + \sum_{p=1}^{\infty}\left(\beta_{pk+q} - \beta_{pk-q}\right)}{1 + 2\sum_{p=1}^{\infty}\alpha_{pk}}$$

in terms of the cosine and sine moments of the base continuous distribution.

Imoto et al. (2020) have given discrete von Mises, cardioid, nonnegative trigonometric sums, wrapped Cauchy, and four-parameter wrapped Cauchy distributions as examples. For the cardioid distribution with pdf

$$f_C(\theta) = \frac{1}{2\pi}\left\{1 + 2\rho\cos(\theta - \mu)\right\}, \quad 0 \le \theta < 2\pi$$

for $0 \le \mu < 2\pi$ (mean direction) and $0 \le \rho \le 1/2$ (mean resultant length) as the base distribution as an example, the corresponding discrete cardioid distribution has pmf

$$p_C(j) = \frac{1}{k}\left\{1 + 2\rho\cos\left(\frac{2\pi j}{k} - \mu\right)\right\}, \quad j = 0, 1, 2, \ldots, k-1.$$

(B.2) Class Probabilities at Class-Marks

Another discretization technique is to posit the class (cell) probability $h_k(j) = F(\pi(2j+1)/k) - F(\pi(2j-1)/k)$ as the pmf of the corresponding class-mark $2\pi j/k$ for $j = 1, 2, 3, \ldots, k-1$, where $F(\cdot)$ is the distribution function of the base continuous distribution. For $j = 0$, we define $h_k(0) = F(\pi/k) + \{1 - F(\pi(2k-1)/k)\}$.

As an example, consider the eight points for wind direction: N, NE, E, SE, S, SW, W, and NW. The probability of N for wind direction is computed as

$$q_{\mathrm{N}} = P(\Theta = \mathrm{N}) = \int_{-\pi/8}^{\pi/8} f(\theta)d\theta = \int_0^{\pi/8} f(\theta)d\theta + \int_{15\pi/8}^{2\pi} f(\theta)d\theta,$$

the probability of NE as

$$q_{\mathrm{NE}} = P(\Theta = \mathrm{NE}) = \int_{\pi/8}^{3\pi/8} f(\theta)d\theta,$$

and so on.

This method can be applied to any circular continuous, e.g. cardioid, von Mises, etc., distribution. As an example, we have the discrete cardioid distribution given by,

$$h(j) = \frac{1}{k}\left\{1 + 2\rho \frac{\sin(\pi/k)}{\pi/k} \cos\left(\frac{2\pi j}{k} - \mu\right)\right\}, \quad j = 0, 1, 2, \ldots, k-1.$$

All the above methods of discretization employed on a circular distribution can naturally be applied also to circular distributions obtained from linear continuous distributions, e.g. derived by the methods of wrapping, projection, and conditioning as presented in Chapter 3.

Remarks.

Remark 6.1 It is easily seen that the wrapping $\mathfrak{W}$ and discretization $\mathfrak{D}$ operators are not commutative, i.e. $\mathfrak{W}(\mathfrak{D}(g(\theta)) \neq \mathfrak{D}(\mathfrak{W}(g(\theta))$. For example, wrapped discrete Laplace distribution obtained by wrapping of a discrete Laplace distribution is not the same distribution as a discrete wrapped Laplace distribution obtained by discretizing (by any of the methods discussed above) a wrapped Laplace distribution.

Remark 6.2 As a method of discretization from a continuous distribution with distribution function $F(y)$ on the real line to a discrete distribution with pmf h_j on integers, there exists a mean preserving procedure

$$H(x) = \sum_{j=0}^{x} h_j = \int_{x}^{x+1} F(y)dy, \quad x = 0, 1, 2, \ldots$$

with the distribution function $H(x)$ of the discrete distribution. But we do not know about a mean direction preserving procedure from a continuous distribution to a discrete distribution on the circle.

6.3 Compound Distributions

A compound distribution as defined earlier will, in general, have a complicated form. However for some special cases, e.g. with geometric or Poisson compounding, this may take elegant forms and also may lead to very useful results. Further for the special case of geometric compounding as defined in Section 6.2.2, we have the following:

Result 6.2
(i) $M_S(t) = M_N[\ln M_X(t)]$,
(ii) $\phi_S(t) = (1-p)/(1-p\phi(t))$, where $M(\cdot)$ and $\phi(\cdot)$ represent m.g.f. and c.f., respectively.
Result 6.2 (ii) easily generalizes to circular rvs, wherein t is replaced by p'.

6.3.1 Geometrically Infinitely Divisibility of Circular Distributions

In the context of compound distributions, we next take up geometrically infinitely divisibility.

Recall from Section 6.1.2 that a random variable X (and its probability distribution) is said to be geometrically infinitely divisible (GID) (Kotz et al., 2001, p. 48) if for any $p \in (0, 1)$ it satisfies the relation

$$X \stackrel{\mathrm{d}}{=} \sum_{j=1}^{\nu_p} X_p^{(j)},$$

where ν_p is a geometric random variable with mean $1/p$, whose probability mass function is $P(\nu_p = k) = pq^{k-1},\ k = 1, 2, 3, \ldots,$ with $q = 1 - p$, the random variables $X_p^{(j)}$ are independent and identically distributed for each p, and ν_p and $(X_p^{(j)})$ are independent.

An exponential distribution $\mathrm{Exp}(\lambda)$ with mean $1/\lambda$ is geometrically infinitely divisible with $X_p^{(j)} \sim \mathrm{Exp}(\lambda/p)$ for any p. In fact, the probability generating function of the geometric distribution is

$$E(t^{\nu_p}) = \frac{pt}{1 - qt}, \qquad |qt| < 1,$$

and if $X_p^{(j)}$ is independent and identically distributed as the exponential distribution $\mathrm{Exp}(\lambda/p)$ with characteristic function $E[\exp\{itX_p^{(j)}\}] = 1/(1 - ipt/\lambda)$, then the characteristic function of $\sum_{j=1}^{\nu_p} X_p^{(j)}$ is

$$\begin{aligned} E\left\{e^{it\sum_{j=1}^{\nu_p} X_p^{(j)}}\right\} &= E\left[E\left\{e^{it\sum_{j=1}^{\nu_p} X_p^{(j)}} \mid \nu_p\right\}\right] \\ &= \left(p\,\frac{1}{1 - ipt/\lambda}\right) \Big/ \left(1 - q\,\frac{1}{1 - ipt/\lambda}\right) \\ &= \frac{1}{1 - it/\lambda}, \end{aligned}$$

which is the characteristic function of $\mathrm{Exp}(\lambda)$.

An asymmetric Laplace distribution $L(\lambda, \kappa)$ is geometrically infinitely divisible with $X_p^{(j)} \sim L(\lambda/\sqrt{p}, \kappa_p)$ for any p, where κ_p is the unique solution of the equation

$$\frac{1}{\kappa_p} - \kappa_p = \sqrt{p}\left(\frac{1}{\kappa} - \kappa\right).$$

For circular random variables and probability distributions, the definition of geometric infinite divisibility is obvious with linear summand variables replaced by circular ones and the sum is taken naturally using mod 2π.

von Mises–Fisher, and in particular von Mises, distribution has been proven (Kent, 1977) to be ID for all $\kappa \in (0, \infty)$ and for all dimensions.

6.3.2 GID of Wrapped Exponential and Laplace Distributions

From the geometric infinite divisibility of the exponential distribution Exp(λ) and the asymmetric Laplace distribution $L(\lambda, \kappa)$, we have the geometric infinite divisibility of the wrapped exponential distribution WE(λ) with $\Theta_j \sim \mathrm{WE}(\lambda/p)$ and that of the wrapped Laplace distribution WL(λ, κ) with $\Theta_j \sim \mathrm{WL}(\lambda/\sqrt{p}, \kappa_p)$, where κ_p satisfies the same equation as in the case of $L(\lambda, \kappa)$.

6.3.3 GID of Wrapped Skew Linnik Distribution

We show below that the wrapped skew or asymmetric Linnik (AL) rv Θ admits of a compound distribution representation, where the primary distribution is geometric and the secondary distribution is wrapped skew Linnik, and hence Θ is GID.

Let Θ_j be i.i.d. wrapped AL ($\alpha, \theta, \sigma = 1$) and v_p be geometric random variables with c.f. and pmf given, respectively, by,

$$\varphi(p') = (1 + e^{-i\theta \ \mathrm{sgn}(p')}|p'|^{\alpha})^{-1} \quad \text{for } \alpha \in (0, 2],\ |\theta| \le \min(\pi\alpha/2, \pi - \pi\alpha/2)$$

$$P(v_p = r) = p(1-p)^{r-1}, \quad r = 1, 2, 3, \dots .$$

Assume that v_p is independent of the Θ_j's. Then,

$$\begin{aligned} E[e^{ip'p^{1/\alpha}(\Theta_1+\Theta_2+\cdots+\Theta_{v_p})}] &= \sum_{r=1}^{\infty} E[e^{ip'p^{1/\alpha}(\Theta_1+\Theta_2+\cdots+\Theta_r)}]p(1-p)^{r-1} \\ &= \sum_{r=1}^{\infty}\left(\frac{1}{1+e^{-i\theta \ \mathrm{sgn}(p^{1/\alpha}p')}p|p'|^{\alpha}}\right)^{r} p(1-p)^{r-1} \\ &= \frac{1}{1+e^{-i\theta \ \mathrm{sgn}(p')}|p'|^{\alpha}}, \end{aligned}$$

which is the c.f. of the wrapped asymmetric Linnik ($\alpha, \theta, \sigma = 1$) distribution. Hence the wrapped asymmetric Linnik family of distributions is GID with closure property.

$\theta = 0$ yields the special case of wrapped symmetric Linnik distribution with c.f. given by

$$\varphi(p') = \frac{1}{1 + |p'|^{\alpha}},$$

where $0 < \alpha \le 2$.

Linnik distribution, defined on $\mathbb{R}$, has been investigated in detail in Pakes (1998) and Erdoğan and Ostrovskii (1998), and it reduces to the Laplace distribution for $\alpha = 2$. The GID property of the (linear) asymmetric Linnik distribution has been established by SenGupta et al. (2026) following an approach similar to the above, and Lin (1998) has also presented an alternative proof of the above result. However, the former one gives a direct proof and establishes the closure property too.

6.3.4 GID of Wrapped Geometric Distribution

Let $\Theta_1, \Theta_2, \Theta_3, \dots$ be independent and identically distributed wrapped geometric random variables with parameter p. Let v_p, which is independent of Θ_j $(j = 1, 2, 3, \dots)$, have the geometric distribution with pmf $P(v_p = r) = pq^{r-1}$ $(r = 1, 2, 3, \dots;\ 0 < p = 1 - q < 1)$, Then the c.f. of the sum $\Theta = \Theta_1 + \cdots + \Theta_{v_p}$ is represented by

$$\begin{aligned}
E(e^{ip'\Theta}) &= \sum_{r=1}^{\infty} E\{e^{ip'(\Theta_1+\Theta_2+\cdots+\Theta_r)}\}pq^{r-1} \\
&= \sum_{r=1}^{\infty} \left(\frac{P}{1 - Qe^{2\pi ip'/k}}\right)^r pq^{r-1} \\
&= \frac{A}{1 - Be^{2\pi ip'/k}},
\end{aligned}$$

which is the c.f. of the wrapped geometric distribution with parameter $A = pP/(1 - Pq)$ $(B = 1 - A = Q/(1 - Pq))$. This shows that the wrapped geometric distribution is geometrically infinitely divisible with closure property.

6.3.5 GID of Wrapped Discrete Skew Laplace Distribution

Denote by v_p the same geometric random variable as the above. Then the characteristic function of the sum $\Theta_1 + \cdots + \Theta_{v_p}$ of wrapped discrete skew Laplace random variables is

$$\begin{aligned}
E\left\{e^{ip'(\Theta_1+\cdots+\Theta_{v_p})}\right\} &= \sum_{r=1}^{\infty} E\left\{e^{ip'(\Theta_1+\cdots+\Theta_r)}\right\} pq^{r-1} \\
&= \sum_{r=1}^{\infty} \left\{\frac{(1-p^*)(1-q^*)}{(1-p^*e^{2\pi ip'/k})(1-q^*e^{-2\pi ip'/k})}\right\}^r pq^{r-1} \\
&= \frac{(1-p^*)(1-q^*)p}{(1-p^*e^{2\pi ip'/k})(1-q^*e^{-2\pi ip'/k}) - (1-p^*)(1-q^*)q}.
\end{aligned}$$

It is shown in Jayakumar and Jakob (2012) that the last equation is of the form

$$\frac{(1-A)(1-B)}{(1-Ae^{2\pi ip'/k})(1-Be^{-2\pi ip'/k})}, \quad 0 < A = 1 - B < 1$$

and hence the wrapped discrete skew Laplace distribution is geometrically infinitely divisible with closure property.

6.4 Miscellaneous Distributions

6.4.1 Perturbed and Mixture Distributions

We now turn to some special cases where original distributions need to be adjusted or perturbed. In practice, usual probability distributions may fail to give good fit at certain points in the discrete case, or at certain intervals in the continuous case. Such a situation frequently encountered with discrete data as that of inflated or deflated frequency at the starting point zero. Let the original pmf $f_Y(y)$ be needed to be inflated at $\theta = 2\pi m/k$, for a given $m \in \{1, \ldots, k\}$. Then $f_Y(y)$ is perturbed to $g_Y(y)$ as,

$$g_Y(y) = \pi_1 + (1-\pi_1)f_Y(m),\ y = m; = (1-\pi_1)f_Y(y),\ y = 0, 1, \ldots, m-1, m+1, \ldots, k,$$

where $\pi_1 \in (0, 1)$. Taking $m = 0$ results in inflated pmf at the starting point 0. Note that this perturbation constrains the probability masses assigned to all the other points to get deflated accordingly. Deflated distributions can be obtained also by appropriate modifications of $f_Y(y)$.

Perturbation can also be envisaged for the continuous case. Let $f_0(\theta)$ be a circular pdf defined on the arc $[0, \theta_0)$. The resulting perturbed pdf $g_\Theta(\theta)$ of the original pdf $f_\Theta(\theta)$ is then given by

$$g_\Theta(\theta) = \pi_1 f_0(\theta) + (1-\pi_1)f_\Theta(\theta), \quad \theta \in [0, 2\pi).$$

In contrast to the discrete case, $f_0(\theta) >$ or $< f_\Theta(\theta)$, will, respectively, yield an inflated or deflated probability for the interval $[0, \theta_0)$. The above perturbation may be adopted to produce inflated or deflated probabilities at several arcs $[\theta_i, \theta_{i+1}),\ i = 1, 2, \ldots,$ on the circle.

The perturbation models above were motivated by the need to modify certain specific parts of the original distribution. However, there is often a need based on the observed data to modify the shape of the original distribution as a whole. Toward this goal, we next discuss circular mixture distributions as defined in Section 6.1.2. The component distributions $F_j,\ j = 1, \ldots, k$, could come from the same family or from different families. For example, letting $k = 2$, we may have a mixture of two von Mises or a mixture of a von Mises and a wrapped Cauchy distribution. We may also construct circular discrete mixture distributions. As an example, note that the pmf of the wrapped GDAL given in (6.1) corresponds to that of a mixture distribution with its components as wrapped discrete geometric and negative geometric pmfs.

Let F_j be unimodal symmetric distributions with the modes $M_j,\ j = 1, \ldots, k$. If M_js are the same, say equal to M_0, then the mixture distribution G will also be unimodal symmetric with the same mode M_0, but with different tails than the components. If further, M_js are different, then G will be possibly multimodal. In general, one may take F_js to be skewed, flat-topped, highly peaked, etc.

distributions. Recall from Chapter 5 that the Inverse and Power Batschelet families of distributions provide rich classes of flat-topped and highly peaked symmetric distributions. A mixture of distributions from these families can then provide multi-modality also. Such mixtures of peaked Power Batschelet distributions as discussed were proposed by Mulder et al. (2020).

However, just as in the linear case, all circular mixture distributions are not necessarily multi-modal. Consider the family of "mirror-image" distributions given by

$$g(\theta;\mu) = \frac{1}{2}\,\left[f(\theta;\mu) + f(\theta;-\mu)\right], \quad \theta,\mu \in (-\pi,\pi],$$

where $f(\theta;\mu)$ is symmetric about μ. Then, taking $f(\cdot)$ as the pdf of the CN(μ,κ), it is known that the resulting $g(\theta;\mu,\kappa)$ is unimodal and platykurtic for $\mu \le \arccos[-1 + (1+4\kappa^2)^{1/2}/(2\kappa)]$. Bartels (1984) has thus referred to this density as the bidirectional, rather than the bimodal, von Mises density.

Trigonometric moments of the mixture distribution can be easily obtained from those of the components. For example,

$$E_G(\cos(p\Theta)) \equiv \alpha_G(p) = \sum_{j=1}^{k}\pi_j\alpha_{F_j}(p); \quad E_G(\sin(p\Theta)) \equiv \beta_G(p) = \sum_{j=1}^{k}\pi_j\beta_{F_j}(p).$$

From the above result, the circular mean and dispersion for mixture distribution $G(\Theta)$ can be easily determined. This facilitates the method of trigonometric moments for estimation of the parameters of a circular mixture distribution.

There is a unique mixture distribution for the circular case as compared to the linear rvs. A linear mixture distribution with a component as uniform distribution on $\mathbb{R}$, is not meaningful. However, the circular uniform distribution can be very useful as a component of a circular mixture distribution. Consider the case of a two-component mixture distribution: $g(\theta) = (1-\pi_1)f_1(\theta) + \pi_1 f_2(\theta)$, $\pi_1 \in [0,1]$. When π_1 is small, $f_2(\theta)$ is interpreted as a contaminating distribution on $f_1(\theta)$. While $f_2(\theta)$ in general can change the shape of the original distribution $f_1(\theta)$, use of the circular uniform (CU) as the contaminating distribution will retain the general shape and can affect the tails of $f_1(\theta)$ more prominently. Further, in this case, the trigonometric moments are easily obtained as,

$$\alpha_G(p) = (1-\pi_1)\alpha_{F_1}(p) + \pi_1 I_{(p=0)}; \quad \beta_G(p) = (1-\pi_1)\beta_{F_1}(p), \quad p \ge 0,$$

where $I_{(p=0)}$ is the indicator function given by $I_{(p=0)} = 1,\ p = 0; = 0,\ p > 0$. Thus the CU distribution has been quite popular as a contaminating distribution - see, e.g. the CN–CU mixture used by Guttorp and Lockhart (1988). The symmetric Wrapped Stable–Circular Uniform (SWS–CU) distribution proposed by SenGupta and Pal (2001) as

$$g(\theta) = (1-\pi_1)\left[\frac{1}{2\pi}\left\{1 + 2\sum_{p=1}^{\infty}\rho^{p^{\alpha}}\cos p(\theta-\mu)\right\}\right] + \pi_1\,\frac{1}{2\pi},$$

$\pi_1 \in [0,1]$, $\rho \in [0,1]$, $\alpha \in [0,2]$, $\mu \in [0,2\pi)$, is a useful, flexible, unimodal, symmetric family of circular mixture distributions, which includes CU, WN, and WC as its contaminating distributions. Similarly, one may consider the symmetric Wrapped Linnik–Circular Uniform (SWLn–CU) family of distributions, given by

$$g(\theta) = (1-\pi_1)\left[\frac{1}{2\pi}\left\{1 + 2\sum_{p=1}^{\infty}(1+(p\sigma)^{\alpha})^{(-1)}\cos p(\theta-\mu)\right\}\right] + \pi_1\frac{1}{2\pi},$$

where $\pi_1 \in [0,1]$, $\alpha \in [0,2]$, $\sigma \in (0,\infty)$, $\mu \in [0,2\pi)$. This is also a useful, highly flexible, unimodal, symmetric family of circular mixture distributions, and includes the CU and wrapped Laplace (for ($\alpha = 2$)), (WLp) as its contaminating distributions. These mixture distributions are specially appealing since for small π_1 they serve as probability models for the flexible SWS or SWLn with a uniform contamination, while for large π_1 they model isotropy with contamination from a member of these rich families.

As noted above, a major property of the mixture distributions is their ability to model multi-modal distributions. Nevertheless, this comes at the expense of the increase in the number of parameters ($\boldsymbol{\pi}, \boldsymbol{\zeta}$) with the number of components (possible modes) k, and also the sacrifice of the exponential family property. This has led to the search for one-component circular distributions in the regular exponential family which can model multi-modality with much fewer number of parameters. Such distributions do exist – as examples, recall the distributions of Kim and SenGupta (2013, 2021) given in Chapter 4.

6.4.1.1 Identifiability of Circular Mixture Distributions

In Section 6.1.2, some of the components of $\boldsymbol{\rho}$ may have support on $\mathbb{R}^+$ in order to accommodate scale parameters and location-scale families. Also, an example of a popular (univariate) circular distribution with $s > 1$ and $r > 1$, is provided by the generalized von Mises distribution ($s = r = 2$).

Fraser et al. (1981) established the identifiability of finite mixtures of von Mises distributions and this result was extended by Kent (1983) to some generalized von Mises distributions.

The following gives another general result which can be sometimes conveniently applied to establish identifiability of finite mixtures of a family of distributions through tail conditions on the corresponding c.f.s.

Result 6.3 Consider a two-parameter location-scale family of distributions $f((x-\mu)/\sigma)$, $\sigma \in \mathbb{R}^+$. Let the c.f. $\phi(t)$ of $f(\cdot)$ satisfy $\lim_{t\to\infty}\phi(\sigma_2 t)/\phi(\sigma_1 t) = 0$, $\sigma_2 > \sigma_1$. Then the finite mixtures of $f(\cdot)$ are identifiable.

For the proof of the above result see the discussions related to equation (7) in Holzmann et al. (2004).

Recall that the c.f. $\phi_\Theta(p)$ of the generated circular wrapped variable is simply the c.f. $\phi_X(t)$ of the generating linear rv X with $t \in \mathbb{R}$ replaced by $p \in \mathbb{Z}$. The identifiability of circular wrapped distributions of identifiable linear location-scale families is then immediate. As an example, note that for the c.f. of a standard symmetric stable distribution given by $\phi_X(t) = \exp(-|t|^\alpha)$, $\alpha \in (0, 2]$, this condition is easily seen to hold. Thus identifiability of finite mixtures also holds for the circular wrapped stable family of distributions, and in particular for the wrapped normal and wrapped Cauchy distributions. However, the condition in **Result 6.3** may not be convenient to apply to the linear location-scale Linnik family of distributions.

Results for the linear case need not carry over to the circular case. Holzmann et al. (2004) provide counter examples – while finite mixtures of Poisson distribution are well known to be identifiable, finite mixtures of wrapped Poisson distributions are not. Another counter example is provided by considering the location families of pdfs $f_\mu(x) = f(x - \mu)$ on the real line, with $f(\cdot)$ being the triangular distribution (Jammalamadaka and SenGupta, 2001, p. 34), whose finite mixtures are known to be identifiable. However, for the wrapped densities, the particular mixture (1/2) $[f(\theta - \mu) + f(\theta - \mu - \pi)]$ gives the uniform distribution on the circle for any μ, and hence it is not identifiable.

6.4.2 Mixed Distributions

We start with a motivating and well-known example of a mixed distribution in the linear case. Recall the GID property of the $\mathcal{E}(\lambda)$. Let us consider the distribution of $S = \sum_{j=1}^{N} X_{(j)}$ where N as before is a geometric rv $\mathcal{GE}(p)$, $X_1, \dots, X_n$, given $N = n$ are now i.i.d. random variables, each following $\mathcal{E}(\lambda)$ but now with (λ) free of p, and $X_{(j)}$ and N are independent. Then the compound distribution of S is a mixed distribution with a probability mass of p at 0 and a continuous component as that of $\mathcal{E}(p\lambda)$ weighted by $1 - p$. The proof follows easily by recalling **Result 6.2** (ii) in Section 6.3.

We now define a mixed distribution for a circular rv. Let $f(\theta), \theta \in [-\pi, \pi)$ be a pdf for a continuous circular rv. Let, w.l.o.g., p_0 be a point mass assigned at $\theta = -\pi$. Then the probability function for the resulting rv Θ is the mixed distribution given by,

$$g_\Theta(\theta) = p_0, \quad \theta = -\pi; \quad = (1 - p_0)f(\theta), \quad \theta \in (-\pi, \pi).$$

It will be interesting to establish that, similar to the case of the linear rv $\mathcal{E}(\lambda)$ and under the similar setup, the compound geometric–$\mathcal{WE}(\lambda)$ distribution is also a circular mixed distribution.

An interesting situation occurs when modeling, e.g. wind directions. In general the daily wind directions may be modeled by using a circular rv $\Theta \in [0, 2\pi)$ as usual. However, "calm" or "no wind" may be encountered during the day, which is a qualitative (categorical) variable, and one needs to be careful in incorporating it also in the probability distribution model.

6.4.3 Distributions Based on Hidden Truncation Models

The hidden truncation method for linear distributions by Arnold and Beaver (2000) is applicable to bivariate circular distributions to generate asymmetric circular distributions. Kim and SenGupta (2013) obtained a distribution by using this method.

First, consider the circular normal conditionals density given by

$$\begin{aligned} f(\theta,\phi) \propto \exp(&a_1\cos\theta + a_2\sin\theta + a_3\cos\phi + a_4\sin\phi \\ &+ a_5\cos\theta\cos\phi + a_6\cos\theta\sin\phi + a_7\sin\theta\cos\phi + a_8\sin\theta\sin\phi) \end{aligned}$$

for $-\pi \le \theta, \phi < \pi$, where a_j $(j = 1, \ldots, 8)$ are real constants. Its name is taken by the fact that its conditional distributions are circular normal (von Mises) distributions. In fact, the marginal density of Φ is

$$f(\phi) \propto \exp(a_3\cos\phi + a_4\sin\phi)$$

and the conditional density of Θ given $\Phi = \phi$ is

$$f(\theta|\phi) \propto \exp\{(a_1 + a_5\cos\phi + a_6\sin\phi)\cos\theta + (a_2 + a_7\cos\phi + a_8\sin\phi)\sin\theta\}.$$

Second, we consider the truncated density of Θ given $\Phi > b$ for some $b \in (-\pi, \pi)$. The resulting density is

$$\begin{aligned} f(\theta|\phi > b) &= \frac{f(\theta, \phi > b)}{P(\Phi > b)} \\ &= \frac{\exp(a_1\cos\theta + a_2\sin\theta)\int_b \exp[h_1(\theta)\cos\{\phi - h_2(\theta)\}]d\phi}{\int_b\int \exp[a_1\cos\theta + a_2\sin\theta + h_1(\theta)\cos\{\phi - h_2(\theta)\}]d\theta d\phi}, \end{aligned}$$

where

$$h_1(\theta) = \sqrt{(a_3 + a_5\cos\theta + a_7\sin\theta)^2 + (a_4 + a_6\cos\theta + a_8\sin\theta)^2}$$

and

$$h_2(\theta) = \tan^{-1}\left\{\frac{a_4 + a_6\cos\theta + a_8\sin\theta}{a_3 + a_5\cos\theta + a_7\sin\theta}\right\}.$$

The distribution is called the hidden truncation circular normal (HTCN) distribution (Kim and SenGupta, 2013), analogous to the hidden truncation normal density (Arnold and Beaver, 2000) obtained by applying the method to the bivariate normal distribution. The HTCN distribution is possibly asymmetric and bimodal. The trigonometric moments, asymptotic inference about the truncation parameter, simulation study, and real data example can be seen in Kim and SenGupta (2013).

6.5 Statistical Inference

For the maximum likelihood estimation in the discrete model (Section 6.2.2.1) by Imoto et al. (2020), the log-likelihood function of the discrete distribution with pmf $p_k(j)$ is

$$L(\psi) = \sum_{j=0}^{k-1} f_j \log p_k(j;\psi) = \sum_{j=0}^{k-1} f_j \log f(2\pi j/k;\psi) + n \log C_k(\psi).$$

Here ψ denotes the (vector) parameter of the distribution, f_j $(j = 0, 1, 2, \dots, k-1)$ is the frequency observed at the point $2\pi j/k$, and $n = \sum_{j=0}^{k-1} f_j$ is the size of a sample. The notation $p_k(j;\psi)$, $f(2\pi j/k;\psi)$, $C_k(\psi)$ is used for $p_k(j)$, $f(2\pi j/k)$, C_k in the log-likelihood function as a function of the (vector) parameter. Notice that the log-likelihood function of the discrete distribution is equal to the log-likelihood function of the base continuous distribution subtracting the penalty term $n \log C_k(\psi)$ caused by the discretization. As a result, the maximum likelihood estimates of the discrete distribution are not the same as the maximum likelihood estimates of the base continuous distribution in general. However, in the case of the cardioid distribution in Section 6.2.2.1, the penalty term is a constant, independent of the parameters μ and ρ, if k is greater than or equal to two, because the pth cosine moment of the cardioid distribution is $\alpha_p = \rho$ for $p = 1$ and $\alpha_p = 0$ for $p = 2, 3, 4, \dots$. Thus, in that case, the maximum likelihood estimates of μ and ρ for the discrete cardioid distribution are identical with those for the cardioid distribution.

For circular mixture models, as pointed out, trigonometric moments can provide simplicity in estimation, and the estimators will be consistent. The EM algorithm is an obvious choice when likelihood-based estimation is preferred. Mulder et al. (2020) provide a R-package to estimate the parameters of their Power Batschelet mixture distributions based on the EM algorithm. However, convergence of the EM algorithm needs to be usually established, e.g. when the components are not members of the regular exponential family, which is often the case. Convergence of EM algorithm has been established (SenGupta and Roy, 2021; SenGupta et al., 2021) for certain toroidal and cylindrical mixture distributions when the components come from curved exponential families.

6.6 Applications

For discrete distributions on the circle, the number of points, k, is determined in general by the domain knowledge. For example, $k = 24$ for pre-fixed equidistant points is commonly used for hourly data such as gun crime data (Gill and

Hangartner, 2010; Mastrantonio et al., 2015) and intensive care unit arrival times on the 24-hour clock, while $k = 8$ and 16 are used for wind directions at a monitoring site.

Imoto et al. (2020) used $k = 9, 18, 36$ for the Clostridium and Sulfolobus data for structural comparisons of paired prokaryotic genomes in synthetic and evolutionary biology. Kim and SenGupta (2013) illustrate the goodness of fit of their distribution in Section 6.4.3 using orientations of the nests of 50 noisy scrub birds along the bank of a creek bed in Fisher (1993), B.19.

Mixed circular distributions occur in practice, e.g. when we consider the distribution of arrival times at the ticket counter of a popular music festival where there is already a queue waiting before its opening time. This is similar to the situation in life-time modeling in reliability theory where there is a positive probability of instantaneous (compared to that over an interval) failure at a given time-point, say at the outset t_0, of the object under study. This distribution naturally is in contrast to a continuous distribution with a "spike" at t_0.

CN–CU mixture has been used by Guttorp and Lockhart (1988) in a Bayesian framework to obtain the location of a downed aircraft from the distress signals, which could possibly be disoriented due to atmospheric and geographical interventions. The unique feature of the SWS–CU mixture distribution to modify the tails without increasing the number of parameters within a two-component mixture family was exploited by SenGupta and Pal (2001) to model the well-known Jander's ant data and demonstrate that it gives a better fit than the CU or the SWS family of distributions. Some organisms, however, do not target the light or heat source, but rather travel at a chosen angle, say μ, to the source (or 0°). The bidirectional von Mises mixture distribution was suggested by Bartels (1984) to model the orientation of some strains of *Dictyostelium* slugs in response to light or heat, where the slugs were assumed to make a random choice to travel toward the right or left of the source of the emanating rays at the diametrically opposite end of their starting point.

Mixture models play a pivotal role in Statistical Machine Learning – more specifically, in model-based clustering for linear data. This aspect extends to circular and, more generally, to directional data. CN mixtures have been widely used to cluster microarray gene expression data, while WN mixtures have been used for clustering signal directions in communication theory. In eye-tracking research, a crucial role is played by the saccade direction, the angle between two consecutive fixations. Due to the high kurtosis and also pronounced skewness in such data, mixtures of peaked power Batschelet distributions have been enhanced by Mulder et al. (2020) to cluster saccade directions. Clustering techniques based on mixtures of toroidal, spherical, and cylindrical distributions have been recently developed and applied to real-life microarray gene and environmental data sets – see, e.g. SenGupta and Roy (2021, 2023a,b) and SenGupta et al. (2021).

References

Arnold, B. C. and Beaver, R. J. (2000). Hidden truncation models, *Sankhyā: The Indian Journal of Statistics, Series A* (1961–2002), 62, 23–35.

Arnold, B. C. and SenGupta, A. (2025). Wrapping generalized Laplace and generalized discrete Laplace distributions, *Axioms*, 14, 529, 1–17.

Atienza, N., Garcia-Heras, J. and Mu noz-Pichardo, J. M. (2006). A new condition for identifiability of finite mixture distributions, *Metrika*, 63, 215–221.

Ball, F. and Blackwell, P. (1992). A finite form for the wrapped Poisson distribution, *Advances in Applied Probability*, 24, 221–222.

Bartels, R. (1984). Estimation in a bidirectional mixture of von Mises distributions, *Biometrics*, 40, 777–784.

Erdoğan, M. B. and Ostrovskii, I. V. (1998). Analytic and asymptotic properties of generalized Linnik probability densities, *Journal of Mathematical Analysis and Applications*, 217, 555–578.

Fisher, N. I. (1993). *Statistical Analysis of Circular Data*, Cambridge University Press, Cambridge.

Fraser, M. D., Hsu, Y. S. and Walker, J. J. (1981). Identifiability of finite mixtures of von Mises distributions, *The Annals of Statistics*, 9, 1130–1131.

Gill, J. and Hangartner, D. (2010). Circular data in political science and how to handle it, *Political Analysis*, 18, 316–336.

Guttorp, P. and Lockhart, R. A. (1988). Finding the location of a signal: A Bayesian analysis, *Journal of the American Statistical Association*, 83, 322–330.

Holzmann, H., Munk, A. and Stratmann, B. (2004). Identifiability of finite mixtures - with applications to circular distributions, *Sankhyā: The Indian Journal of Statistics* (2003–2007), 66, 440–449.

Imoto, T., Shieh, G. S. and Shimizu, K. (2020). Discrete circular distributions with applications to shared orthologs of paired circular genomes, *Computer Modeling in Engineering & Sciences*, 123, 1131–1149.

Jacob, S. and Jayakumar, K. (2013). Wrapped geometric distribution: A new probability model for circular data, *Journal of Statistical Theory and Applications*, 12, 348–355.

Jammalamadaka, S. R. and SenGupta, A. (2001). *Topics in Circular Statistics*, World Scientific, Singapore.

Jayakumar, K. and Jacob, S. (2012). Wrapped skew Laplace distribution on integers: A new probability model for circular data, *Open Journal of Statistics*, 2, 106–114.

Kent, J. T. (1977). The infinite divisibility of the von Mises–Fisher distribution for all values of the parameter in all dimensions, *Proceedings of the London Mathematical Society*, 35, 359–384.

Kent, J. T. (1983). Identifiability of finite mixtures for directional data, *The Annals of Statistics*, 11, 984–988.

Kim, S. and SenGupta, A. (2013). A three-parameter generalized von Mises distribution, *Statistical Papers*, 54, 685–693.

Kim, S. and SenGupta, A. (2021). Multimodal exponential families of circular distributions with application to daily peak hours of PM2.5 level in a large city, *Journal of Applied Statistics*, 48, 3193–3207.

Kotz, S., Kozubowski, T. J. and Podgórski, K. (2001). *The Laplace Distribution and Generalizations: A Revisit with Applications to Communications, Economics, Engineering, and Finance*, Birkhäuser, Boston, MA.

Lin, G. D. (1998). A note on the Linnik distributions, *Journal of Mathematical Analysis and Applications*, 217, 701–706.

Mardia, K. V. and Jupp, P. E. (2000). *Directional Statistics*, Wiley, Chichester.

Mardia, K. V. and Sriram, K. (2022). Families of discrete circular distributions with some novel applications, *Sankhyā A: The Indian Journal of Statistics*, https://doi.org/10.1007/s13171-022-00298-z.

Mastrantonio, G., Maruotti, A. and Jona-Lasinio, G. (2015). Bayesian hidden Markov modelling using circular-linear general projected normal distribution, *Environmetrics*, 26, 145–158.

Mulder, K., Klugkist, I., van Renswoude, D. and Visser, I. (2020). Mixtures of peaked power Batschelet distributions for circular data with application to saccade directions, *Journal of Mathematical Psychology*, 95, 102309.

Pakes, A. G. (1998). Mixture representations for symmetric generalized Linnik laws, *Statistics & Probability Letters*, 37, 213–221.

SenGupta, A. and Pal, C. (2001). On optimal tests for isotropy against the symmetric wrapped stable - circular uniform mixture family, *Journal of Applied Statistics*, 28, 129–143.

SenGupta, A. and Roy, M. (2021). Clustering on the torus, Special Issue: Celebrating the Centenary of Professor C. R. Rao, (Eds. Khattree, R., Jammalamadaka, S. R. and Rao, M. B.), *Journal of Statistical Theory and Practice*, 15, Article number: 58.

SenGupta, A. and Roy, M. (2023a). A characteristic function based circular distribution family and its goodness of fit: The flexible wrapped Linnik, *Journal of Applied Statistics*, https://doi.org/10.1080/02664763.2023.2283689. Also in *50 Years of Journal of Applied Statistics: Directional Statistics: Advances in Theory and Emerging Applications*, Editor: A. SenGupta. Taylor and Francis, 1–18.

SenGupta, A. and Roy, M. (2023b). Clustering on the cylinder, sphere and torus - statistical machine learning approaches, In *Proceedings of the International Conference COMDATA 2021*, Kuala Lumpur, Malaysia, *AIP Conf. Proc. 2756, 020001*, https://doi.org/10.1063/5.0141859.

SenGupta, A., Roy, M. and Chattopadhyay, A. K. (2021). Model-based clustering for cylindrical data, In *Advances in Statistics - Theory and Applications: Honoring the Contributions of Barry C. Arnold in Statistical Science*, (Eds. Ghosh, I., Balakrishnan, N. and Ng, H. K. T.), Springer, New York, 347–364.

SenGupta, A., Ozalp, M. A. and Yildirak, K. (2026). Compound Geo-Linnik distribution and application to COVID regime, To Appear.

7

Axial Distributions

An axis is an undirected line for which the starting and ending points cannot be distinguished (see also Fisher 1993, p. xvii). Axial data on the unit circle have a range of possible values $[0, \pi)$, and each direction θ ($\in [0, \pi)$) is considered as equivalent to the diametrically opposite direction $\theta + \pi$ about the origin. Examples of such data relate to, e.g. a fracture in a rock exposure, face-cleat in a coal seam, long-axis orientations of feldspar laths in basalt, horizontal axes of outwash pebbles, orientations of rock cores (Fisher 1993). In this chapter, we first deal with univariate axial distributions and then proceed to bivariate and multivariate axial distributions.

7.1 Preliminary Notions

The theoretical probability density function (pdf) of a continuous axial random variable Θ which takes values in $[0, \pi)$ is defined by $f(\theta)$ which satisfies the following properties:

(a) for any $\theta \in [0, \pi), f(\theta) \geq 0$,

(b) $\displaystyle\int_0^{\pi} f(\theta)d\theta = 1$, and

(c) for any integer $k, f(\theta) = f(\theta + k\pi)$.

Since the domain of $f(\theta)$ is extended to the whole real line using the periodicity (c), the distribution function $F(\theta)$ of Θ, which is obtained by $F(\theta) = \int_0^{\theta} f(\xi)d\xi$ ($\theta \in [0, \pi)$), satisfies the relation

$$F(\theta + (k+1)\pi) - F(\theta + k\pi) = 1, \quad k = 0, \pm 1, \pm 2, \dots .$$

Thus,

$$\lim_{\theta \to -\infty} F(\theta) = -\infty \quad \text{and} \quad \lim_{\theta \to \infty} F(\theta) = \infty.$$

Note that a distribution restricted on $[0, \pi)$ which is derived by truncating a circular distribution on $[0, 2\pi)$ does not, in general, give an axial distribution in the above sense.

7.2 General Methods of Construction for Axial Distributions

We now discuss several methods of deriving axial distributions.

7.2.1 Doubling Method

Any circular density can be used to produce an axial density by noting that the angle corresponding to a circular distribution is double of that for an axial distribution; i.e. simple technique of *doubling*. Thus, if $g(\theta)$ is a circular density with support $[0, 2\pi)$, the corresponding axial density will be of the form

$$f(\theta) = 2g(2\theta)\ I(0 \leq \theta < \pi). \tag{7.1}$$

The above may be referred to as that of "Doubling" method.

Note that the argument 2θ will traverse the entire circle from $[0, 2\pi)$. Further, since $g(\theta)$ is 2π-periodic, $f(\theta)$ will be π-periodic. This yields the very useful

Result 7.1 The shape of the axial distribution $f(\cdot)$ resembles that of the generating circular distribution $g(\cdot)$ from which it is obtained by the Doubling method.

By virtue of Result 7.1, properties pertaining to membership in exponential family, symmetry, modality, peakedness, flat-top, tail behavior, etc., of $g(\cdot)$ will be preserved in $f(\cdot)$. Thus, the $f(\cdot)$ required to fit a desired shape can be easily obtained from any $g(\cdot)$ known to possess that shape. Further, the effect of the parameters on the shape of the axial density is similar to that for the circular density and, hence, may be inferred from the latter. Of course, the corresponding critical angles will naturally differ.

An exponential family of symmetric axial densities can be constructed (Arnold and SenGupta 2026) by doubling a von Mises random variable, which yields

$$f_\Theta(\theta; \mu, \kappa) \propto \exp[\kappa \cos\{2(\theta - \mu)\}]\ I(0 \leq \theta < \pi). \tag{7.2}$$

We will refer to this distribution as axial von Mises Type I, AvM-I(μ, κ) distribution.

A natural extension of (7.2) for an axial variable was given by Arnold and SenGupta (2026) as the axial power cardioid distribution,

$$f_\Theta(\theta; \alpha, \beta, \gamma) \propto \{1 + \alpha\cos(2\theta) + \beta\sin(2\theta)\}^\gamma\ I(0 \leq \theta < \pi), \tag{7.3}$$

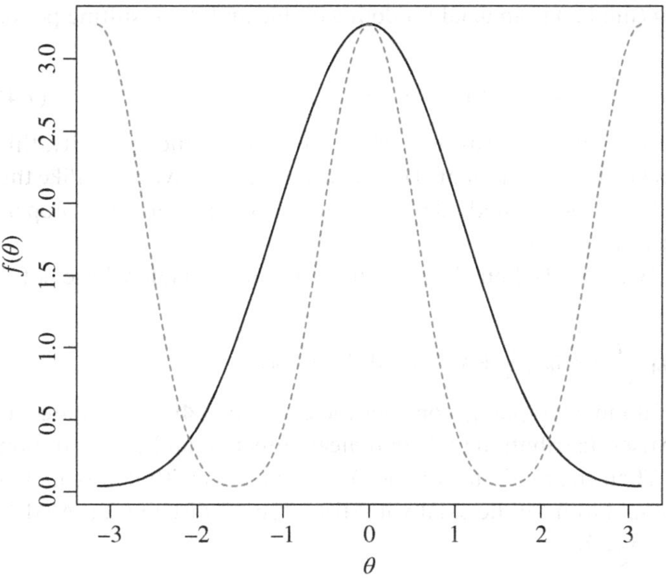

Figure 7.1 Power cardioid unnormalized circular (—) and axial (· · ·) pdfs by Doubling.

where $\gamma \in (-\infty, \infty)$ and $\alpha^2 + \beta^2 < 1$. The similarity in the shapes of the circular and axial power cardioid densities is depicted in Figure 7.1, where the location parameter is taken to be 0 (implying $\beta = 0$), $\alpha = 0.8$ and $\gamma = 2$. Note that in the figure, since the circular random variable has support $(-\pi, \pi)$, the corresponding axial random variable has support $(-\pi/2, \pi/2)$. Thus, the shape of the axial pdf over $(-\pi/2, \pi/2)$ should be compared to that of the circular pdf over $(-\pi, \pi)$.

7.2.2 Wrapping Method

In Section 3.3, we introduced the wrapping method to construct a circular distribution from a distribution on the real line. Similarly, axial distributions can also be constructed by wrapping a linear distribution. Thus, for V, a real-valued random variable, we may define an axial variable as $\Theta = V \bmod (\pi)$, and derive its axial distribution $f_\Theta(\theta)$ corresponding to the linear distribution $g_V(v)$ of V, using the same technique of wrapping:

$$f_\Theta(\theta) = \sum_{k=-\infty}^{\infty} g_V(\theta + k\pi).$$

Arnold and SenGupta (2006, 2011) proposed wrapping of a circular distribution to yield an axial distribution. A further wrapping technique may be used by wrapping circular density. Let Φ be a circular random variable with pdf $f_\Phi(\phi)$ on

$[0, 2\pi)$. Then $\Theta = \Phi \pmod{\pi}$ is an axial random variable and the resulting pdf of Θ becomes

$$f_\Theta(\theta) = \{f_\Phi(\theta) + f_\Phi(\theta + \pi)\}\, I(0 \le \theta < \pi). \tag{7.4}$$

In case the circular density $f_\Phi(\theta)$ was derived by wrapping a linear density, this technique may then be viewed as that of "Double wrapping." However, unlike the doubling method, the wrapping method will not necessarily preserve the properties of the generating distribution.

If we define $V = 2\Theta$, in (7.4) then V is a circular random variable whose pdf is given by

$$f_V(v) = \frac{1}{2}\left\{f_\Phi\left(\frac{v}{2}\right) + f_\Phi\left(\frac{v}{2} + \pi\right)\right\}, \quad 0 \le v < 2\pi.$$

As an example of double wrapping, consider the case when Φ is distributed as VM$(0, \kappa)$, the von Mises distribution with zero mean direction and concentration parameter κ (≥ 0). Then the random variable $\Theta = \Phi \pmod{\pi}$ with Φ distributed as VM(μ, κ) will be said to have the axial von Mises Type II distribution AvM-II (μ, κ), whose pdf is given by

$$f_\Theta(\theta) = \frac{1}{\pi I_0(\kappa)} \cosh\{\kappa \cos(\theta - \mu)\}, \quad 0 \le \theta < \pi. \tag{7.5}$$

AvM-II(μ, κ) is symmetric about μ and reduces to the axial uniform distribution, i.e. uniform distribution on the semicircle $[0, \pi)$, when $\kappa = 0$. This family of axial normal densities, which are symmetric, though arising naturally from circular normal variables, suffers from the fact that it does not correspond to an exponential family.

For p even, the pth cosine and sine moments of AvM-II(μ, κ) are

$$E\{\cos(p\Theta)\} = E\{\cos(p\Phi)\} = B_p(\kappa)\cos(p\mu)$$

and

$$E\{\sin(p\Theta)\} = E\{\sin(p\Phi)\} = B_p(\kappa)\sin(p\mu)$$

with notation $B_p(\kappa) = I_p(\kappa)/I_0(\kappa)$. For p odd,

$$E\{\cos(p\Theta)\} = \int_0^\pi \cos(p\phi) f_\Phi(\phi) d\phi - \int_\pi^{2\pi} \cos(p\phi) f_\Phi(\phi) d\phi$$

and

$$E\{\sin(p\Theta)\} = \int_0^\pi \sin(p\phi) f_\Phi(\phi) d\phi - \int_\pi^{2\pi} \sin(p\phi) f_\Phi(\phi) d\phi,$$

where $f_\Phi(\phi)$ denotes the pdf of Φ.

The pdf of AvM-II (μ, κ) can also be written

$$f_\Theta(\theta) = \frac{1}{\pi I_0(\kappa)} \cosh(a \sin\theta + b\cos\theta), \quad 0 \le \theta < \pi,$$

where $a = \kappa \sin \mu$ and $b = \kappa \cos \mu$ $(a, b \in \mathbb{R})$. The axial von Mises distributions do not constitute an exponential family.

7.2.3 Angular Transformation on Bivariate Linear Variables

Axial distributions may be derived using angular transformation on bivariate linear random variables $(X, Y)'$, as discussed in Arnold and SenGupta (2026). Let $(X, Y)'$ have an absolutely continuous distribution on $\mathbb{R}^2$, with mean vector $(0, 0)'$ and arbitrary variance–covariance matrix. Define,

$$\Theta = \tan^{-1}\left(\frac{Y}{X}\right) + \frac{\pi}{2}, \tag{7.6}$$

where $\theta = \tan^{-1}(u)$ has range $(-\pi/2, \pi/2)$. Thus, Θ is an axial random variable, and its distribution can be derived from that of $(X, Y)'$ using the above angular transformation.

The central Gaussian axial density results by the above method when $(X, Y)'$ follows a bivariate normal distribution with the null mean vector. Interestingly, this leads to the same density as one obtained by doubling the wrapped Cauchy density, i.e.

$$f_\Theta(\theta; \alpha, \beta) \propto \{1 + \alpha \cos(2\theta) + \beta \sin(2\theta)\}^{-1}\, I(0 \leq \theta < \pi), \tag{7.7}$$

where $\alpha^2 + \beta^2 < 1$. The density (7.7) is a special case of (7.3).

In (7.6), the random variable Y/X has a Cauchy distribution. Other choices for the distribution of Y/X can also be rewarding and will be discussed below in the context of asymmetric distributions.

7.2.4 By-products of Distributions of Linear Statistics

In presenting the distribution of the classical t-distribution, Rao (1973, p. 171) obtained a distribution of a random angle Θ defined on $[0, \pi)$. By considering fractional degrees of freedom in the definition of the t-distribution, the distribution of Θ may be parameterized by a positive real-valued parameter, say α. Then the resulting family of densities is of the form

$$f_\Theta(\theta; \alpha) = [B(1/2, \alpha/2)]^{-1}\{\cos(\theta - \pi/2)\}^{\alpha-1}\, I(0 \leq \theta < \pi), \tag{7.8}$$

where $\alpha \in (0, \infty)$. Observe that this distribution however is not π-periodic.

7.3 Skew Distributions

All of the techniques for deriving asymmetric circular distributions may be adopted, with obvious modifications, to derive asymmetric axial distributions. Thus, an exhaustive treatment for it is not required. However, we present below some such methods as for illustration and further insights.

7.3.1 Wrapping Method

If the distribution of a circular random variable Φ is skew, then the corresponding axial distribution of $\Theta = \Phi \,(\mathrm{mod}\, \pi)$ by wrapping will be skew.

Wrapping a circular pdf $f_C(\phi)$ on $[-\pi, \pi)$ symmetric about $\phi = 0$ gives $f_0(\theta) = f_C(\theta) + f_C(\theta + \pi)$, and $f_0(\theta)$ becomes an axial pdf on $[-\pi/2, \pi/2)$ symmetric about $\theta = 0$. Denoting an axial random variable with pdf $f_0(\theta)$ on $[-\pi/2, \pi/2)$ by Θ_0, we have $E\{\sin(p\Theta_0)\} = 0$ for $p = 0, \pm 1, \pm 2, \ldots$. Consider an axial pdf $g(\theta)$ on $[-\pi/2, \pi/2)$ symmetric about $\theta = 0$ and let $G(\theta) = \int_{-\pi/2}^{\theta} g(t)dt$ be its distribution function. If an odd and periodic function $w(\theta)$ on $[-\pi/2, \pi/2)$ satisfies $|w(\theta)| \leq \pi/2$, then the function

$$f^*(\theta) = 2G\,(w(\theta))\,f_0(\theta) \tag{7.9}$$

is an axial pdf on $[-\pi/2, \pi/2)$, which is generally skew. An extension with location parameter μ $(-\pi/2 \leq \mu < \pi/2)$ is easily defined by replacing θ in (7.9) with $\theta - \mu$. Such families of distributions as defined by $f^*(\theta)$ above may be termed Skew-symmetric families of axial distributions. There are various choices of the skewing function $w(\theta)$. Below we consider one such function as a special case.

7.3.2 Skew-Symmetric Functions

Consider the case where $G(\theta)$ in Section 7.3.1 is $G(\theta) = (\theta + \pi/2)/\pi$, the distribution function of the axial uniform distribution on $[-\pi/2, \pi/2)$, and $w(\theta) = (\pi\lambda/2)\sin(k\theta)$ with parameter λ $(-1 \leq \lambda \leq 1)$ and k $(k = 1, 2, 3, \ldots)$. Then, (7.9) leads to the sine-skewed axial pdf

$$f(\theta) = \{1 + \lambda\sin(k\theta)\}f_0(\theta), \quad -\frac{\pi}{2} \leq \theta < \frac{\pi}{2}, \tag{7.10}$$

which has properties $f(0) = f_0(0)$ and $f(-\pi/k) = f_0(-\pi/k)$ for all λ $(-1 \leq \lambda \leq 1)$. The distribution with pdf (7.10) is generally skewed, but obviously, it is symmetric about $\theta = 0$ if $\lambda = 0$. Hereafter, we shall study the case where $k = 2$.

Trigonometric moments of the sine-skewed axial random variable Θ^* with pdf (7.10) and $k = 2$ are calculated by

$$\begin{aligned}
E\{\cos(p\Theta^*)\} &= \int_{-\pi/2}^{\pi/2} \cos(p\theta)f(\theta)d\theta = \int_{-\pi/2}^{\pi/2} \cos(p\theta)\,\{1 + \lambda\sin(2\theta)\}f_0(\theta)d\theta \\
&= \int_{-\pi/2}^{\pi/2} \cos(p\theta)f_0(\theta)d\theta = E\{\cos(p\Theta_0)\}, \\
E\{\sin(p\Theta^*)\} &= \int_{-\pi/2}^{\pi/2} \sin(p\theta)f(\theta)d\theta = \int_{-\pi/2}^{\pi/2} \sin(p\theta)\,\{1 + \lambda\sin(2\theta)\}f_0(\theta)d\theta \\
&= \frac{\lambda}{2}\int_{-\pi/2}^{\pi/2} [\cos\{(p-2)\theta\} - \cos\{(p+2)\theta\}]f_0(\theta)d\theta \\
&= \frac{\lambda}{2}\left(E[\cos\{(p-2)\Theta_0\}] - E[\cos\{(p+2)\Theta_0\}]\right)
\end{aligned}$$

for $p = 0, \pm 1, \pm 2, \ldots$. Here Θ_0 has axial pdf $f_0(\theta) = f_C(\theta) + f_C(\theta + \pi)$ on $[-\pi/2, \pi/2)$ symmetric about $\theta = 0$, i.e. $\Theta_0 = \Phi \pmod{\pi}$ when Φ is a circular random variable with pdf $f_C(\phi)$ symmetric about $\phi = 0$. Hence, for p even, the trigonometric moments of the sine-skewed axial distribution with the random variable Θ^* are expressed in terms of $\alpha_{p,0} = E\{\cos(p\Phi)\}$, the cosine moments of the circular random variable Φ, as

$$E\{\cos(p\Theta^*)\} = \alpha_{p,0} \quad \text{and} \quad E\{\sin(p\Theta^*)\} = \lambda(\alpha_{p-2,0} - \alpha_{p+2,0})/2.$$

We now present some distributions obtained as special cases of the above technique.

7.3.2.1 Sine-Skewed Axial von Mises Distributions

Consider the von Mises distribution VM$(0, \kappa)$ as a special case of (7.14). Then the sine-skewed axial von Mises distribution has pdf

$$f(\theta) = \frac{1}{\pi I_0(\kappa)} \{1 + \lambda \sin(2\theta)\} \cosh(\kappa \cos\theta), -\frac{\pi}{2} \leq \theta < \frac{\pi}{2}; \ \kappa \geq 0, -1 \leq \lambda \leq 1. \tag{7.11}$$

The sine-skewed axial von Mises distribution with location parameter μ $(-\pi/2 \leq \mu < \pi/2)$ is obtained by replacing θ in (7.11) with $\theta - \mu$.

Another version of axial sine-skewed von Mises distribution is provided by the Doubling method as

$$f(\theta) = \frac{1}{2\pi I_0(\kappa)} \{1 + \lambda \sin(2\theta)\} \exp\{\kappa \cos(2\theta)\}, -\frac{\pi}{2} \leq \theta < \frac{\pi}{2}; \ \kappa \geq 0, -1 \leq \lambda \leq 1. \tag{7.12}$$

This Doubling method used in (7.12) may be applied in a similar fashion to other distributions discussed below.

For p even, if Θ^* has pdf (7.11), then the pth cosine and sine moments of Θ^* are

$$E\{\cos(p\Theta^*)\} = B_p(\kappa) \quad \text{and} \quad E\{\sin(p\Theta^*)\} = \frac{\lambda}{2} \left\{B_{p-2}(\kappa) - B_{p+2}(\kappa)\right\}.$$

7.3.2.2 Sine-Skewed Axial Wrapped Cauchy Distribution

If the wrapped Cauchy distribution WC$(0, \rho)$ with pdf

$$f_C(\phi) = \frac{1}{2\pi} \frac{1 - \rho^2}{1 + \rho^2 - 2\rho\cos\phi}, \qquad -\pi \leq \phi < \pi; \ 0 \leq \rho \leq 1,$$

as a special case of (7.14) is applied to (7.10) with $k = 2$, then the sine-skewed axial wrapped Cauchy distribution has pdf

$$f(\theta) = \frac{1}{\pi} \{1 + \lambda \sin(2\theta)\} \frac{1 - \rho^4}{1 + \rho^4 - 2\rho^2 \cos(2\theta)},$$
$$-\frac{\pi}{2} \leq \theta < \frac{\pi}{2}; \ 0 \leq \rho \leq 1, -1 \leq \lambda \leq 1. \tag{7.13}$$

If $\lambda = 0$, then (7.13) gives the symmetric axial wrapped Cauchy distribution as a special case. The sine-skewed axial wrapped Cauchy distribution with location parameter μ $(-\pi/2 \leq \mu < \pi/2)$ is obtained by replacing θ in (7.13) with $\theta - \mu$. From the distribution function of the sine-skewed wrapped Cauchy distribution on the circle in Abe and Pewsey (2011), the distribution function of (7.13) is expressed as

$$F(\theta) = \frac{1}{2} + \frac{1}{\pi} \arctan\left(\frac{1+\rho^2}{1-\rho^2} \tan\theta\right) + \frac{\lambda(1-\rho^4)}{4\pi\rho^2} \log\left\{\frac{1+\rho^4 - 2\rho^2\cos(2\theta)}{(1+\rho^2)^2}\right\}$$

for $-\pi/2 \leq \theta < \pi/2$.

Since the sine-skewed wrapped Cauchy distributions on the circle are always unimodal (Abe and Pewsey 2011), (7.13) is unimodal, independent of the values of ρ and λ as shown below. In what follows, we assume that $0 \leq \rho < 1$ because the density is concentrated at the point $\theta = 0$ when $\rho = 1$. Differentiation of (7.13) with respect to θ yields

$$f'(\theta) = \frac{\sqrt{1-a^2}\sqrt{\lambda^2+a^2}}{\pi\{1 - a\cos(2\theta)\}^2} \{\cos(2\theta + \xi) - \zeta\}.$$

Here $a = 2\rho^2/(1+\rho^4)$ $(\in [0,1))$, $\zeta = a\lambda/(\sqrt{\lambda^2+a^2})$ $(\in [-1/\sqrt{2}, 1/\sqrt{2}])$ and $\xi = \arg(\lambda + ia)$. From this derivative, we see that (7.13) is unimodal. It is possible to obtain the mode and antimode as follows: Note that $\xi \in [0,\pi]$ from $a \geq 0$. For $0 < \lambda \leq 1$, we have $\xi \in [0, \pi/2)$ and the mode and antimode occur at

$$\theta^* = \frac{-\xi \pm \cos^{-1}\zeta}{2}$$

with density values

$$f(\theta^*) = \frac{1-\rho^4}{\pi(1+\rho^4)} \frac{\sqrt{\lambda^2 + a^2 - (a\lambda)^2} \pm \lambda^2}{\sqrt{\lambda^2 + a^2 - (a\lambda)^2} \mp a^2}.$$

For $-1 \leq \lambda < 0$, we have $\xi \in [\pi/2, \pi]$ and the mode and antimode occur at $\theta^* = (-\xi + \cos^{-1}\zeta)/2$ and $\theta^* = -(\xi + \cos^{-1}\zeta)/2 + \pi$ with the same density values as those for $0 < \lambda \leq 1$.

For p even, the pth cosine and sine moments of Θ^* with pdf (7.13) are

$$E\{\cos(p\Theta^*)\} = \rho^p \quad \text{and} \quad E\{\sin(p\Theta^*)\} = \frac{\lambda}{2}\left(\rho^{p-2} - \rho^{p+2}\right).$$

7.3.2.3 Sine-Skewed Axial Power Cardioid Distribution

Consider the power cardioid distribution on the circle (Eq. (4.39) in Section 4.6.3.1) with pdf

$$f_C(\phi) = \frac{\cosh^{1/\psi}(\kappa\psi)\{1 + \tanh(\kappa\psi)\cos\phi\}^{1/\psi}}{2\pi P_{1/\psi}(\cosh(\kappa\psi))}, \quad -\pi \leq \phi < \pi, \qquad (7.14)$$

where κ (≥ 0) and ψ ($-\infty < \psi < \infty$) are parameters, and $P_{1/\psi}$ denotes the associated Legendre function of the first kind of degree $1/\psi$ and order zero. Wrapping (7.14) gives an axial pdf, symmetric about $\theta = 0$,

$$f_0(\theta) = \frac{\cosh^{1/\psi}(\kappa\psi)}{2\pi P_{1/\psi}(\cosh(\kappa\psi))} \times \left[\{1 + \tanh(\kappa\psi)\cos\theta\}^{1/\psi} + \{1 - \tanh(\kappa\psi)\cos\theta\}^{1/\psi}\right], -\frac{\pi}{2} \leq \theta < \frac{\pi}{2}, \tag{7.15}$$

which has mode at $\theta = 0$ if $\psi < 1$ and $\theta = -\pi$ if $\psi > 1$ under the condition $\kappa > 0$. Note that (7.15) reduces to the axial von Mises distribution when $\psi = 0$ by continuity and axial wrapped Cauchy distribution when $\psi = -1$ ($\rho = \tanh(\kappa/2)$) as special cases. For $\psi = 1$ or $\psi \to \infty$, (7.15) reduces to the axial uniform distribution and has no modal point.

The symmetric axial power cardioid distribution with pdf (7.15) has property $f_0(0) > f_C(0)$. As for its flat-topped and sharply peaked property in terms of the curvature of $f_0(\theta)$ at the mode $\theta = 0$ for $\psi < 1$, defined by

$$\frac{f_0''(0)}{[1 + \{f_0'(0)\}^2]^{3/2}} = f_0''(0)$$

since $f_0'(0) = 0$, the symmetric axial power cardioid distribution for $\psi < 1$ is more flat-topped than the symmetric circular power cardioid distribution at $\theta = 0$, i.e. $f_0''(0) > f_C''(0)$. In fact, since

$$f_0''(\theta) = f_C''(\theta) + \frac{\cosh^{1/\psi}(\kappa\psi)}{2\pi P_{1/\psi}(\cosh(\kappa\psi))} \frac{\tanh(\kappa\psi)}{\psi} \times \left\{\cos\theta - \tanh(\kappa\psi)\cos^2\theta - \frac{\tanh(\kappa\psi)}{\psi}\sin^2\theta\right\} \times \{1 - \tanh(\kappa\psi)\cos\theta\}^{1/\psi-2},$$

we have

$$f_0''(0) = f_C''(0) + \frac{1}{2\pi P_{1/\psi}(\cosh(\kappa\psi))} \frac{\tanh(\kappa\psi)}{\psi} \{1 - \tanh(\kappa\psi)\}^{1/\psi-1} > f_C''(0).$$

Density plots of the sine-skewed axial power cardioid pdf $f(\theta) = \{1 + \lambda\sin(2\theta)\}f_0(\theta)$ with $f_0(\theta)$ given by (7.15) are shown as Figure 1 in Abe et al. (2012b). Graphically, the distributional shape for $\psi \geq 1$ is not so affected by a change in the values of κ.

Another general method for constructing skew axial distributions is to use skewing functions on symmetric axial distributions, as is used to derive various skew distributions by following Azzalini's (2005) idea. For example, like sine-skewed distributions on the circle in Section 5.11, sine-skewed axial distributions may be derived (see, e.g., Abe et al., 2012b).

7.3.3 Angular Projection Method

The angular projection method (Arnold and SenGupta 2026) of generating an axial density is as follows: Start with a given random vector $(X, Y)'$ and consider

$$\Theta = \tan^{-1}\left(\frac{Y}{X}\right) + \frac{\pi}{2} \tag{7.16}$$

(where $\tan^{-1}$ has range $(-\pi/2, \pi/2)$).

For the special case, where $(X, Y)'$ has a centered bivariate normal density, Y/X has Cauchy density. Then the model (7.16) can be represented as

$$\Theta = \tan^{-1}(\mu + \sigma Z) + \pi/2 \tag{7.17}$$

where Z has a standard *Cauchy*$(0, 1)$ distribution.

Z in (7.17) may be replaced by a random variable W having some non-Cauchy distribution. Even if W has a symmetric density, the density of Θ can have quite a rich variety of shapes.

7.3.4 By-Products of Noncentral Distributions of Linear Statistics

An asymmetric version of the density obtained as by-product of the t-distribution discussed above, was may be obtained (see Rao 1973, p. 172) as

$$\begin{aligned} f_\Theta(\theta; \alpha, \delta) = e^{-\delta^2/2}\{\cos(\theta - \pi/2)\}^{\alpha-1} \sum_{j=0}^{\infty} B(\alpha/2, (j+1)/2) \\ \times \{2\delta \sin(\theta - \pi/2)\}^j \, I(0 \le \theta < \pi), \end{aligned} \tag{7.18}$$

where $\alpha \in (0, \infty)$ and $\delta \in (-\infty, \infty)$. For $\delta = 0$, as expected (7.18) reduces to the density corresponding to the central t-distribution (7.8). Also, for $\alpha = 1$, the resulting density is uniform on $[0, \pi)$.

7.3.5 Flat-Topped, Highly Peaked Family

The general methods of wrapping and doubling may be used with the families of asymmetric, flat-topped, and highly peaked circular distributions of Papakonstantinou (1979) and Batschelet (1981), and wrapped stable and wrapped Linnik distributions discussed in Chapter 4. As an example, Figure 7.2 shows the flat-topped symmetric shape is preserved in the axial distribution, derived by the Doubling method applied to the corresponding circular distribution in (15.7.3) of Batschelet (1981) with the same parameter values, $\kappa = 0.8$, $\mu = 0$ and $\nu = -1$, given by

$$f(x) \propto \exp\left(\kappa \cos[2(\theta - \mu) - \nu \sin\{2(\theta - \mu)\}]\right).$$

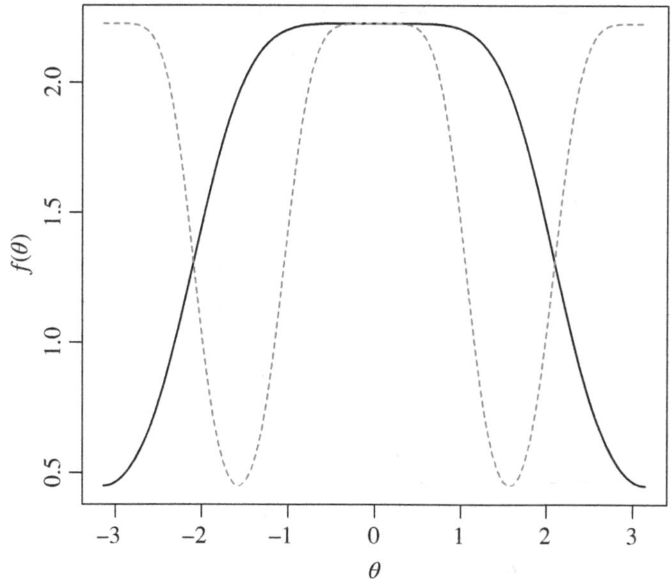

Figure 7.2 Flat-topped unnormalized circular (−) and axial (· · ·) pdfs by Doubling.

7.3.6 Flexible Multimodal Asymmetric Exponential Family

To capture multimodality in axial data, one may use the usual approach of adopting mixture distributions with number of components equaling that of the number of modes and parameters of the components chosen accordingly. However, it is desirable to derive a distribution with parameters which are interpretable and are smaller in number than that required for mixture distributions. It will be even more rewarding if that family is highly flexible in the sense that it can capture various degrees of asymmetry and skewness also. Finally, it is highly desirable that those distributions be also members of the regular exponential family since this will facilitate optimal statistical inference. Such families of circular distributions have been recently presented by Kim and SenGupta (2021) (see also Kim and SenGupta 2013, for the bimodal case) as discussed in Chapter 5.

Again, the general methods of deriving axial distributions discussed above may be applied to these families also. The doubling method applied to the multimodal circular distribution of Kim and SenGupta (2021) yields the corresponding axial pdf as

$$f(x) \propto \exp[\kappa_1 \cos\{2(\theta - \mu)\} + \kappa_2 \sin\{2q(\theta - \mu)\}].$$

As an example, Figure 7.3 shows the multi(tri)modal, asymmetric shape is preserved in the axial distribution derived by the Doubling method when applied to

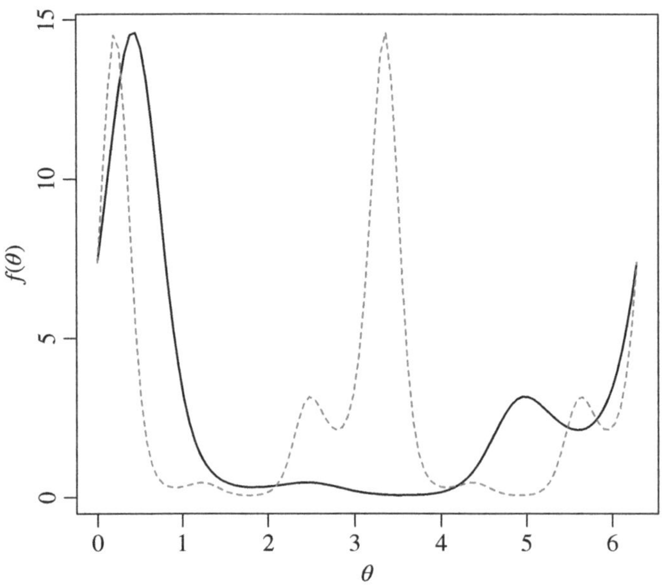

Figure 7.3 Multimodal unnormalized circular (—) and axial (· · ·) pdfs by Doubling.

the corresponding circular distribution with the same parameter values, $\kappa_1 = 2.0$, $\kappa_2 = 0.9$, $\mu = 0$, and $q = 3$.

This family is also highly flexible as it includes symmetric, asymmetric, unimodal, and multimodal distributions with varying degrees of kurtosis with just four parameters. The parameters are also easily interpretable: $\mu \in [-\pi/2, \pi/2)$, $\kappa_1 \geq 0$, $\kappa_2 \in [-1, 1]$ and $q \in \mathbb{I}^*$, a finite set of positive integers, can be interpreted as location, concentration, skewness, and modality parameters, respectively. This density has at most q modes – it is unimodal (AvM-I) for $\kappa_2 = 0$. Being a member of the regular exponential family, it is also amenable to optimal statistical inference procedures.

7.4 Bivariate Axial Distributions

Dependency can be introduced between two axial random variables in several innovative ways, some of which are specific to the axial and circular cases. Also, interest lies in the construction of such joint distributions as having preferred or specified marginals in some cases and specified conditionals in some others. Arnold and SenGupta (2006, 2011, 2026) presented various types of bivariate axial distributions with specified marginals and conditionals. We consider some general methods and their special cases below.

7.4.1 Marginal Angular Projection Method

Let $(X_1, X_2)' \in \mathbb{R}^2$. Using angular projection separately on each component, define

$$\Theta_j = \tan^{-1}(\mu_j + \sigma_j X_j) + \frac{\pi}{2}, \quad j = 1, 2,$$

where $-\infty < \mu_j < \infty$ and $\sigma_j > 0$. Then the random vector $(\Theta_1, \Theta_2)'$ provides a bivariate axial distribution on $[0, \pi)^2$.

As an example, let $(X_1, X_2)'$ have a bivariate Cauchy distribution with standard Cauchy marginal, whose joint pdf is given by

$$f(\boldsymbol{x}) = \frac{1}{2\pi\ |R|^{1/2}}\ (1 + \boldsymbol{x}' R^{-1} \boldsymbol{x})^{-3/2}, \quad \boldsymbol{x} \in \mathbb{R}^2$$

with correlation matrix R. This method is applicable to other bivariate distributions on the plane and multivariate distributions in p-dimensional Euclidean space.

7.4.2 Marginal Wrappings

The method of wrapping used for the univariate case can be extended to the construction of bivariate axial models by wrapping each variable separately. Begin with a suitable bivariate model or a distribution on the torus for a circular random vector $(\Phi_1, \Phi_2)'$ and then define a related bivariate axial random vector $(\Theta_1, \Theta_2)'$ by $\Theta_1 = \Phi_1 \pmod{\pi}$ and $\Theta_2 = \Phi_2 \pmod{\pi}$.

For example, we begin with a bivariate distribution having von Mises conditionals (SenGupta 2004), whose joint pdf is given by

$$f_{\Phi_1,\Phi_2}(\phi_1, \phi_2) = \exp\{(1, \sin\phi_1, \cos\phi_1) M (1, \sin\phi_2, \cos\phi_2)'\}, \quad (\phi_1, \phi_2)' \in [0, 2\pi)^2,$$

where M is a 3×3 matrix of parameters. Then the axial random vector $(\Theta_1, \Theta_2)'$ has pdf

$$\begin{aligned} f_{\Theta_1,\Theta_2}(\theta_1, \theta_2) = &\ f_{\Phi_1,\Phi_2}(\phi_1, \phi_2) + f_{\Phi_1,\Phi_2}(\phi_1, \phi_2 + \pi) \\ &+ f_{\Phi_1,\Phi_2}(\phi_1 + \pi, \phi_2) + f_{\Phi_1,\Phi_2}(\phi_1 + \pi, \phi_2 + \pi), \\ &(\phi_1, \phi_2)' \in [0, \pi)^2. \end{aligned}$$

Such a distribution fails to have axial von Mises marginals or conditionals. Of course, random variables from the above joint distribution of the form may be easy to simulate provided that holds also for $(X, Y)'$.

7.4.3 Axial Copula

We introduce here Axial Copula based on Arnold and SenGupta (2026). Let F_1 and F_2 be two given axial cdfs with corresponding F_1^{-1} and F_2^{-1}, respectively. Let

$(U_1, U_2)'$ be a bivariate random variable with Uniform(0,1) marginal distributions. To derive a bivariate axial distribution with marginal densities given by F_1^{-1} and F_2^{-1}, define

$$(\Theta_1, \Theta_2)' = (F_1^{-1}(U_1), F_2^{-1}(U_2))'. \tag{7.19}$$

Joint distributions with Uniform(0,1) marginals are often called *copulas*. Replacement of this (Uniform) copula by different distributions will yield different bivariate axial distributions in (7.19). Usefulness of this method depends on the availability of simple forms of cdfs. This is, however, not usually the case with axial distributions. Notable exceptions include the cardioid, four-parameter generalized cardioid model (Kato and Jones 2015), wrapped stable and wrapped Linnik distributions.

7.4.4 Specified Marginals Using Wehrly and Johnson's Family

The axial version of Wehrly and Johnson's (1980) type distribution with specified marginals, whose pdfs are $f_j(\theta_j),\ j = 1, 2$, is given by

$$f(\theta_1, \theta_2) = \pi f_1(\theta_1) f_2(\theta_2) g[\pi\{F_1(\theta_1) \pm F_2(\theta_2)\}], \quad (\theta_1, \theta_2)' \in [0, \pi)^2,$$

where g is an axial pdf and $F_j(\theta_j)$ are the distribution functions of $f_j(\theta_j)$.

7.4.5 Conditionally Specified Bivariate Family

It is often easier and meaningful to specify the conditionals of bivariate distributions, which are termed conditionally specified bivariate distributions. If the specified conditionals are both members of the regular exponential family, then Arnold and Strauss (1991) has characterized the class of all bivariate distributions with such conditionals as also one in a regular exponential family and has given an explicit form for it. However, for nonexponential family conditionals, such a characterization is not available in general. Based on the above facts, we now present several conditionally specified bivariate axial distributions.

(i) **Axial von Mises Type I Conditionals**
If we desire to have axial von Mises Type I conditionals (see (7.2)), i.e. conditionals of the form

$$f_\Theta(\theta) \propto \exp\{\alpha \cos(2\theta) + \beta \sin(2\theta)\}, \quad 0 \le \theta < \pi$$

with parameters α and β, then these are members of the regular exponential family. Following the circular case given in SenGupta (2004), such bivariate axial pdfs are readily derived as

$$f_{\Theta_1,\Theta_2}(\theta_1, \theta_2) \propto \exp\{(1, \sin(2\theta_1), \cos(2\theta_1))A(1, \sin(2\theta_2), \cos(2\theta_2))'\},$$
$$(\theta_1, \theta_2)' \in [0, \pi)^2,$$

where $A = (a_{jk})$ is a 3×3 matrix of parameters with a_{11}, a function of the other elements, being a normalizing constant.

(ii) **Axial von Mises Type II Conditionals**

Consider bivariate axial pdfs of the form

$$f_{\Theta_1,\Theta_2}(\theta_1, \theta_2) = \frac{1}{\pi^2 I_0(\sqrt{A^2 + B^2})} \cosh(A \sin\theta_1 + B\cos\theta_1), \quad (\theta_1, \theta_2)' \in [0, \pi)^2 \tag{7.20}$$

with

$$A = a_{11} \sin\theta_2 + a_{12} \cos\theta_2 \quad \text{and} \quad B = a_{21} \sin\theta_2 + a_{22} \cos\theta_2.$$

Here a_{11}, a_{12}, a_{21}, and a_{22} are real numbers. The conditional distribution of Θ_1 given $\Theta_2 = \theta_2 \in [0, \pi)$, obtained from (7.20), is an axial von Mises Type II distribution (see (7.5)) with pdf

$$f_{\Theta_1|\Theta_2}(\theta_1|\theta_2) = \frac{1}{\pi I_0(\sqrt{A^2 + B^2})} \cosh(A \sin\theta_1 + B\cos\theta_1), \quad 0 \le \theta_1 < \pi.$$

Observe that the marginal distribution of Θ_2 is axial uniform on $[0, \pi)$.

(iii) **Axial Power Cardioid Conditionals**

A joint pdf of the form

$$f_{\Theta_1,\Theta_2}(\theta_1, \theta_2) \propto \{(1, \sin(2\theta_1), \cos(2\theta_1))A(1, \sin(2\theta_2), \cos(2\theta_2))'\}^{\rho},$$
$$(\theta_1, \theta_2)' \in [0, \pi)^2$$

with a 3×3 matrix A of parameters and an additional parameter ρ has conditionals of the form (see (7.3))

$$f_{\Theta}(\theta) \propto \{1 + \gamma \sin(2\theta) + \delta \cos(2\theta)\}^{\rho}, \quad 0 \le \theta < \pi$$

with parameters γ and δ. The entries in the parameter matrix A must be constrained to ensure that $f_{\Theta_1,\Theta_2}(\theta_1, \theta_2)$ is non-negative over the region $(0, \pi)^2$, but otherwise these may be quite arbitrary.

(iv) **Axial Multimodal Conditionals**

Multimodal, exponential family axial distributions of Kim and SenGupta (2021) may also be used as conditionals. The corresponding bivariate axial pdfs follow from SenGupta (2004) and in the same lines of (i) as

$$f_{\Theta_1,\Theta_2}(\theta_1, \theta_2) \propto \exp\{(1, \sin(2\theta_1), \cos(2\theta_1), \sin(2q\theta_1))A(1, \sin(2\theta_2),$$
$$\cos(2\theta_2), \sin(2q\theta_2))'\}, \quad (\theta_1, \theta_2)' \in [0, \pi)^2,$$

where $A = (a_{jk})$ is a 4×4 matrix of parameters with a_{11}, a function of the other elements, being a normalizing constant.

While the family in items (i) and (iv) provide all bivariate axial densities with the corresponding conditionals since these are members of the regular exponential family, the same need not be true for the families in (ii) and (iii).

7.4.6 Additive and Ratio Components

Another easy way (Arnold and SenGupta 2026) to introduce dependency and also facilitate simulation from the resulting bivariate distribution is the use of a common component. Consider three axial random variables Λ_1, Λ_2, and Λ_3 which, for simplicity, are usually taken to be independent. By the method of additive component, one then defines a bivariate axial variable by

$$(\Theta_1, \Theta_2)' = ((\Lambda_1 + \Lambda_3) \bmod (\pi), (\Lambda_2 + \Lambda_3) \bmod (\pi))'. \tag{7.21}$$

Similarly, by the method of common component in the ratio, one defines bivariate axial variable by

$$(\Theta_1, \Theta_2)' = ((\Lambda_1/\Lambda_3) \bmod (\pi), (\Lambda_2/\Lambda_3) \bmod (\pi))'. \tag{7.22}$$

The challenge here is to identify such distributions using (7.21) or (7.22) for Λ_is as will lead to elegant analytical forms for the bivariate axial distribution.

7.4.7 Method of Mixing

Another easy way of introducing dependency is that of parameter mixing, as shown by Ong and SenGupta (2022). Starting with independent axial random variables, a common new random variable is mixed appropriately with the parameters of the two distributions and then integrated out from the product density. This results in a new bivariate axial density.

The above method is best explained by an example. The concentration or precision mixing is a convenient source of bivariate (and potentially, multivariate) circular and axial cardioid models.

Let Θ_1 and Θ_2 be conditionally independent given $X = x \in \mathbb{R}$, each having an axial cardioid density. Thus,

$$\begin{aligned} f_{\Theta_1,\Theta_2|X=x}(\theta_1, \theta_2|x) &= \frac{1}{2\pi} [1 + \kappa_1 x \cos\{2(\theta_1 - \mu_1)\}] \\ &\times \frac{1}{2\pi} [1 + \kappa_2 x \cos\{2(\theta_2 - \mu_2)\}]\, I((\theta_1, \theta_2)' \in (0, 2\pi)^2), \end{aligned} \tag{7.23}$$

where $(\kappa_1, \kappa_2)' \in (-1, 1)^2$ and $(\mu_1, \mu_2)' \in (0, \pi)^2$.

Assume (Arnold and SenGupta 2026) that X has a quite arbitrary distribution function $G(x)$ (generalizing the assumption of beta distribution in Ong and SenGupta 2022) on the interval $(0, 1)$ with necessarily finite first and second moments $E_G(X)$ and $E_G(X^2)$. Then the joint density of $(\Theta_1, \Theta_2)'$ is obtained from (7.23) as follows:

$$\begin{aligned} &f_{\Theta_1,\Theta_2}(\theta_1, \theta_2; \kappa_1, \kappa_2, \mu_1, \mu_2, G) \\ &= \int_0^1 \frac{1}{\pi} [1 + \kappa_1 x \cos\{2(\theta_1 - \mu_1)\}] \frac{1}{\pi} [1 + \kappa_2 x \cos\{2(\theta_2 - \mu_2)\}] \\ &\times dG(x)\, I((\theta_1, \theta_2)' \in [0, \pi)^2) \end{aligned}$$

$$= (1/\pi)^2 \left(1 + E_G(X) \left[\kappa_1 \cos\{2(\theta_1 - \mu_1)\} + \kappa_2 \cos\{2(\theta_2 - \mu_2)\}\right] \right.$$
$$\left. + E_G(X^2)\kappa_1\kappa_2 \cos\{2(\theta_1 - \mu_1)\} \cos\{2(\theta_2 - \mu_2)\}\right) I((\theta_1, \theta_2)' \in [0, \pi)^2).$$

It is clear that this joint density has marginal densities of the axial cardioid form. The conditional densities have, however, complicated forms.

Following Ong and SenGupta (2022) and Arnold and SenGupta (2026) considered the case in which a two-dimensional mixing variable $(X, Y)'$ is utilized. Whereas the former considered various bivariate beta densities for the mixing distribution, the latter demonstrated that it is possible to assume a more general mixing distribution, with certain moment constraints, and still preserve the closure property, as also presented below.

Let the joint conditional density of $(\theta_1, \theta_2)'$ given $(X, Y)' = (x, y)'$ be

$$f_{\Theta_1,\Theta_2|X=x,Y=y}(\theta_1, \theta_2|x, y) = \frac{1}{\pi} \left[1 + \kappa_1 x \cos\{2(\theta_1 - \mu_1)\}\right]$$
$$\times \frac{1}{\pi} \left[1 + \kappa_2 y \cos\{2(\theta_2 - \mu_2)\}\right] I((\theta_1, \theta_2)' \in [0, \pi)^2), \tag{7.24}$$

where $(\kappa_1, \kappa_2)' \in (-1, 1)^2$ and $(\mu_1, \mu_2)' \in (0, \pi)^2$. Assume that $(X, Y)'$ has an arbitrary joint distribution $H(x, y)$ on the set $(0, 1)^2$ with necessarily finite moments $E_H(X), E_H(Y)$ and $E_H(XY)$. The unconditional joint density of $(\theta_1, \theta_2)'$ is then, from (7.24), given by

$$f_{\Theta_1,\Theta_2}(\theta_1, \theta_2; \kappa_1, \kappa_2, \mu_1, \mu_2, H)$$
$$= \int_0^1 \int_0^1 \frac{1}{\pi} \left[1 + \kappa_1 x \cos\{2(\theta_1 - \mu_1)\}\right] \frac{1}{\pi} \left[1 + \kappa_2 y \cos\{2(\theta_2 - \mu_2)\}\right]$$
$$\times dH(x, y) \, I((\theta_1, \theta_2)' \in (0, \pi)^2)$$
$$= (1/\pi)^2 [1 + E_H(X)\kappa_1 \cos\{2(\theta_1 - \mu_1)\} + E_H(Y)\kappa_2 \cos\{2(\theta_2 - \mu_2)\}$$
$$+ E_H(XY)\kappa_1\kappa_2 \cos\{2(\theta_1 - \mu_1)\} \cos\{2(\theta_2 - \mu_2)\}] \, I((\theta_1, \theta_2)' \in [0, \pi)^2).$$

It can be easily verified that this joint density also has marginal densities of the axial cardioid form, while the conditional densities have complicated forms.

As shown above, any arbitrary mixing distribution $H(X, Y)$ may be chosen as long as the moments $E_H(X), E_H(Y)$, and $E_H(XY)$ exist. Ong and SenGupta (2022) presented a spectrum of examples in which various bivariate beta models were used for the mixing distribution. Further generalizations and discussions on the models related to the method of mixing are given in Arnold and SenGupta (2026).

7.5 Multivariate Axial Distributions

Models for axial data can be seen in Mardia and Jupp (2000, pp. 180–183). Since the vectors $\boldsymbol{x}$ and $-\boldsymbol{x}$ on the $(p-1)$-dimensional unit sphere S^{p-1} in p-dimensional

Euclidean space are indistinguishable, pdfs for $\boldsymbol{x}$ have property $f(-\boldsymbol{x}) = f(\boldsymbol{x})$. The Dimroth–Scheidegger–Watson distribution has pdf

$$f(\pm\boldsymbol{x};\boldsymbol{\mu},\kappa) = \left\{{}_1F_1(1/2;p/2;\kappa)\right\}^{-1} \exp\left\{\kappa(\boldsymbol{\mu}'\boldsymbol{x})^2\right\}, \quad \|\boldsymbol{x}\| = 1$$

for $-\infty < \kappa < \infty$, which is rotationally symmetric about $\boldsymbol{\mu}$. The pdf of the Bingham distribution is proportional to

$$f(\pm\boldsymbol{x};A) \propto \exp(\boldsymbol{x}'A\boldsymbol{x}),$$

where tr $A = 0$, without loss of generality, whose normalizing constant $\int_{S^{p-1}} \exp(\boldsymbol{x}'A\boldsymbol{x})d\boldsymbol{x}$ is expressed using the hypergeometric function ${}_1F_1$ of matrix argument. The projected normal distribution $\text{PN}_p(\boldsymbol{0},\Sigma)$ in later on Section 8.4.2 can be viewed as an axial model.

7.6 Statistical Inference

Statistical inference for the axial von Mises Type II distribution AvM-II(μ,κ) has been studied in Arnold and SenGupta (2006) and Abe et al. (2012b).

Suppose that a random sample $\theta_1,\dots,\theta_n$ of size n (≥ 2) is from AvM-II(μ,κ) and denote the sample pth cosine and sine moments by

$$C_p = \frac{1}{n}\sum_{j=1}^{n}\cos(p\theta_j) \quad \text{and} \quad S_p = \frac{1}{n}\sum_{j=1}^{n}\sin(p\theta_j).$$

Then the second trigonometric-moment estimates $\tilde{\mu}$ and $\tilde{\kappa}$ for μ and κ are given by solving the following equations:

$$B_2(\kappa)\cos(2\mu) = C_2 \quad \text{and} \quad B_2(\kappa)\sin(2\mu) = S_2.$$

If $C_2^2 + S_2^2 > 0$, the estimates are explicitly expressed as

$$\tilde{\mu} = \frac{1}{2}\arg(C_2 + i\,S_2) \quad \text{and} \quad \tilde{\kappa} = B_2^{-1}\left(\sqrt{C_2^2+S_2^2}\right) \quad (>0).$$

Note that the inverse function of B_2 exists because $B_2(\kappa) = A_4(\kappa)A_2(\kappa)$, where $A_p(\kappa) = I_{p/2}(\kappa)/I_{p/2-1}(\kappa)$ (Section 8.4.3), is an increasing function, $B_2(0) = 0$, and $B_2(\kappa) \to 1$ as $\kappa \to \infty$ (cf. Kato and Shimizu 2008, Appendix A).

The method of second trigonometric moments for μ may be viewed as an approximation of the maximum likelihood estimation for μ in the following sense. The likelihood equation is

$$\frac{\partial\ell(\mu,\kappa)}{\partial\mu} = \kappa\sum_{j=1}^{n}\tanh\{\kappa\cos(\theta_j-\mu)\}\sin(\theta_j-\mu) = 0,$$

where $\ell(\mu, \kappa)$ denotes the log likelihood function. From an approximation $\tanh x = x - x^3/3 + \cdots$ ($|x| < \pi/2$), the above equation is approximated by

$$\sum_{j=1}^{n} \cos(\theta_j - \mu)\sin(\theta_j - \mu) = \frac{1}{2}\{S_2\cos(2\mu) - C_2\sin(2\mu)\} = 0,$$

from which we have $\tilde{\mu} = \arg(C_2 + i\, S_2)/2$.

We briefly take up statistical tests for axial uniformity for the AvM-II(μ, κ) distribution. The optimal locally most mean powerful unbiased test for the multiparameter hypothesis $H_0 : a = 0$ and $b = 0$ can be derived from the construction in SenGupta and Vermeire (1986). However, a likelihood ratio test for an equivalent hypothesis $H_0 : \kappa = 0$ is presented here. Note that H_0 specifies the value of the parameter κ on the boundary. The critical region is given by

$$T_1 \equiv -n\log I_0(\hat{\kappa}) + \sum_{j=1}^{n} \log\cosh\{\hat{\kappa}\cos(\theta_j - \hat{\mu})\} > K_1$$

with the maximum likelihood estimates $\hat{\mu}$ and $\hat{\kappa}$ of μ and κ, and the cut-off point K_1 to be determined from the specified size of the test. The exact distribution of T_1 is nontrivial to derive, though T_1 is easy to compute. Moreover, the standard result on the asymptotic χ^2 distribution of the $-2\log$ (likelihood ratio statistics) need not hold because of the nonregular situation of testing for the value of the parameter κ on the boundary of the parameter space. However, the exact cut-off points can be conveniently determined through simulation.

For testing the hypothesis $H_0 : \mu = \mu_0$ for a specified μ_0, the usual likelihood ratio test is applicable. For the unconstrained maximum likelihood estimate $\hat{\mu}$ and $\hat{\kappa}$ of μ and κ, and the maximum likelihood estimate $\hat{\kappa}_0$ of κ under H_0, we have the critical region

$$\begin{aligned} T_2 \equiv &-n\{\log I_0(\hat{\kappa}_0) - \log I_0(\hat{\kappa})\} \\ &+\sum_{j=1}^{n} \left[\log\cosh\{\hat{\kappa}_0\cos(\theta_j - \mu_0)\} - \log\cosh\{\hat{\kappa}\cos(\theta_j - \hat{\mu})\}\right] < K_2, \end{aligned}$$

where K_2 is the cut-off point. In large samples, $-2\log T_2$ follows a χ^2 distribution with 1 degree of freedom.

7.7 Applications

Arnold and SenGupta (2006) use measurements of long-axis orientations of 60 feldspar laths in basalt, listed in degrees in Appendix B5 of Fisher's (1993) book. Another example in the same paper refers to 63 measurements of median directions of face-cleats taken at 20 m intervals along a tunnel in Wallsend Borehole

Colliery, NSW, Australia. The data are listed in degrees in Appendix B22 in Fisher (1993). The fitted axial von Mises pdfs by using maximum likelihood estimates are shown as Figures 2 and 4 in Arnold and SenGupta (2006).

Abe et al. (2012b) used a data set consisting of the direction of 552 fallen logs, measured with a compass, in 5° classes. Study area (68°18′N, 30°22′E) was located within the northern boreal zone in the Kazkim River area in the Laplandsky Forest-conservation area, Murmansk Province, Russia, and all dominated by Norway spruce (*Picea abies*). The orientations as well as coordinates and length in the (40 m)×(400 m) study area are shown as Figure 2 in Abe et al. (2012b). It may be possible to view the observations as circular or cylindrical data, but in this study, the data are treated as undirected lines. The sine-skewed axial power cardioid distribution as well as sine-skewed axial von Mises, sine-skewed axial wrapped Cauchy, symmetric axial power cardioid, symmetric axial von Mises, and symmetric axial wrapped Cauchy distributions as submodels are fitted to the data using the maximum likelihood estimates. The sine-skewed axial wrapped Cauchy was selected as an optimal model among the six models above in the sense of minimum Akaike information criterion (AIC), with p-value 0.198 for the chi-squared goodness-of-fit statistic. Analysis of fallen logs using circular distributions can be seen in Abe et al. (2012a).

References

Abe, T. and Pewsey, A. (2011). Sine-skewed circular distributions. *Statistical Papers* 52: 683–707.

Abe, T., Kubota, Y., Shimatani, K. et al. (2012a). Circular distributions of fallen logs as an indicator of forest disturbance regimes. *Ecological Indicators* 18: 559–566.

Abe, T., Shimizu, K., Kuuluvainen, T. and Aakala, T. (2012b), Sine-skewed axial distributions with an application for fallen tree data. *Environmental and Ecological Statistics* 19: 295–307.

Arnold, B. C. and SenGupta, A. (2006). Probability distributions and statistical inference for axial data. *Environmental and Ecological Statistics* 13: 271–285.

Arnold, B. C. and SenGupta, A. (2011). Models for axial data. In: *Advances in Directional and Linear Statistics* (Editors: M. T. Wells and A. SenGupta), 1–9. Berlin: Springer-Verlag.

Arnold, B. C. and SenGupta, A. (2026). Symmetric and asymmetric axial and directional models and bivariate extensions thereof. In *Quantitative Frontiers Exploring Applications of Mathematics and Statistics* (Editors: A. Adhikari, E. Trell and A. SenGupta), 221–237. Berlin: Springer-Verlag.

Arnold, B. C. and Strauss, D. J. (1991). Bivariate distributions with conditionals in prescribed exponential families. *Journal of the Royal Statistical Society* B53: 365–375.

Azzalini, A. (2005). The skew-normal distribution and related multivariate families. *Scandinavian Journal of Statistics* 32, 159–200.

Batschelet, E. (1981). *Circular Statistics in Biology*. London: Academic Press.

Fisher, N. I. (1993). *Statistical Analysis of Circular Data*. Cambridge: Cambridge University Press.

Kato, S. and Jones, M. C. (2015). A tractable and interpretable four-parameter family of unimodal distributions on the circle. *Biometrika* 102: 181–190.

Kato, S. and Shimizu, K. (2008). Dependent models for observations which include angular ones. *Journal of Statistical Planning and Inference* 138: 3538–3549.

Kim, S. and SenGupta, A. (2013). A three-parameter generalized von Mises distribution. *Statistical Papers* 54: 685–693.

Kim, S. and SenGupta, A. (2021). Multimodal exponential families of circular distributions with application to daily peak hours of PM2.5 level in a large city. *Journal of Applied Statistics* 48: 3193–3207.

Mardia, K. V. and Jupp, P. E. (2000). *Directional Statistics*. Chichester: Wiley.

Ong, S. H. and SenGupta, A. (2022). Bivariate cardioid distributions. In: *Directional Statistics for Innovative Applications: A Bicentennial Tribute to Florence Nightingale* (ed. A. SenGupta and B. C. Arnold). Springer. 245–262.

Papakonstantinou, V. (1979). Beiträge zur zirkulären Statistik. PhD dissertation. Switzerland: University of Zurich.

Rao, C. R. (1973). *Linear Statistical Inference and Its Applications*, 2e. New York: Wiley.

SenGupta, A. (2004). On the constructions of probability distributions for directional data. *Bulletin of Calcutta Mathematical Society* 96: 139–154.

SenGupta, A. and Vermeire, L. (1986). Locally optimal tests for multiparameter hypotheses. *Journal of the American Statistical Association* 81: 819–825.

Wehrly, T. E. and Johnson, R. A. (1980). Bivariate models for dependence of angular observations and a related Markov process. *Biometrika* 67: 255–256.

8

Distributions on the Sphere, Hemi-Sphere and Hyper-Sphere

In this chapter we consider distributions on the (hemi-) sphere and the hyper-(hemi) sphere. A circle may be viewed as a one-dimensional sphere in the plane, the two-dimensional Euclidean space, and distributions on the circle or circular distributions have already been introduced in Chapter 2. Thus the main topic in this chapter is related to higher dimensional spheres. It is possible to visualize a distribution on the two-dimensional sphere in the three-dimensional Euclidean space. The distribution may be a statistical model for events which occur on the surface of the earth, though the earth is not a perfect globe. However, it is difficult to visualize hyper-spherical distributions physically.

In Sections 8.1–8.5, we start with preliminary notions for spherical and hemispherical distributions and proceed to several probability distributions such as uniform, von Mises–Fisher, and generalized cardioid distributions on these manifolds.

8.1 Preliminary Notions for Spherical, Hemi-Spherical and Hyper-Spherical Distributions

8.1.1 Terminologies and Manifolds

A sphere is an important manifold different from torus, cylinder and disc. Recall that, in mathematics, a manifold is a geometric object which locally has the structure of a finite-dimensional real vector space $\mathbb{R}^p$, but globally it might not. Geometrically a sphere is a perfectly round object in $\mathbb{R}^3$; arithmetically the unit sphere with center at the origin, which is our main concern, is the set of points which are situated at the same unit distance (the radius of the sphere) from the origin (the center of the sphere). Earthquake epicenters may be viewed approximately as data on the sphere. In general, the unit hyper-sphere with center at the origin is a smooth embedding from $\mathbb{R}^p$ to S^{p-1}, i.e. $S^{p-1} = \{\boldsymbol{x} = (x_1, x_2, \dots, x_p)' \in \mathbb{R}^p \mid \|\boldsymbol{x}\| = \sqrt{\sum_{j=1}^{p} x_j^2} = 1\}$.

A torus is the surface of revolution generated by revolving a circle in $\mathbb{R}^3$, i.e. $S^1 \times S^1$, and a cylinder is the surface generated by the points at the same distance from a given segment, i.e. $S^1 \times \mathbb{R}$. We use these geometrical terminologies. However, it may be easy to understand the notions when we think of data sets which consist of two wind directions (data on the torus), and wind direction and speed (data on the cylinder). The unit disc is the part of the plane bounded by the unit circle and containing the origin as its center, i.e. $\{\boldsymbol{x} = (x_1, x_2)' \in \mathbb{R}^2 \mid x_1^2 + x_2^2 < 1\}$. These other manifolds will be taken up in Chapters 9, 10 and 11.

Chikuse (2003) discusses statistics on the two special manifolds, the Stiefel manifold and the Grassmann manifold. The Stiefel manifold $V_{k,m}$ is represented by the set of $m \times k$ matrices X such that $X'X = I_k$, where I_k denotes the $k \times k$ identity matrix, i.e. $V_{k,m} = \{X\,(m \times k) \mid X'X = I_k\}$. The special case $k = 1$ is regarded as the unit hyper-sphere. The Grassman manifold $G_r(\mathbb{R}^p)$ is the set of r-dimensional subspaces in $\mathbb{R}^p$, and each subspace V corresponds to a $p \times p$ orthogonal projection matrix idempotent of rank r onto V. For the special case $r = 1$, the Grassman manifold reduces to the real projective space which consists of all lines through the origin. However, these two general manifolds will not be further discussed in our book.

Let the surface of the globe be viewed approximately as the surface of a sphere, i.e. a spherical surface. Then the distribution of certain random events, say the locations of earthquakes, on the globe can be modeled as a probability distribution on the sphere, i.e. as a spherical distribution. As an alternative example, a 3-D scatter plot on the sphere for 107 measurements of magnetic remanence from specimens of Precambrian volcanics (Fisher et al., 1993, B6 Data) is shown in Figure 8.1.

8.1.2 Polar Transformation

A spherical distribution here will mean a distribution on the unit sphere in the three-dimensional Euclidean space. Similarly, a hyper-spherical distribution is defined as a distribution on the unit hyper-sphere in a higher-dimensional Euclidean space.

The joint pdf $f(\boldsymbol{x})$ of a spherical distribution satisfies the properties that $f(\boldsymbol{x}) \geq 0$ for all $\boldsymbol{x} \in S^2$ and $\int_{\boldsymbol{x}\in S^2} f(\boldsymbol{x}) d\boldsymbol{x} = 1$, where $S^2 = \{\boldsymbol{x} = (x_1, x_2, x_3)' \in \mathbb{R}^3 \mid \|\boldsymbol{x}\| = \sqrt{x_1^2 + x_2^2 + x_3^2} = 1\}$, i.e. the unit sphere surface. From the polar transformation (Figure 8.2)

$$\begin{cases} x_1 &= r \sin\theta \cos\phi, \\ x_2 &= r \sin\theta \sin\phi, \\ x_3 &= r \cos\theta, \end{cases}$$

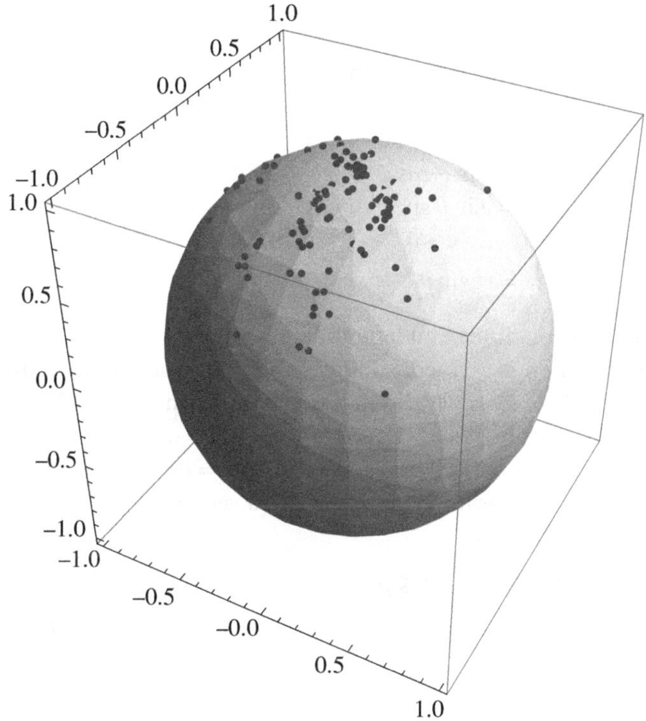

Figure 8.1 3-D scatter plot on the sphere for 107 measurements of magnetic remnants from specimens of Precambrian volcanics *Source:* Data from (Fisher et al., 1993, B6 Data).

Figure 8.2 Polar transformation in the three-dimensional space.

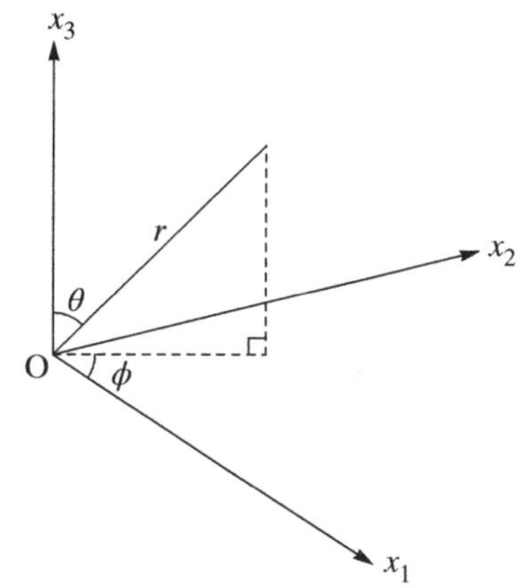

with $r = 1$, $0 \le \theta < \pi$, $0 \le \phi < 2\pi$, the angular representation $g(\theta, \phi)$ of $f(\boldsymbol{x})$ is written as

$$g(\theta, \phi) = f(\sin\theta\cos\phi, \sin\theta\sin\phi, \cos\theta)\sin\theta$$

since the Jacobian is

$$\begin{vmatrix} \partial x_1/\partial r & \partial x_1/\partial\theta & \partial x_1/\partial\phi \\ \partial x_2/\partial r & \partial x_2/\partial\theta & \partial x_2/\partial\phi \\ \partial x_3/\partial r & \partial x_3/\partial\theta & \partial x_3/\partial\phi \end{vmatrix} = \begin{vmatrix} \sin\theta\cos\phi & r\cos\theta\cos\phi & -r\sin\theta\sin\phi \\ \sin\theta\sin\phi & r\cos\theta\sin\phi & r\sin\theta\cos\phi \\ \cos\theta & -r\sin\theta & 0 \end{vmatrix}$$
$$= r^2\sin\theta.$$

The angles θ and ϕ stand for colatitude and longitude, respectively.

A hyper-spherical distribution is defined, similarly to a spherical distribution, on the unit hyper-sphere S^{p-1}. A general polar transformation is as follows. Let $\boldsymbol{X} = (X_1, \dots, X_p)' = R\boldsymbol{Y}$, where $R = \| \boldsymbol{X} \|$ is the length of $\boldsymbol{X}$ and $\boldsymbol{Y} = (Y_1, \dots, Y_p)' = \boldsymbol{e}(\boldsymbol{T})$ is the direction cosine of $\boldsymbol{X}$. Here $\boldsymbol{T} = (\Theta_1, \dots, \Theta_{p-1})'$ is the direction of $\boldsymbol{X}$. The polar transformation is applied to a vector $(x_1, \dots, x_p)'$:

$$\begin{cases} x_j = r\left(\prod_{k=1}^{j-1}\sin\theta_k\right)\cos\theta_j, \quad 1 \le j \le p-1, \\ x_p = r\left(\prod_{k=1}^{p-2}\sin\theta_k\right)\sin\theta_{p-1}, \end{cases} \tag{8.1}$$

where $r > 0$, $\theta_k \in [0, \pi)$ for $k = 1, \dots, p-2$, $\theta_{p-1} \in [0,2\pi)$ and we interpret $\prod_{k=1}^{0}\sin\theta_k$ as 1. Then the Jacobian of the transformation from $x_1, \dots, x_p$ to $r, \theta_1, \dots, \theta_{p-1}$ given by (8.1) is

$$\frac{\partial(x_1, \dots, x_p)}{\partial(r, \theta_1, \dots, \theta_{p-1})} = r^{p-1}\prod_{k=1}^{p-2}\sin^{p-k-1}\theta_k.$$

8.1.3 Tangent–Normal Decomposition

Let $\varepsilon_1, \dots, \varepsilon_p$ denote a set of orthonormal vectors in $\mathbb{R}^p$. For $\boldsymbol{x} \in S^{p-1}$, the tangent–normal decomposition (cf. Watson, 1983, p. 44; Mardia and Jupp, 2000, p. 161) is $\boldsymbol{x} = t\varepsilon_p + \sqrt{1-t^2}\xi_{p-1}$ $(-1 \le t \le 1)$, where $t = \boldsymbol{x}'\varepsilon_p$, and ξ_{p-1} is a unit vector in the space spanned by $\varepsilon_1, \dots, \varepsilon_{p-1}$, i.e. $\xi_{p-1} \in S^{p-2} = \{(x_1, \dots, x_{p-1}, 0)' \in \mathbb{R}^p \mid \sqrt{\sum_{j=1}^{p-1} x_j^2} = 1\}$.

Let $d\omega_p$ denote the area element of S^{p-1}. Then

$$d\omega_p = (1-t^2)^{(p-3)/2}\, dt d\omega_{p-1}$$

and

$$\omega_p = \omega_{p-1} \int_{-1}^{1} (1-t^2)^{(p-3)/2}\,dt = 2^{p-2} B\left(\frac{p-1}{2}, \frac{p-1}{2}\right) \omega_{p-1}$$
$$= B\left(\frac{1}{2}, \frac{p-1}{2}\right) \omega_{p-1}, \tag{8.2}$$

where B denotes the beta function defined by (B.5). To establish the last equation in (8.2), we can use the duplication formula of the gamma function (B.2). From the recursion relation in (8.2), the area of the unit hyper-sphere surface ω_p is

$$\omega_p = \frac{2\pi^{p/2}}{\Gamma(p/2)},$$

which takes a maximum at $p = 7$.

Equation (8.2) leads to the distribution of t in the tangent–normal decomposition, whose pdf is expressed by

$$\frac{\omega_{p-1}}{\omega_p}(1-t^2)^{(p-3)/2} = \frac{1}{B(1/2,(p-1)/2)}(1-t^2)^{(p-3)/2}, -1 < t < 1. \tag{8.3}$$

The right side of (8.3) represents the pdf of a beta distribution on the interval $(-1,1)$ with parameters $((p-1)/2, (p-1)/2)$. For $p = 2$, the pdf is symmetric about zero and U-shaped, and for $p = 3$, it is constant. If $p > 3$, then the pdf is symmetric about zero and unimodal with mode at zero.

8.1.4 Rotational Symmetry

The notion of rotational symmetry for a spherical distribution on S^2 or a hyper-spherical distribution on S^{p-1} parallels that of symmetry for a circular distribution.

Let $\boldsymbol{\mu}$ denote a fixed direction of a spherical pdf $f(\boldsymbol{x})$. Then the distribution is called rotationally symmetric about $\boldsymbol{\mu}$ if $f(\boldsymbol{x})$ is of the form

$$f(\boldsymbol{x}) = g(\boldsymbol{\mu}'\boldsymbol{x}),$$

where g is a known function. From the tangent–normal decomposition $\boldsymbol{x} = t\boldsymbol{\mu} + (1-t^2)^{1/2}\boldsymbol{\xi}$, where $\boldsymbol{\mu}'\boldsymbol{\xi} = 0$, rotational symmetry means that the distribution of the unit tangent $\boldsymbol{\xi}$ at $\boldsymbol{\mu}$ is uniformly distributed on S^{p-2} and is independent of t.

8.2 General Methods of Construction

8.2.1 Direct Substitution Method

Observe that for circular distributions, the pdf of a symmetric circular distribution is written in terms of $(\cos\theta, \sin\theta)$. However, it can also be written equivalently in terms of (x_1, x_2), where $x_1^2 + x_2^2 = 1$. Thus $\cos(\theta - \mu)$ can be represented as $\boldsymbol{x}'\boldsymbol{\eta}$. This fact can be utilized to generalize pdf of a distribution on the circle to that of a corresponding one on a sphere. The von Mises–Fisher and the distributions in the generalized power cardioid family given later in this chapter are simple examples of this fact. Similarly, all the pdfs of the wrapped stable family on the circle lend them to elegant generalizations on the sphere, though not necessarily with any such "wrapping" interpretation. The asymmetric pdfs on the circle may also be generalized by an extension of this method as has been done by SenGupta and Kim (2016) to derive the generalized AGvM distribution given below. The main hurdle often for such a distribution is the unavailability of its normalizing constant in a closed analytical form.

8.2.2 Spherical Harmonics and Exponential Functions

Just as the pdf of a circular random variable admits of its representation in terms of Fourier coefficients, it is natural to expect a similar representation for a distribution on S^{p-1} in terms of spherical harmonics. Watson (1983) builds up this idea to derive an exponential family of distributions on S^{p-1}. Watson (1983, p. 82) suggests the following density as the natural generalization to S^{p-1} of the exponential of a finite Fourier series on the circle:

$$C(\boldsymbol{\beta}) \exp\left(\sum_{j=1}^{m} \sum_{k=1}^{N_{jp}} \beta_{jk} S_{jk}(\boldsymbol{y}) \right),$$

where $C(\boldsymbol{\beta})$ is the normalizing constant, $\{S_{jk}(\boldsymbol{y})\}$ is a complete set of orthogonal spherical harmonics of degree j on S^{p-1}, and $\boldsymbol{y} = \boldsymbol{H}\boldsymbol{x}$ is the rotation of $\boldsymbol{x}$ through an orthogonal matrix $\boldsymbol{H}$.

The above approach also leads to "linear" (in terms of $\mathbf{x}$) distributions on S^{p-1}, as may be given by generalizing, e.g. the symmetric wrapped stable family on the circle to the sphere.

8.2.3 Conditioning Method

The conditioning method to derive spherical distributions is similar to that for the circular case. After the polar transformation (8.1) is applied to a random vector $\boldsymbol{Z}$ in $\mathbb{R}^p$ $(p \geq 2)$ such that $P(\boldsymbol{Z} = \boldsymbol{0}) = 0$ and the joint pdf of $(R, \boldsymbol{T}')'$ is obtained, the conditional distribution of $\boldsymbol{T}$ or $\boldsymbol{e}(\boldsymbol{T})$ given $R = r$ gives a distribution on the unit hyper-sphere. For example, conditioning of scale mixtures of normal produces a fruitful spherical family, since distributions such as Student t, logistic, Laplace or double-exponential, stable family, and exponential power family can all be constructed as scale mixtures of normals.

8.2.4 Projections: Stereographic, Tangent and Lambert

As in the circular case, the projected distribution of a random vector $\boldsymbol{X}$ in $\mathbb{R}^p$ ($p \geq 2$) such that $P(\boldsymbol{X} = \boldsymbol{0}) = 0$ is defined by the directional distribution of the unit random vector $\boldsymbol{X}/ \parallel \boldsymbol{X} \parallel$. It is obvious that $a\boldsymbol{X}/ \parallel a\boldsymbol{X} \parallel = \boldsymbol{X}/ \parallel \boldsymbol{X} \parallel$ for any positive constant a. This means that the projected distribution is invariant to scale transformation of X. If $\boldsymbol{X}$ has distribution $N_p(\boldsymbol{\mu}, \Sigma)$, then the corresponding projected distribution is called the projected normal distribution and denoted by $\mathrm{PN}_p(\boldsymbol{\mu}, \Sigma)$. On transforming the joint pdf of $\boldsymbol{X}$ to that of $(R, \boldsymbol{T}')'$ from the polar transformation (8.1) and integrating it with respect to r, we can obtain the pdf of $\boldsymbol{T}$.

Two variants of stereographic projection are considered now. Let $\boldsymbol{x} \in \mathbb{R}^2$ and (θ, ϕ), as before map a point on a unit sphere. The first type of projection, to be termed Type I, is given by,

$$x = \tan(\phi/2)\cos\theta,\ y = \tan(\phi/2)\sin\theta,\ |J| = \frac{\sin\phi}{(1+\cos\phi)^2},$$

where $|J|$ is the Jacobian. In the univariate case, this also corresponds to the Möbius transformation relating the angle, θ, of an unit circle to its corresponding point x on the real line.

The second type of projection, to be termed Type II, is given by,

$$x = \tan(\pi/4 + \theta/2),\ y = \tan\phi,\ |J| = (1/2)\mathrm{sec}^2(\pi/4 + \theta/2)\mathrm{sec}^2\phi.$$

These two projections are depicted graphically in Figure 8.3.

Various bivariate distributions on the plane, $\mathbb{R}^2$, can be used to produce distributions on the sphere by these projections, to be termed as spherical (corresponding) Type I and Type II distributions.

Figure 8.4a displays a great circle which goes through the north pole N, a point P, the south pole S, and a point Q. By the Lambert projection, the points P and Q, not equal to N, are, respectively, transformed to P′ and Q′ in a disc on the tangent plane of S (Figure 8.4b). Figure 8.4c shows that the point P on the northern hemisphere may be transformed to a point P′ in a disc on the tangent plane of N, while the

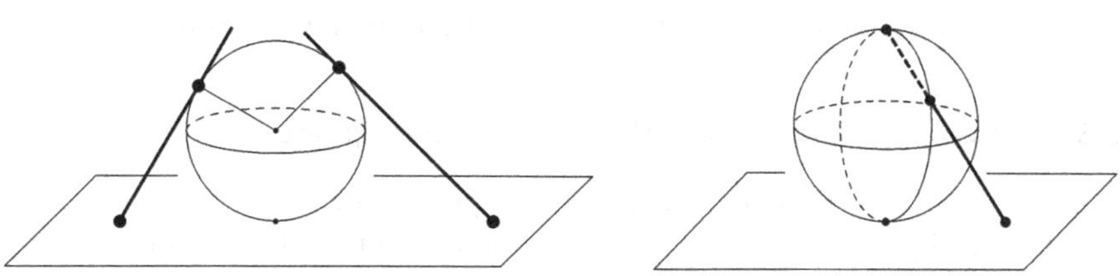

(a) (b)

Figure 8.3 (a) Tangent projection and (b) Stereographic projection.

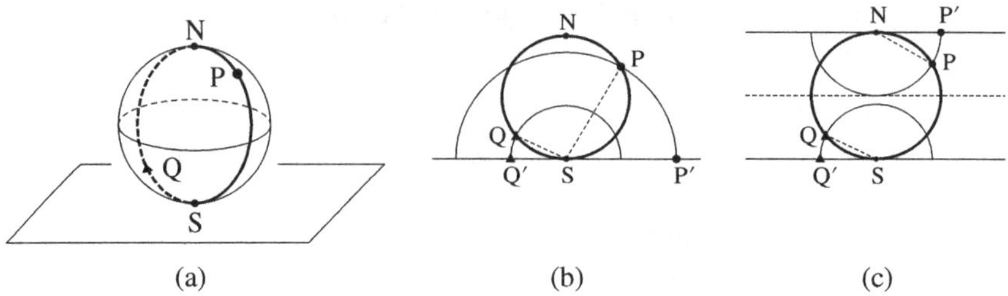

Figure 8.4 Lambert azimuthal equal-area projection. (a) Great circle which goes through points P and Q on the unit sphere, (b) Projection of P and Q to P′ and Q′, and (c) Projection of P and Q to P′ and Q′.

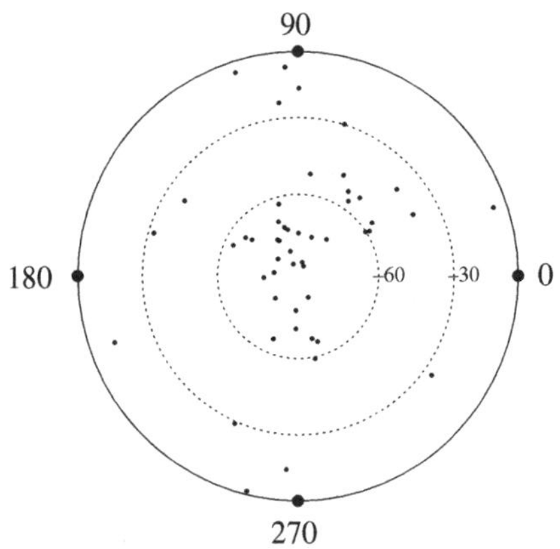

Figure 8.5 Lambert azimuthal equal-area projection of 50 pole positions (Fisher et al., 1993, B1 Data).

point Q on the southern hemisphere is transformed to Q′ in an alternative disc on the tangent plane of S.

An example of the Lambert projection is given in Figure 8.5 for the data set B1 in the book by Fisher et al. (1993). The projection is useful not only as a visualization technique but also as a method for giving the relationship between the corresponding distributions on the disc and on the sphere. It is easy to see that this method may also be used to produce probability distributions on the hemisphere.

8.2.5 Stereographic Projection Method for Hyper-Sphere

A three-dimensional version of the stereographic projection which preserves angles is given by Watson (1983). However, when we apply it for the construction of spherical distributions, the Jacobian does not take a simple form. Thus, here we propose a projection which maps the exterior point $\boldsymbol{x} \in \mathbb{R}^{p-1}$ into the southern hemisphere of the p-dimensional sphere of radius v, S_p^v, and the interior point $\boldsymbol{x} \in \mathbb{R}^{p-1}$ into the northern hemisphere of S_p^v. A graphical representation of the proposed projection is provided as Figure 4 in Abe et al. (2010).

Let $p \geq 3, v > 0$, and $S_p^v = \{\boldsymbol{\xi} \in \mathbb{R}^p \mid \parallel \boldsymbol{\xi} \parallel = v\}$. For $-\pi \leq \phi < \pi$ and $0 \leq \theta_j < \pi$ $(j = 1, \dots, p-2)$, Abe et al. (2010) used the polar coordinates

$$\begin{pmatrix} \xi_1 \\ \xi_2 \\ \vdots \\ \xi_{p-1} \\ \xi_p \end{pmatrix} = \begin{pmatrix} v \sin\theta_1 \cdots \sin\theta_{p-2} \cos\phi \\ v \sin\theta_1 \cdots \sin\theta_{p-2} \sin\phi \\ \vdots \\ v \sin\theta_{p-2} \cos\theta_{p-3} \\ v \cos\theta_{p-2} \end{pmatrix}. \tag{8.4}$$

Then, for any $\boldsymbol{x} = (x_1, \dots, x_{p-1})' \in \mathbb{R}^{p-1}$, there exists a unique point $\boldsymbol{\xi} = (\xi_1, \dots, \xi_p)' \in S_p^v$ such that

$$\begin{aligned}(x_1, \dots, x_{p-1})' &= \left(\frac{v\xi_1}{v + \xi_p}, \dots, \frac{v\xi_{p-1}}{v + \xi_p} \right)' \\ &= \left(\frac{v \sin\theta_1 \dots \sin\theta_{p-2} \cos\phi}{1 + \cos\theta_{p-2}}, \dots, \frac{v \sin\theta_{p-2} \cos\theta_{p-3}}{1 + \cos\theta_{p-2}} \right)',\end{aligned}$$

where $\xi_p \neq -v$. The Jacobian is

$$\begin{aligned}\left| \frac{\partial(x_1, x_2, \dots, x_{p-1})}{\partial(\phi, \theta_1, \dots, \theta_{p-2})} \right| &= \frac{v \sin^{p-2}\theta_{p-2}}{(1 + \cos\theta_{p-2})^{p-1}} \left| \frac{\partial(x_1, x_2, \dots, x_{p-1})}{\partial(v, \phi, \theta_1, \dots, \theta_{p-3})} \right| \\ &= \left(\frac{v}{1 + \cos\theta_{p-2}} \right)^{p-1} \sin\theta_1 \cdots \sin^{p-2}\theta_{p-2} \, (\equiv |J|, \text{ say}).\end{aligned}$$

The discussion above leads to the following assertion. For $j = 1, \dots, p-2$, suppose Φ and Θ_j are randomly chosen angles in the intervals $(-\pi, \pi)$ and $(0, \pi)$, respectively. Let $g(\phi, \theta_1, \dots, \theta_{p-2})$ denote the pdf of $(\Phi, \Theta_1, \dots, \Theta_{p-2})$. Then (8.4) is a random point on the sphere of radius v. Let $\boldsymbol{X} = (x_1, \dots, x_{p-1})' \in \mathbb{R}^{p-1}$ and $f(\boldsymbol{x})$ denote the pdf of $\boldsymbol{X}$. Then, g can be written in terms of f by

$$\begin{aligned}g\,(\phi, \theta_1, \dots, \theta_{p-2}) &= f(\boldsymbol{x}) \left| \frac{\partial(x_1, x_2, \dots, x_{p-1})}{\partial(\phi, \theta_1, \dots, \theta_{p-2})} \right| \\ &= f\left(\frac{v \sin\theta_1 \cdots \sin\theta_{p-2} \cos\phi}{1 + \cos\theta_{p-2}}, \dots, \frac{v \sin\theta_{p-2} \cos\theta_{p-3}}{1 + \cos\theta_{p-2}} \right) |J|. \end{aligned} \tag{8.5}$$

8.3 Distributions on the Sphere

8.3.1 Uniform Distribution

Using the probability density element for a unit sphere, we get the pdf having the Uniform distribution on the sphere as,

$$f(\theta,\phi)=\frac{\sin\theta}{4\pi},\quad 0\le\theta<\pi,\ 0\le\phi<2\pi.$$

Note that the marginal pdf of ϕ is circular Uniform, but that of θ is not (axial) uniform. The various pdfs on the sphere as defined below reduce to this isotropic density for the appropriate choices of their parameters.

8.3.2 Spherical Analogs to Linear Distributions by Projections

Spherical Normal, t and Cauchy – Type I and Type II Distributions For the bivariate Cauchy distribution with parameters (x_0, y_0, γ), the pdf of the corresponding S^{3-1} spherical Cauchy Type I distribution is given by,

$$\begin{aligned}f(\theta,\phi)&=\gamma\sin\phi[2\pi(1+\cos\phi)^2\\&\quad\times\{\tan^2(\phi/2)-2r_0\tan(\phi/2)(\cos\theta+\sin\theta)+x_0^2+y_0^2+\gamma^2\}^{1.5}]^{-1},\end{aligned}$$

where $x_0=r_0\cos\theta_0$ and $y_0=r_0\sin\theta_0$.

The pdf of the S^{3-1} spherical Cauchy Type II distribution is given by,

$$\begin{aligned}f(\theta,\phi)&=\gamma[4\pi\cos^2(\pi/4+\theta/2)\cos^2\phi\\&\quad\times[\{\tan^2(\pi/4+\theta/2)-x_0\}^2+(\tan\phi-y_0)^2+\gamma^2]^{1.5}]^{-1}.\end{aligned}$$

Similarly, the pdfs for the S^{3-1} spherical normal and t Type I and Type II distributions can be written down explicitly in closed forms.

8.3.3 Bi-modal Distributions

8.3.3.1 Wood Distribution

Wood (1982) proposed a bimodal distribution on the sphere whose density is given by

$$f(\theta,\phi)=\frac{\kappa}{4\pi\sinh\kappa}\exp[\kappa\{\cos\alpha\cos\theta+\sin\alpha\sin\theta\cos(2\phi-\beta)\}]\sin\theta \quad (8.6)$$

for colatitude θ and longitude ϕ as the angular representation, where $0\le\alpha\le\pi/2$, $0\le\beta<2\pi$ and $\kappa\ge0$. The difference from the Fisher density is that $\cos(2\phi-\beta)$ replaces $\cos(\phi-\beta)$. When $\kappa=0$, (8.6) reduces to the spherical uniform distribution.

8.3.3.2 Kent and Generalized AGvM Distributions

The Fisher or von Mises–Fisher distribution on the sphere is the basic model for spherical data, just as the von Mises distribution is for circular data. The Fisher distribution is capable of modeling unimodal and symmetric spherical data, and was shown by the direct substitution method to be a simple generalization of the von Mises distribution. In fact, these two distributions are, respectively, the circular and the spherical members ($p = 2$ and $p = 3$) of a family of distributions on the p-sphere called Langevin distributions. The vectorial form of the Fisher distribution, in terms of the directional cosines, is given by

$$f(\boldsymbol{x}) \propto \exp\{\kappa(\boldsymbol{x}'\boldsymbol{\zeta})\},$$

where $\boldsymbol{x} = (\sin\theta\cos\phi, \sin\theta\sin\phi, \cos\theta)'$ and $\boldsymbol{\zeta} = (\sin\nu\cos\mu, \sin\nu\sin\mu, \cos\nu)'$. In terms of polar coordinates, the Fisher distribution is given by

$$f(\theta,\phi) \propto \exp[\kappa\{\sin\theta\sin\nu\cos(\phi-\mu) + \cos\theta\cos\nu\}]\sin\theta,$$
$$0 \le \theta < \pi,\ 0 \le \phi < 2\pi,$$

where ν and μ are location parameters, $\mu \in (0,\pi)$, $\nu \in [0,2\pi)$, and κ is the concentration parameter toward the direction $(\nu,\mu)'$, $\kappa \ge 0$. One can note that the Fisher distribution is reduced to the von Mises distribution, when $\theta = \nu = \pi/2$.

A shortcoming of the Fisher distribution is that its shape is always symmetric and unimodal. We next present two spherical distributions (Kim and SenGupta, 2018) which are direct extensions of the 4-parameter generalized von Mises (GvM) distribution (Yfantis and Borgman, 1982) and the 3-parameter asymmetric generalized von Mises (AGvM) distribution (Kim and SenGupta, 2013). Like the GvM and the AGvM distributions on the circle, these new spherical distributions (Kim and SenGupta, 2018) are suitable to model unimodal or bimodal, and symmetric or asymmetric data on the sphere.

8.3.3.3 A Generalization to Spherical Model of the GvM Distribution

The density form of the GvM distribution, in terms of polar coordinates, is shown below:

$$f(\phi) \propto \exp[\kappa_1\cos(\phi-\mu_1) + \kappa_2\cos\{2(\phi-\mu_2)\}],$$

where μ_1 and μ_2 are location parameters, $\mu_1, \mu_2 \in [0,2\pi)$, and κ_1 and κ_2 are shape parameters, $\kappa_1 \ge 0$ and $\kappa_2 \in \mathbb{R}$. Letting $\boldsymbol{y} = (\cos\phi, \sin\phi)'$, $\boldsymbol{\eta}_1 = (\cos\mu_1, \sin\mu_1)'$, $\boldsymbol{\eta}_2 = (\cos\mu_2, \sin\mu_2)'$ and $\boldsymbol{\eta}_3 = (-\sin\mu_2, \cos\mu_2)'$, the vectorial form of the GvM distribution is given by

$$f(\boldsymbol{y}) \propto \exp[\kappa_1\boldsymbol{y}'\boldsymbol{\eta}_1 + \kappa_2\{(\boldsymbol{y}'\boldsymbol{\eta}_2)^2 - (\boldsymbol{y}'\boldsymbol{\eta}_3)^2\}].$$

The density form of a generalization to spherical model of the GvM distribution, in terms of directional cosines, is given by

$$f(\boldsymbol{x}) \propto \exp[\kappa_1(\boldsymbol{x}'\boldsymbol{\zeta}_1) + \kappa_2\{(\boldsymbol{x}'\boldsymbol{\zeta}_2)^2 - (\boldsymbol{x}'\boldsymbol{\zeta}_3)^2\}],$$

where

$$\boldsymbol{x} = (\sin\theta\cos\phi, \sin\theta\sin\phi, \cos\theta)', \ \zeta_1 = (\sin\nu\cos\mu_1, \sin\nu\sin\mu_1, \cos\nu)',$$

$$\zeta_2 = (\sin\nu\cos\mu_2, \sin\nu\sin\mu_2, \cos\nu)' \quad \text{and} \quad \zeta_3 = (-\sin\mu_2, \cos\mu_2, 0)'.$$

In terms of polar coordinates, we have

$$f(\theta,\phi) \propto \exp\big(\kappa_1\{\sin\theta\sin\nu\cos(\phi-\mu_1)+\cos\theta\cos\nu\} \\ + \kappa_2[\{\sin\theta\sin\nu\cos(\phi-\mu_2)+\cos\theta\cos\nu\}^2 - \sin^2\theta\sin^2(\phi-\mu_2)]\big)\sin\theta,$$

where μ_1, μ_2, and ν are location parameters, $\mu_1, \mu_2 \in [0,2\pi)$, and $\nu \in [0,\pi)$, and κ_1 and κ_2 are shape parameters. In general this distribution represents the possibly asymmetric and possibly bimodal distribution proposed by Kent (1982). When $\kappa_2 = 0$, it reduces to the Fisher density. For $\theta = \nu = \pi/2$, it reduces to the GvM distribution.

8.3.3.4 A Generalization to Spherical Model of the AGvM Distribution

In this section, we present yet another generalization to spherical model of the AGvM distribution, whose density form is given by

$$f(\phi) \propto \exp[\kappa_1\cos(\phi-\mu_1)+\kappa_2\sin\{2(\phi-\mu_2)\}],$$

where μ_1 and μ_2 are location parameters, $\mu_1, \mu_2 \in [0,2\pi)$, and κ_1 and κ_2 are shape parameters, where $\kappa_1 \geq 0$ and $\kappa_2 \in \mathbb{R}$. Letting $\boldsymbol{y} = (\cos\phi, \sin\phi)'$, $\boldsymbol{\eta}_1 = (\cos\mu_1, \sin\mu_1)'$, $\boldsymbol{\eta}_2 = (\cos\mu_2, \sin\mu_2)'$ and $\boldsymbol{\eta}_3 = (-\sin\mu_2, \cos\mu_2)'$, the vectorial form of the AGvM distribution is given by

$$f(\boldsymbol{y}) \propto \exp\{\kappa_1\boldsymbol{y}'\boldsymbol{\eta}_1 + 2\kappa_2(\boldsymbol{y}'\boldsymbol{\eta}_2)(\boldsymbol{y}'\boldsymbol{\eta}_3)\}.$$

The pdf of this generalization to spherical model of the AGvM distribution, in terms of directional cosines, is given by

$$f(\boldsymbol{x}) \propto \exp\{\kappa_1(\boldsymbol{x}'\zeta_1) + 2\kappa_2(\boldsymbol{x}'\zeta_2)(\boldsymbol{x}'\zeta_3)\},$$

where

$$\boldsymbol{x} = (\sin\theta\cos\phi, \sin\theta\sin\phi, \cos\theta)', \ \zeta_1 = (\sin\nu\cos\mu_1, \sin\nu\sin\mu_1, \cos\nu)',$$

$$\zeta_2 = (\sin\nu\cos\mu_2, \sin\nu\sin\mu_2, \cos\nu)' \quad \text{and} \quad \zeta_3 = (-\sin\mu_2, \cos\mu_2, 0)'.$$

In terms of polar coordinates, we have

$$f(\theta,\phi) \propto \exp\big[\kappa_1\{\sin\theta\sin\nu\cos(\phi-\mu_1)+\cos\theta\cos\nu\} \\ +2\kappa_2\{\sin\theta\sin\nu\cos(\phi-\mu_2)+\cos\theta\cos\nu\}\sin\theta\sin(\phi-\mu_2)\big]\sin\theta,$$

where μ_1, μ_2 and ν are location parameters, $\mu_1, \mu_2 \in [0,2\pi)$ and $\nu \in [0,\pi)$, and κ_1 and κ_2 are shape parameters. Note that, when $\theta = \nu = \pi/2$, the above density reduces to the form of the AGvM distribution and when $\kappa_2 = 0$, it reduces to the von Mises–Fisher density.

8.3.4 Bivariate Spherical Distributions

Correlated distributions on the sphere may be constructed by using the general methods of conditional specification and maximum entropy as given in SenGupta (2004). Let $\boldsymbol{X}_j \in S^2$ $(j = 1,2)$ be two correlated spherical random variables. By the maximum entropy method, the easiest way to obtain a joint distribution is to specify the first-order marginal and joint moments, $E(\boldsymbol{X}_1)$, $E(\boldsymbol{X}_2)$ and $E(\boldsymbol{X}_1\boldsymbol{X}_2')$. In this case, the maximum entropy density from a general Theorem in Section 9.2.2 is

$$C\ \exp(\boldsymbol{\eta}_1'\boldsymbol{x}_1 + \boldsymbol{\eta}_2'\boldsymbol{x}_2 + \boldsymbol{\xi}_1'\boldsymbol{x}_1\boldsymbol{x}_2'\boldsymbol{\xi}_2),$$

where C is the normalizing constant and $\boldsymbol{\eta}_j$, $\boldsymbol{\xi}_j$ $(j = 1,2)$ are parameter vectors. In particular, this method leads to a simple form for a bivariate von Mises–Fisher distribution.

When the spherical distribution is a member of the regular exponential family, the conditional specification method can be conveniently exploited. For example, following the general method this leads to the Bivariate Conditional von Mises–Fisher distribution with pdf as,

$$C\ \exp\{(1, \boldsymbol{x}_1')M(1, \boldsymbol{x}_2')'\},$$

where C is the normalizing constant and M is the parameter matrix. The number of parameters can be, as usual, greatly reduced by choosing the elements of M appropriately. Recall also that this model admits the independence model which results by taking the lower diagonal submatrix of order 3×3 of M as the null matrix. The corresponding form of this distribution in terms of the polar coordinates can be trivially written down.

These two methods generalize easily to yield multivariate spherical distributions.

8.4 Distributions on the Hyper-Sphere

We next present various probability distributions on the hyper-sphere. Note that all these distributions reduce to their counterparts as spherical distributions by taking $p = 3$.

8.4.1 Uniform Distribution

8.4.1.1 Definition and Properties

Just as the uniform distribution is the basic one among the distributions on the circle, the hyper-spherical uniform distribution is the most fundamental one among distributions on the unit sphere surface S^{p-1}. This distribution also plays an essential role in the theory of spherically symmetric (or simply spherical) and elliptically symmetric (or simply elliptical) distributions (Fang et al., 1990). We shall denote a random vector which follows the uniform distribution on the unit sphere by $\boldsymbol{U}^{(p)} = (U_1, \dots, U_p)'$.

We now derive the pdf of $\boldsymbol{U}^{(p)}$.

Theorem 8.1 The probability density function of $\boldsymbol{U}^{(p)}$ is given by

$$f(\boldsymbol{u}^{(p)}) = 1/\omega_p = \frac{\Gamma(p/2)}{2\pi^{p/2}},$$

where ω_p denotes the surface area of S^{p-1}.

Proof: It is obvious that the pdf is constant and is obtained by the reciprocal of the area of S^{p-1}.

The area can be calculated in several ways. The tangent–normal decomposition was used in Section 8.1. Here we introduce an alternative method of using the polar transformation (8.1). The Jacobian of the transformation is given in Section 8.1. Then the following result holds:

Result 8.1 ***(Fang et al., 1990, Theorem 2.11, p. 37; see also Muirhead, 1982, Theorem 1.5.5, p. 36)*** For a spherical random vector $\boldsymbol{X} \in \mathbb{R}^p$ whose pdf is $g(\boldsymbol{x})$ as a function of $\boldsymbol{x}'\boldsymbol{x}$, the transformed variables $R, \Theta_1, \dots, \Theta_{p-1}$ by using the polar transformation (8.1) are mutually independent and, respectively, have the pdfs

$$\begin{cases} h_r(r) = \dfrac{2\pi^{p/2}}{\Gamma(p/2)} r^{p-1} g(r^2), & r \ge 0, \\ h_{\theta_k}(\theta_k) = \dfrac{1}{B(1/2, (p-k)/2)} \sin^{p-k-1}\theta_k, & 0 \le \theta_k < \pi,\ k = 1, \dots, p-2, \\ h_{\theta_{p-1}}(\theta_{p-1}) = \dfrac{1}{2\pi}, & 0 \le \theta_{p-1} < 2\pi, \end{cases}$$

where B denotes the beta function.

Taking $g(\boldsymbol{x}'\boldsymbol{x}) = c$, $\boldsymbol{x}'\boldsymbol{x} \le a^2; = 0$, otherwise, we have

$$\int_{\boldsymbol{x}'\boldsymbol{x} \le a^2} c \prod_{j=1}^{p} dx_j = c \int_0^a \frac{2\pi^{p/2}}{\Gamma(p/2)} r^{p-1} dr = \frac{2\pi^{p/2} a^p}{p\Gamma(p/2)} c = \frac{\pi^{p/2} a^p}{\Gamma(p/2+1)} c = 1,$$

and the volume of the inside of the sphere with radius a and center at the origin is

$$\int_{\boldsymbol{x}'\boldsymbol{x} \le a^2} \prod_{j=1}^{p} dx_j = \frac{\pi^{p/2} a^p}{\Gamma(p/2+1)}.$$

The required volume then is obtained by setting $a = 1$, and it takes a maximum at $p = 5$. The area of the p-dimensional unit surface is obtained by differentiating the volume with respect to the radius a and putting $a = 1$, i.e.

$$\omega_p = \frac{d}{da}\left(\frac{2\pi^{p/2}a^p}{p\Gamma(p/2)}\right)\Bigg|_{a=1} = \frac{2\pi^{p/2}}{\Gamma(p/2)}.$$

Like the above result, the polar transformation (8.1) is important as well as tangent–normal decomposition and both will be used repeatedly.

As for the volume, however, the Liouville integral immediately gives a formula

$$\int_{\mathbb{R}^p} f\left(\sum_{j=1}^{p} x_j^2\right)\prod_{j=1}^{p} dx_j = \frac{\pi^{p/2}}{\Gamma(p/2)}\int_0^\infty y^{p/2-1}f(y)dy,$$

from which the volume is obtained by putting $f(y) = 1,\ y \le a^2; = 0$, otherwise.

Mixed Moments For any nonnegative integers $m_1, \dots, m_p$ with $m = \sum_{j=1}^{p} m_j$, the mixed moments of $\boldsymbol{U}^{(p)}$ are expressed (Fang et al., 1990, Theorem 3.3, p. 72) as

$$E\left(\prod_{j=1}^{p} U_j^{m_j}\right) = \begin{cases} \dfrac{1}{(p/2)_\ell}\displaystyle\prod_{j=1}^{p}\dfrac{(2\ell_j)!}{4^{\ell_j}(\ell_j)!}, & \text{if } m_j = 2\ell_j \text{ are even,} \\ & j = 1, \dots, p,\ m = 2\ell, \\ 0, & \text{if at least one of the } m_j \text{ is odd,} \end{cases}$$

where $(x)_\ell$ denotes Pochhammer's symbol (see (B.8)).

Marginal and Conditional Distributions The marginal distributions of $\boldsymbol{U}^{(p)}$ are not uniform distributions. Indeed, we have

Theorem 8.2 The distribution of $(U_1, \dots, U_k)'$, $1 \le k < p$, has pdf

$$\frac{\Gamma(p/2)}{\Gamma((p-k)/2)\pi^{k/2}}\left(1 - \sum_{j=1}^{k} u_j^2\right)^{(p-k)/2-1}, \quad \sum_{j=1}^{k} u_j^2 < 1. \tag{8.7}$$

Proof: Let $\boldsymbol{X} = (X_1, \dots, X_p)'$ denote a random vector having the p-dimensional standard normal distribution $N_p(\boldsymbol{0}, I)$. Since $N_p(\boldsymbol{0}, I)$ is a spherically symmetric distribution, we have $\boldsymbol{X}/\parallel \boldsymbol{X} \parallel \stackrel{d}{=} \boldsymbol{U}^{(p)}$ and $(X_1^2/\parallel \boldsymbol{X}\parallel^2, \dots, X_p^2/\parallel \boldsymbol{X}\parallel^2)'$ has a Dirichlet distribution with pdf

$$\frac{\Gamma(p/2)}{\{\Gamma(1/2)\}^p}\prod_{j=1}^{p} y_j^{1/2-1},\ y_j = |x_j|/\parallel \boldsymbol{x} \parallel,\quad \sum_{j=1}^{p} y_j = 1,\ y_j > 0\,(j = 1, \ldots, p).$$

From the partitioning property of the Dirichlet distribution, $(X_1^2/\parallel \boldsymbol{X}\parallel^2, \ldots, X_k^2/\parallel \boldsymbol{X}\parallel^2)'$ has a Dirichlet distribution with pdf

$$\frac{\Gamma(p/2)}{\{\Gamma(1/2)\}^k\Gamma((p-k)/2)}\prod_{j=1}^{k}\left(\frac{x_j^2}{\parallel \boldsymbol{x}\parallel^2}\right)^{1/2-1}\left(1-\frac{\sum_{j=1}^{k} x_j^2}{\parallel \boldsymbol{x}\parallel^2}\right)^{(p-k)/2-1}\prod_{j=1}^{k} d\left(\frac{x_j^2}{\parallel \boldsymbol{x}\parallel^2}\right)$$

for $1 \le k < p$. From

$$d\left(\frac{|x_j|}{\parallel \boldsymbol{x} \parallel}\right) = \frac{1}{2\sqrt{x_j^2/\parallel \boldsymbol{x}\parallel^2}}\, d\left(\frac{x_j^2}{\parallel \boldsymbol{x}\parallel^2}\right),$$

the pdf of $(y_1, \ldots, y_k)'$, where $y_j = |x_j|/\parallel \boldsymbol{x} \parallel\ (1 \le j \le k) = |u_j|$, is given by

$$\frac{\Gamma(p/2)2^k}{\Gamma((p-k)/2)\pi^{k/2}}\left(1-\sum_{j=1}^{k} y_j^2\right)^{(p-k)/2-1}\prod_{j=1}^{k} dy_j,\quad y_j \ge 0\,(1 \le j \le k),\ \sum_{j=1}^{k} y_j^2 < 1.$$

Additionally, since the joint pdf of $(u_1, \ldots, u_k)'$ is a function of $\sum_{j=1}^{k} u_j^2$, $(u_1, \ldots, u_k)'$ is spherical and decomposed into $(u_1, \ldots, u_k)' \overset{d}{=} r\boldsymbol{U}^{(k)}$, where r is a positive random variable and independent of the random vector $\boldsymbol{U}^{(k)}$ having the uniform distribution on S^{k-1}. The pdf of r is

$$f(r) = \frac{2}{B(k/2, (p-k)/2)}\, r^{k-1}(1-r^2)^{(p-k)/2-1},\quad 0 < r < 1$$

and r^2 has a beta distribution with parameters $(k/2, (p-k)/2)$.

Each univariate marginal distribution of $\boldsymbol{U}^{(k)}$, say that of U_1, is given by the fact that $T = (U_1 + 1)/2$ follows a Beta distribution with both parameters as $(p-1)/2$.

From (8.7), the conditional distribution of $\boldsymbol{U}_{(1)}$ given $\boldsymbol{U}_{(2)} = \boldsymbol{a}$, where $\boldsymbol{U}^{(p)} = (\boldsymbol{U}_{(1)}', \boldsymbol{U}_{(2)}')'$, $\boldsymbol{U}_{(1)}$ is $q \times 1$ and $\boldsymbol{U}_{(2)}$ is $(p-q) \times 1$, is a uniform distribution on the sphere with radius $(1-\parallel \boldsymbol{a}\parallel^2)^{1/2}$ and center at the origin.

Characteristic Function Similar to the Fourier series representation of a function on the circle in terms of the Fourier coefficients, there also exists a Fourier-series-type representation of a function on the sphere in terms of the coefficients termed spherical harmonics. Hence, the characteristic function of a spherical representation may also admit of such a representation. In the following however we give a representation of the characteristic function directly in terms of a generalized hypergeometric function.

Theorem 8.3 The characteristic function of $\boldsymbol{U}^{(p)}$ is given by

$$E\left(e^{i\boldsymbol{v}'\boldsymbol{U}^{(p)}}\right) = {}_0F_1\left(\frac{p}{2}; -\frac{1}{4} \parallel \boldsymbol{v}\parallel^2\right), \tag{8.8}$$

where $\boldsymbol{v} \in \mathbb{R}^p$, $i = \sqrt{-1}$, and ${}_0F_1$ represents a member of the generalized hypergeometric function defined by (B.16).

Proof: The characteristic function of $\boldsymbol{U}^{(p)}$ is defined by

$$E\left(e^{i\boldsymbol{v}'\boldsymbol{U}^{(p)}}\right) = \frac{1}{\omega_p}\int_{S^{p-1}} e^{i\boldsymbol{v}'\boldsymbol{u}^{(p)}} d\omega_p, \quad \boldsymbol{v} \in \mathbb{R}^p.$$

Taking $\boldsymbol{v} = \theta\boldsymbol{a}$, $\theta \in \mathbb{R}$, $\parallel \boldsymbol{a} \parallel = 1$, we may identify $\boldsymbol{a}$ with $\boldsymbol{\varepsilon}_p$. Following Watson (1983, p. 45), we have the characteristic function of $t = \boldsymbol{a}'\boldsymbol{u}^{(p)}$, from the recursion formula $d\omega_p = (1-t^2)^{(p-3)/2} dt d\omega_{p-1}$, as

$$\begin{aligned}
E(e^{i\theta t}) &= \frac{1}{\omega_p}\int_{S^{p-1}} e^{i\theta t} d\omega_p = \frac{\omega_{p-1}}{\omega_p}\int_{-1}^{1} e^{i\theta t}(1-t^2)^{(p-3)/2} dt \\
&= \Gamma\left(\frac{p}{2}\right)\left(\frac{\theta}{2}\right)^{-p/2+1} J_{p/2-1}(\theta) \\
&= {}_0F_1\left(\frac{p}{2}; -\frac{1}{4}\theta^2\right).
\end{aligned}$$

Here we used the two formulae (9.1.69) in p. 362 and (9.1.20) in p. 360 of Abramowitz and Stegun (1972), respectively, given below, with respect to the Bessel function of the first kind:

$$\begin{aligned}
J_\nu(z) &= \frac{2(z/2)^\nu}{\sqrt{\pi}\Gamma(\nu+1/2)}\int_0^1 (1-t^2)^{\nu-1/2}\cos(zt) dt, \quad \operatorname{Re}\nu > -\frac{1}{2} \\
&= \frac{(z/2)^\nu}{\Gamma(\nu+1)}\, {}_0F_1\left(\nu+1; -\frac{1}{4}z^2\right).
\end{aligned}$$

Another direct way of establishing (8.8) stems from the fact that the uniform distribution on the sphere is a spherically symmetric distribution. It holds that $\boldsymbol{v}'\boldsymbol{U}^{(p)} \overset{d}{=} \parallel \boldsymbol{v} \parallel U_1$ and the characteristic function of $\boldsymbol{U}^{(p)}$ is given by

$$\begin{aligned}
E\left(e^{i\boldsymbol{v}'\boldsymbol{U}^{(p)}}\right) &= E\left(e^{i\|\boldsymbol{v}\|U_1}\right) = \int_{-1}^{1} e^{i\|\boldsymbol{v}\|u_1}\frac{1}{B(1/2,(p-1)/2)}(1-u_1^2)^{(p-3)/2} du_1 \\
&= \frac{1}{B(1/2,(p-1)/2)}\int_0^{\pi} e^{i\|\boldsymbol{v}\|\cos\theta}\sin^{p-2}\theta d\theta \\
&= \frac{1}{B(1/2,(p-1)/2)}\sum_{k=0}^{\infty}\frac{(-1)^k \parallel \boldsymbol{v}\parallel^{2k}}{(2k)!} B\left(\frac{p-1}{2}, \frac{2k+1}{2}\right) \\
&= \sum_{k=0}^{\infty}\frac{\Gamma(p/2)}{\Gamma(p/2+k)}\frac{(-1)^k \parallel \boldsymbol{v}\parallel^{2k}}{4^k k!},
\end{aligned}$$

which leads to our desired result on using the duplication formula for the gamma function.

8.4.1.2 General Norm Uniform Distribution

For $\boldsymbol{x} = (x_1, \dots, x_p)' \in \mathbb{R}^p$ and $\alpha \geq 1$, the $(p-1)$-dimensional L_α unit sphere with center at the origin is defined by $S_\alpha^{p-1} = \{\boldsymbol{x} \mid \|\boldsymbol{x}\|_\alpha = (\sum_{j=1}^p |x_j|^\alpha)^{1/\alpha} = 1\}$. Then the random vector $\boldsymbol{U}_{(p)} = (U_1, \dots, U_p)'$ is said to have p-dimensional L_α-norm uniform distribution on the L_α unit sphere (Gupta and Song, 1997; Szabłowski, 1998) if the joint pdf is given by

$$f(u_1, \dots, u_{p-1}) = \left(\frac{\alpha}{2}\right)^{p-1} \frac{\Gamma(p/\alpha)}{\Gamma^p(1/\alpha)} \left(1 - \sum_{j=1}^{p-1} |u_j|^\alpha\right)^{(1-\alpha)/\alpha}, \quad \sum_{j=1}^{p-1} |u_j|^\alpha < 1.$$

From Gupta and Song (1997), if $\boldsymbol{U}_{(p)}$ is partitioned into two parts $\boldsymbol{U}_{(p)} = (\boldsymbol{U}_1', \boldsymbol{U}_2')'$, where $\boldsymbol{U}_j\,(j = 1,2)$ are, respectively, random vectors of p_j component with $p_1 + p_2 = p$, then

$$\begin{pmatrix} \boldsymbol{U}_1 \\ \boldsymbol{U}_2 \end{pmatrix} \stackrel{d}{=} \begin{pmatrix} R\boldsymbol{U}_{(p_1)} \\ (1 - R^\alpha)^{1/\alpha} \boldsymbol{U}_{(p_2)} \end{pmatrix},$$

where R, $\boldsymbol{U}_{(p_1)}$ and $\boldsymbol{U}_{(p_2)}$ are mutually independent, R^α is distributed as a beta distribution with parameters $(p_1/\alpha, p_2/\alpha)$, and $\boldsymbol{U}_{(p_j)}$ are p_j-dimensional L_α-norm uniform distributions. As for the mean vector and variance-covariance matrix of $\boldsymbol{U}_{(p)}$, we have

$$E(\boldsymbol{U}_{(p)}) = \boldsymbol{0} \quad \text{and} \quad V(\boldsymbol{U}_{(p)}) = \frac{\Gamma(p/\alpha)\Gamma(3/\alpha)}{\Gamma(1/\alpha)\Gamma((p+2)/\alpha)} I.$$

8.4.2 Projected Normal Distribution

A general projected normal distribution constructed from $N_p(\boldsymbol{\mu}, \Sigma)$ is denoted by $\mathrm{PN}_p(\boldsymbol{\mu}, \Sigma)$. However, the simplest case when $\boldsymbol{\mu} = \boldsymbol{0}$ in $\mathrm{PN}_p(\boldsymbol{\mu}, \Sigma)$ is considered below.

From the joint pdf $f_{\boldsymbol{X}}(\boldsymbol{x})$ of $\boldsymbol{X}$ which follows $N_p(\boldsymbol{0}, \Sigma)$, we have

$$\begin{aligned} f_{\boldsymbol{X}}(\boldsymbol{x})d\boldsymbol{x} &= \frac{1}{(2\pi)^{p/2}|\Sigma|^{1/2}} \exp\left[-\frac{1}{2}\boldsymbol{x}'\Sigma^{-1}\boldsymbol{x}\right] d\boldsymbol{x} \\ &= \frac{1}{(2\pi)^{p/2}|\Sigma|^{1/2}} \exp\left[-\frac{r^2}{2}\boldsymbol{y}'\Sigma^{-1}\boldsymbol{y}\right] r^{p-1} \prod_{k=1}^{p-2} \sin^{p-k-1}\theta_k \, dr \prod_{k=1}^{p-1} d\theta_k. \end{aligned}$$

Then the pdf of direction $\boldsymbol{T}$ is given by

$$f_T(\boldsymbol{t}) = \int_0^\infty \frac{1}{(2\pi)^{p/2}|\Sigma|^{1/2}} \exp\left[-\frac{r^2}{2}\boldsymbol{y}'\Sigma^{-1}\boldsymbol{y}\right] r^{p-1} \prod_{k=1}^{p-2} \sin^{p-k-1}\theta_k \, dr$$

$$= \frac{1}{(2\pi)^{p/2}|\Sigma|^{1/2}} \prod_{k=1}^{p-2} \sin^{p-k-1}\theta_k \int_0^\infty \left(\frac{2u}{\mathbf{y}'\Sigma^{-1}\mathbf{y}}\right)^{(p-2)/2} e^{-u} \frac{du}{\mathbf{y}'\Sigma^{-1}\mathbf{y}}$$

$$= \frac{\Gamma(p/2)}{2\pi^{p/2}|\Sigma|^{1/2}(\mathbf{y}'\Sigma^{-1}\mathbf{y})^{p/2}} \prod_{k=1}^{p-2} \sin^{p-k-1}\theta_k.$$

The pdf of direction cosine $\boldsymbol{Y} = \boldsymbol{e}(\boldsymbol{T})$ with respect to the uniform distribution on the $(p-1)$-dimensional sphere with unit radius and center at the origin, S^{p-1}, is given by

$$f_Y(\mathbf{y}) = \frac{1}{|\Sigma|^{1/2}(\mathbf{y}'\Sigma^{-1}\mathbf{y})^{p/2}}. \tag{8.9}$$

As a special case, if $\boldsymbol{X}$ has $N_p(\mathbf{0}, \Sigma)$ with $\Sigma = I$ (identity matrix), then $\boldsymbol{Y} = \boldsymbol{X}/\parallel \boldsymbol{X} \parallel$ has the uniform distribution. The above derivation also implies the fact that if $\boldsymbol{X}$ has $N_p(\mathbf{0}, I)$, then $S = \parallel \boldsymbol{X} \parallel^2$ has a χ^2 distribution with p degrees of freedom and $\boldsymbol{X}/\parallel \boldsymbol{X} \parallel$ is independent of S. Moreover, note that (8.9) reduces to the pdf (4.53) of the projected normal distribution in Section 4.10.2 if $p = 2$. As mentioned in Section 4.10.2, if $\boldsymbol{X}$ is spherical, $P(\boldsymbol{X} = \mathbf{0}) = 0$ and $\Sigma\,(p \times p)$ is positive definite, then $\Sigma^{1/2}\boldsymbol{X}/\parallel \Sigma^{1/2}\boldsymbol{X} \parallel$ has the same pdf as (8.9) since $a\Sigma^{1/2}\boldsymbol{X}/\parallel a\Sigma^{1/2}\boldsymbol{X} \parallel \stackrel{d}{=} \Sigma^{1/2}\boldsymbol{X}/\parallel \Sigma^{1/2}\boldsymbol{X} \parallel$ for all $a > 0$. It is clear that if $\boldsymbol{X}$ has the uniform distribution on S^{p-1}, then $\boldsymbol{X}/\parallel \boldsymbol{X} \parallel$ also has the uniform distribution.

8.4.3 von Mises–Fisher Distribution

The von Mises–Fisher or Langevin (1905) distribution $M_p(\boldsymbol{\mu}, \kappa)$ on S^{p-1} has joint pdf with respect to the uniform distribution, given by

$$f(\boldsymbol{x}) = \left(\frac{\kappa}{2}\right)^{p/2-1} \frac{1}{\Gamma(p/2)I_{p/2-1}(\kappa)} \exp(\kappa\boldsymbol{\mu}'\boldsymbol{x}), \quad \boldsymbol{x} \in S^{p-1}, \tag{8.10}$$

where $\kappa \geq 0$, $\boldsymbol{\mu} \in S^{p-1}$ and $I_\nu(\cdot)$ denotes the modified Bessel function of the first kind and order ν. This distribution is rotationally symmetric about $\boldsymbol{\mu}$. Note that $\int_{S^{p-1}} f(\boldsymbol{x})d\boldsymbol{x} = 1$, since

$$\int_{S^{p-1}} \exp(\kappa\boldsymbol{\mu}'\boldsymbol{x})d\omega_p = \omega_{p-1}\int_{-1}^{1} e^{\kappa t}(1-t^2)^{(p-3)/2}dt = \frac{(2\pi)^{p/2}}{\kappa^{p/2-1}} I_{p/2-1}(\kappa)$$

from the recursion formula (8.2) about $d\omega_p$ given in Section 8.1.3 and the formula (B.24).

Equation (8.10) can be rewritten as the pdf of angular variables by using the polar transformation

$$\begin{pmatrix} x_1 \\ x_2 \\ \vdots \\ x_{p-1} \\ x_p \end{pmatrix} = \begin{pmatrix} \cos t_1 \cos t_2 \cdots \cos t_{p-2} \cos t_{p-1} \\ \cos t_1 \cos t_2 \cdots \cos t_{p-2} \sin t_{p-1} \\ \vdots \\ \cos t_1 \sin t_2 \\ \sin t_1 \end{pmatrix} = \boldsymbol{e}(\boldsymbol{t}),$$

where $\boldsymbol{e}(\boldsymbol{t})$ denotes the direction cosine of the vector $\boldsymbol{t} = (t_1, \dots, t_{p-1})'$, whose range is $\omega = \{(t_1, \dots, t_{p-1})' \mid t_1, \dots, t_{p-2} \in [-\pi/2, \pi/2), t_{p-1} \in [-\pi, \pi)\}$. Likewise, we transform $\boldsymbol{\mu}$ into $\boldsymbol{e}(\boldsymbol{\tau})$. Then the angular representation of $M_p(\boldsymbol{\mu}, \kappa)$, the pdf of $\boldsymbol{t}$, is given by

$$f(\boldsymbol{t}) = \frac{\kappa^{p/2-1} \exp(\kappa \langle \boldsymbol{e}(\boldsymbol{t}), \boldsymbol{e}(\boldsymbol{\tau}) \rangle)}{(2\pi)^{p/2} I_{p/2-1}(\kappa)} \prod_{j=1}^{p-2} \cos^{p-j-1} t_j,$$

where $\langle \boldsymbol{e}(\boldsymbol{t}), \boldsymbol{e}(\boldsymbol{\tau}) \rangle$ denotes the inner product between $\boldsymbol{e}(\boldsymbol{t})$ and $\boldsymbol{e}(\boldsymbol{\tau})$, and the value of $\prod_{j=1}^{p-2} \cos^{p-j-1} t_j$ is interpreted as 1 when $p = 2$.

The von Mises–Fisher distribution with pdf given by (8.10) reduces to the uniform distribution if $\kappa \to 0$. It reduces to the von Mises distribution on the circle when $p = 2$ and is called the Fisher distribution on the sphere when $p = 3$ with a simple pdf

$$f(x_1, x_2, x_3) = \frac{\kappa}{\sinh \kappa} e^{\kappa(\mu_1 x_1 + \mu_2 x_2 + \mu_3 x_3)}$$

with respect to the uniform distribution. The normalizing constant is obtained from (B.26). When $p = 3$, if $\boldsymbol{x}$ and $\boldsymbol{\mu}$ are written in polar coordinates as

$$\boldsymbol{x} = (\cos\theta, \sin\theta\cos\phi, \sin\theta\sin\phi)', \ \boldsymbol{\mu} = (\cos\alpha, \sin\alpha\cos\beta, \sin\alpha\sin\beta)'$$

for $0 \le \theta < \pi$, $0 \le \phi < 2\pi$ and $0 \le \alpha < \pi$, $0 \le \beta < 2\pi$, then the joint pdf is

$$f(\theta, \phi) = \frac{\kappa}{4\pi \sinh \kappa} \exp[\kappa\{\cos\alpha\cos\theta + \sin\alpha\sin\theta\cos(\phi - \beta)\}] \sin\theta.$$

Note that though $\kappa = 0$ is not a point of the parameter space, $f(\theta, \phi)$ tends to the uniform distribution with density $\sin\theta/(4\pi)$ as $\kappa \to 0$, since

$$\frac{\kappa}{\sinh \kappa} = \frac{2\kappa}{e^{\kappa} - e^{-\kappa}} \to 1$$

as $\kappa \to 0$.

The marginal distribution of Θ takes a simple form when $\boldsymbol{\mu} = (1, 0, 0)'$, i.e. when $\alpha = 0$, as given by

$$f(\theta) = \frac{\kappa}{2\sinh\kappa} \exp(\kappa\cos\theta)\sin\theta, \quad \theta \in [0, \pi).$$

This axial distribution has received popularity, e.g. in Bioinformatics (see Section 8.7).

The conditional distribution of Φ given $\Theta = \theta$ is a von Mises distribution $\mathrm{VM}(\beta, \kappa\sin\theta\sin\alpha)$.

For $p > 3$, the pdf given by (8.10) is often referred to as the Langevin distribution, since Langevin (1905) contributed significantly to the popularization of this distribution.

The von Mises–Fisher distribution with joint pdf (8.10) can be obtained by using the conditioning method. Let $\boldsymbol{Z}$ be distributed as $N_p(\boldsymbol{\mu}, \kappa^{-1}I)$, where $\kappa > 0$ and $\| \boldsymbol{\mu} \| = 1$. Then the conditional distribution of $\boldsymbol{Z}$ given $\| \boldsymbol{Z} \| = 1$ is $M_p(\boldsymbol{\mu}, \kappa)$.

For a random unit vector $\boldsymbol{X} = (X_1, \dots, X_p)'$, the mean resultant length is defined by

$$L = \left[\sum_{j=1}^{p} \{E(X_j)\}^2\right]^{1/2} = \{E(\mathbf{X})'E(\mathbf{X})\}^{1/2},$$

and when $L > 0$, the mean direction is defined by $E(\boldsymbol{X})/L$. The mean direction of $M_p(\boldsymbol{\mu}, \kappa)$ is $\boldsymbol{\mu}$, and the mean resultant length is given by

$$L = E(\boldsymbol{\mu}'\boldsymbol{X}) = \frac{I_{p/2}(\kappa)}{I_{p/2-1}(\kappa)} \; (= A_p(\kappa), \text{say})$$

since

$$\int_{S^{p-1}} \boldsymbol{\mu}'\boldsymbol{x} \exp(\kappa\boldsymbol{\mu}'\boldsymbol{x}) d\omega_p = \omega_{p-1} \frac{\kappa}{p-1} \int_{-1}^{1} (1-t^2)^{(p-1)/2} e^{\kappa t} dt$$
$$= \frac{(2\pi)^{p/2}}{\kappa^{p/2-1}} I_{p/2}(\kappa).$$

Note that the derivative of $A_p(\kappa)$ appears as the Fisher information of κ for the von Mises–Fisher distribution (cf. Mardia and Jupp, 2000, eq. (10.3.16)):

$$-E\left(\frac{\partial^2}{\partial \kappa^2} \log f\right) = \frac{\partial}{\partial \kappa} A_p(\kappa)$$

for the pdf f given in (8.10).

The parameter κ is called the concentration parameter since for $\kappa \to 0$ the distribution is uniform, whereas for large κ it is clustered around the mean direction $\boldsymbol{\mu}$. When $p = 3$, the mean resultant length has a simple form

$$A_3(\kappa) = \coth \kappa - \frac{1}{\kappa}$$

which follows from (B.26) and (B.27).

For high-dimensional situations $p \to \infty$, Watson (1988) showed that several appealing results in inference and useful applications may be obtained for the special case when κ above is taken as $\kappa p^{1/2}$, i.e. the distribution as $M_p(\boldsymbol{\mu}, \kappa p^{1/2})$.

8.4.4 Fisher–Bingham Distribution

The Fisher–Bingham distribution (see Mardia and Jupp, 2000, p. 174) has pdf

$$f(\boldsymbol{x}; \boldsymbol{\mu}, \kappa, A) = \frac{1}{a(\kappa, A)} \exp(\kappa\boldsymbol{\mu}'\boldsymbol{x} + \boldsymbol{x}'A\boldsymbol{x}), \quad \boldsymbol{x}'\boldsymbol{x} = 1, \tag{8.11}$$

with the normalizing constant $a(\kappa, A)$, where A is a symmetric $p \times p$ matrix and $\mathrm{tr}A = 0$, without loss of generality. Similar to the derivation of the von Mises–Fisher distribution, (8.11) can be obtained by conditioning a p-variate normal distribution $N_p(-\kappa(A + cI_p)^{-1}\boldsymbol{\mu}/2, -(A + cI_p)^{-1}/2)$, where c is such that $A + cI_p$ is negative definite, on $\| \boldsymbol{x} \| = 1$. It is clear that (8.11) reduces to the von Mises–Fisher pdf if $A = \boldsymbol{O}$.

The Kent distribution is obtained from the Fisher–Bingham distribution with pdf (8.11) by imposing the restriction $A\boldsymbol{\mu} = \mathbf{0}$, i.e.

$$f(\boldsymbol{x}; \boldsymbol{\mu}, \kappa, A) = \frac{1}{a(\kappa, A)} \exp(\kappa \boldsymbol{\mu}'\boldsymbol{x} + \boldsymbol{x}'A\boldsymbol{x}),$$

where A is a symmetric $p \times p$ matrix with $\mathrm{tr}\,A = 0$ and $A\boldsymbol{\mu} = \mathbf{0}$.

The Fisher–Bingham family of distributions with six parameters (FB_6) has density (see, e.g. Wood, 1987)

$$f(\boldsymbol{x} : \kappa, \beta, \gamma, \boldsymbol{\mu}) \propto \exp[\kappa\, \boldsymbol{\mu}_3'\, \boldsymbol{x} + \gamma\, (\boldsymbol{\mu}_3'\, \boldsymbol{x})^2 + \beta\{(\boldsymbol{\mu}_1'\, \boldsymbol{x})^2 - (\boldsymbol{\mu}_2'\, \boldsymbol{x})^2\}],$$

$\boldsymbol{x} \in S^2$, $\kappa \geq 0$, $\beta \geq 0$, $\gamma \in (-\infty, \infty)$, where $\boldsymbol{\mu} = [\boldsymbol{\mu}_1\ \boldsymbol{\mu}_2\ \boldsymbol{\mu}_3]$ is a 3×3 orthogonal matrix. This is one of the most popularly used family of spherical distributions that generalizes many special cases of interest, namely, Uniform ($\kappa = \beta = \gamma = 0$), Fisher ($\beta = \gamma = 0$), Bingham ($\kappa = 0$), Dimroth–Watson ($\kappa = \beta = 0$), FB_4 ($\beta = 0$), and FB_5 ($\gamma = 0$). The family easily extends to higher-dimensional hyper-spheres.

8.4.5 Watson Distribution

The Watson distribution, also called the Dimroth–Scheidegger–Watson distribution for axial data, has with respect to the uniform distribution, density

$$f(\pm\boldsymbol{x}; \boldsymbol{\mu}, \kappa) = \left\{M\left(\frac{1}{2}, \frac{p}{2}, \kappa\right)\right\}^{-1} \exp\left\{\kappa\left(\boldsymbol{\mu}'\boldsymbol{x}\right)^2\right\},$$

where M denotes Kummer's function or the confluent hypergeometric function given by (B.9), (B.10) and (B.11). In the S^2 case, the angular representation of the density is

$$\begin{aligned} f(\theta, \phi) = &\left\{4\pi M\left(\frac{1}{2}, \frac{3}{2}, \kappa\right)\right\}^{-1} \\ &\times \exp\left[\kappa\{\sin\theta \sin\alpha \cos(\phi - \beta) + \cos\theta \cos\alpha\}^2\right] \sin\theta, \\ &0 \leq \theta < \pi,\ 0 \leq \phi < 2\pi, \end{aligned}$$

where

$$M\left(\frac{1}{2}, \frac{3}{2}, \kappa\right) = \int_0^1 e^{\kappa u^2} du.$$

For $\kappa > 0$, the distribution is bipolar, and for $\kappa < 0$, it is girdle (equatorial), i.e. concentrated around a great circle (see Figure 8.6). For $\kappa = 0$, the distribution is uniform.

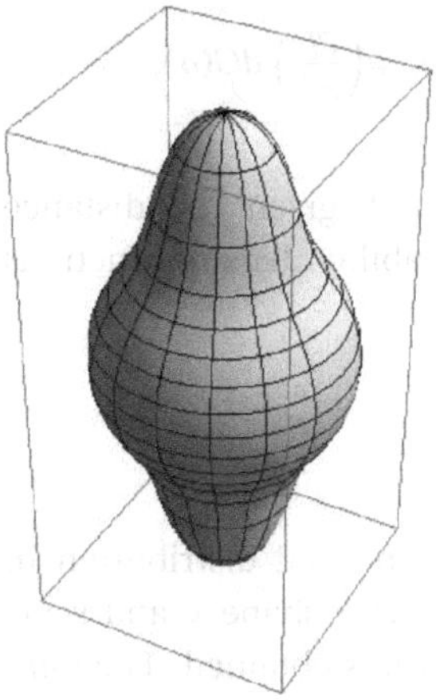

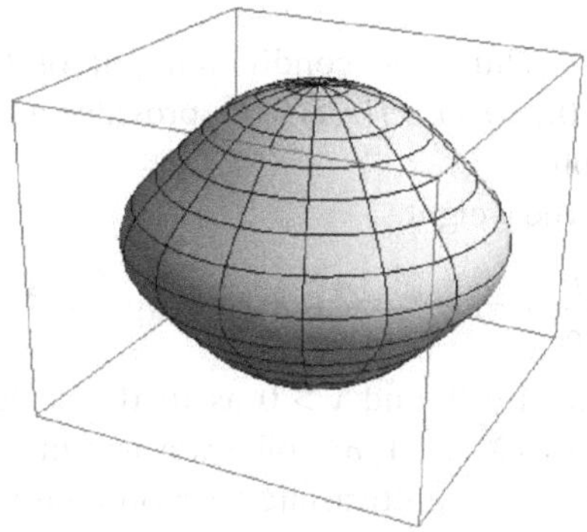

(a) (b)

Figure 8.6 3-D plots of the Watson distribution with $\alpha = 0$. (a) $\kappa = 7$, (b) $\kappa = -7$.

8.4.6 Pearson Type VII Distribution

Consider $N_p(\boldsymbol{\eta}, \sigma^2 I)$, $\sigma > 0$, whose pdf is denoted by $\phi(\boldsymbol{z}; \boldsymbol{\eta}, \sigma^2 I)$, $\boldsymbol{z} \in \mathbb{R}^p$, and let $G(\sigma)$ be a distribution function on the interval $(0, \infty)$. The scale mixture of normal distributions associated with the weight G is defined by a random vector $\boldsymbol{Z}$ with pdf

$$f(\boldsymbol{z}) = \int_0^\infty \phi(\boldsymbol{z}; \boldsymbol{\eta}, \sigma^2 I) dG(\sigma)$$

and forms a class of distributions. Write $\boldsymbol{Z}$ in polar coordinates as $\boldsymbol{Z} = R\boldsymbol{e}(\boldsymbol{T})$, where $R = \| \boldsymbol{Z} \|$ and $\boldsymbol{T} = (T_1, \ldots, T_{p-1})'$. For $\boldsymbol{\eta} = \rho \boldsymbol{e}(\boldsymbol{\tau}) \, (\rho > 0)$, the joint pdf of $(R, \boldsymbol{T}')'$ is expressed as

$$\begin{aligned} f(r, \boldsymbol{t}) = & \int_0^\infty \frac{1}{(2\pi)^{p/2}\sigma^p} \exp\left\{ -\frac{r^2 + \rho^2 - 2\rho r \langle \boldsymbol{e}(\boldsymbol{t}), \boldsymbol{e}(\boldsymbol{\tau}) \rangle}{2\sigma^2} \right\} \\ & \times r^{p-1} \prod_{j=1}^{p-2} \cos^{p-j-1} t_j \, dG(\sigma). \end{aligned} \tag{8.12}$$

Integrating (8.12) with respect to $\boldsymbol{t}$, we obtain the marginal pdf of the distance R as

$$\begin{aligned} f(r) = & \int_0^\infty \left[\frac{r^{p-1}}{(2\pi)^{p/2}\sigma^p} \exp\left(-\frac{r^2 + \rho^2}{2\sigma^2} \right) \right. \\ & \left. \times \int_\omega \exp\left\{ \frac{\rho r \langle \boldsymbol{e}(\boldsymbol{t}), \boldsymbol{e}(\boldsymbol{\tau}) \rangle}{\sigma^2} \right\} \prod_{j=1}^{p-2} \cos^{p-j-1} t_j \, d\boldsymbol{t} \right] dG(\sigma) \end{aligned}$$

$$= \left(\frac{1}{\rho r}\right)^{p/2-1} r^{p-1} \int_0^\infty \frac{1}{\sigma^2} \exp\left(-\frac{r^2+\rho^2}{2\sigma^2}\right) I_{p/2-1}\left(\frac{\rho r}{\sigma^2}\right) dG(\sigma),$$

for $r > 0$. Thus, the conditional pdf of the direction $\boldsymbol{T}$ given the distance $R = r(> 0)$, $f(\boldsymbol{t}|r) = f(r, \boldsymbol{t})/f(r)$, provides a class of probability density functions on the sphere.

Adopt the weight

$$\frac{d}{d\sigma} G(\sigma) = \frac{2^{1-\nu} A^{2\nu}}{\Gamma(\nu)} \sigma^{-1-2\nu} \exp\left(-\frac{A^2}{2\sigma^2}\right)$$

for $\sigma > 0$, $A \neq 0$, and $\nu > 0$ as in the scale mixture of normal distribution in Section 4.10.3, i.e. $1/\sigma^2$ follows a gamma distribution with shape ν and scale $2/A^2$. Then the multivariate Pearson type VII distribution is obtained. The joint pdf of the length R and the direction $\boldsymbol{T}$ is given by

$$f_P(r, \boldsymbol{t}) = \frac{\Gamma(\nu + p/2)}{\Gamma(\nu)} \left(\pi A^2\right)^{-p/2} \left\{1 + \frac{r^2 + \rho^2 - 2\rho r \langle \boldsymbol{e}(\boldsymbol{t}), \boldsymbol{e}(\boldsymbol{\tau})\rangle}{A^2}\right\}^{-\nu-p/2} \times r^{p-1} \prod_{j=1}^{p-2} \cos^{p-j-1} t_j.$$

The marginal pdf of the length R is given by

$$f_P(r) = \frac{2r^{p-1}A^{2\nu}}{B(\nu, p/2)} \left(\frac{1}{r^2 + \rho^2 + A^2}\right)^{\nu+p/2} \times {}_2F_1\left(\frac{\nu}{2} + \frac{p}{4}, \frac{\nu}{2} + \frac{p}{4} + \frac{1}{2}; \frac{p}{2}; \left(\frac{2\rho r}{r^2 + \rho^2 + A^2}\right)^2\right), \quad r > 0,$$

where B and ${}_2F_1$ denote the beta function and the Gauss hypergeometric function, respectively.

Finally we obtain the conditional pdf of the direction $\boldsymbol{T}$ given the distance $R = r(> 0)$ as

$$f_P(\boldsymbol{t}|r) = \frac{\Gamma(p/2)}{2\pi^{p/2}} \left[{}_2F_1\left(\frac{\nu}{2} + \frac{p}{4}, \frac{\nu}{2} + \frac{p}{4} + \frac{1}{2}; \frac{p}{2}; \left\{\frac{2}{A^2} \kappa(A^2, \rho|r)\right\}^2\right)\right]^{-1} \times \left\{1 - \frac{2}{A^2} \kappa(A^2, \rho|r) \langle \boldsymbol{e}(\boldsymbol{t}), \boldsymbol{e}(\boldsymbol{\tau})\rangle\right\}^{-\nu-p/2} \prod_{j=1}^{p-2} \cos^{p-j-1} t_j,$$

where

$$\kappa(A^2, \rho|r) = \frac{\rho r}{1 + (r^2 + \rho^2)/A^2}.$$

The pdf of the distribution with respect to the uniform distribution is explicitly expressed by

$$f_P(\boldsymbol{x}) = \left[{}_2F_1 \left(\frac{\nu}{2} + \frac{p}{4}, \frac{\nu}{2} + \frac{p}{4} + \frac{1}{2}; \frac{p}{2}; \left\{ \frac{2}{A^2} \kappa(A^2, \rho|r) \right\}^2 \right) \right]^{-1}$$
$$\times \left\{ 1 - \frac{2}{A^2} \kappa(A^2, \rho|r) \boldsymbol{\mu}' \boldsymbol{x} \right\}^{-\nu - p/2}.$$

The distribution with pdf $f_P(\boldsymbol{t}|r)$ or $f_P(\boldsymbol{x})$ should be called the Pearson type VII distribution with parameters ν, A, ρ, and τ on the $(p-1)$-dimensional sphere (Shimizu and Iida, 2002). The parameter function $\kappa(A^2, \rho|r)$ may be viewed as a concentration parameter, since the distribution reduces to the uniform distribution when $\rho \to 0$. The distribution is rotationally symmetric about $\boldsymbol{\mu}$, the mean direction is $\boldsymbol{\mu}$, and the mean resultant length is calculated as

$$L = \frac{\nu + p/2}{p} \left\{ \frac{2}{A^2} \kappa(A^2, \rho|r) \right\}$$
$$\times \frac{{}_2F_1 \left(\nu/2 + p/4 + 1/2, \nu/2 + p/4 + 1; p/2 + 1; \left\{ (2/A^2) \kappa(A^2, \rho|r) \right\}^2 \right)}{{}_2F_1 \left(\nu/2 + p/4, \nu/2 + p/4 + 1/2; p/2; \left\{ (2/A^2) \kappa(A^2, \rho|r) \right\}^2 \right)}.$$

The Pearson type VII distribution with $\nu = n/2$ and $A^2 = n$ is the t-distribution with n degrees of freedom, and the distribution whose pdf is given by $f_P(\boldsymbol{t}|r)$ with $\nu = n/2$ and $A^2 = n$ is called the symmetric t-distribution with parameters n (degrees of freedom), ρ and τ on the $(p-1)$-dimensional sphere. As $n \to \infty$, $\kappa(n, \rho|r) = \rho r / \{1 + (r^2 + \rho^2)/n\}$ converges to $\rho r (= \kappa$, say) and the pdf to the von Mises–Fisher pdf with concentration parameter κ.

8.4.7 Generalized Power Cardioid Family

As a generalization of (4.39), the corresponding family on the unit sphere having joint pdf with respect to the uniform distribution is given by (Jones and Pewsey, 2005),

$$f(\boldsymbol{x}) = \frac{|\sinh(\kappa\psi)|^{p/2-1}}{2^{p/2-1}\Gamma(p/2)} \frac{\{\cosh(\kappa\psi) + \sinh(\kappa\psi)\boldsymbol{x}'\boldsymbol{\mu}\}^{1/\psi}}{P^{1-p/2}_{1/\psi+p/2-1}(\cosh(\kappa\psi))}, \quad \boldsymbol{x} \in S^{p-1}. \quad (8.13)$$

Here $\kappa \geq 0, -\infty < \psi < \infty \, (\psi \neq 0)$, $\boldsymbol{\mu} \in S^{p-1}$ and P^{μ}_{ν} stands for the associated Legendre function expressed by

$$P^{\mu}_{\nu}(x) = \frac{1}{\Gamma(1-\mu)} \left(\frac{x+1}{x-1} \right)^{\mu/2} {}_2F_1 \left(-\nu, \nu+1; 1-\mu; \frac{1-x}{2} \right), -1 \leq x \leq 1.$$

In the circular case, note that this distribution provides an extension of the symmetric t-distribution introduced by Shimizu and Iida (2002) in the sense that the parameter space is enlarged and that the normalizing constant is also expressed in terms of the Gauss hypergeometric function. Likewise (8.13) has an expression

$$f(\boldsymbol{x}) = \frac{\left\{1+\tanh(\kappa\psi)\boldsymbol{x}'\boldsymbol{\mu}\right\}^{1/\psi}}{{}_2F_1\left(-1/(2\psi), -1/(2\psi)+1/2; p/2; \tanh^2(\kappa\psi)\right)}, \quad \boldsymbol{x} \in S^{p-1}. \tag{8.14}$$

Indeed, using the formulae (B.37), (B.36), (B.14), and (B.15), we obtain

$$\begin{aligned} P^{1-p/2}_{1/\psi+p/2-1}(\cosh(\kappa\psi)) &= P^{1-p/2}_{-1/\psi-p/2}(\cosh(\kappa\psi)) \\ &= \frac{2^{1-p/2}}{\Gamma(p/2)} \frac{\{\cosh(\kappa\psi)\}^{1/\psi}}{\{\sinh^2(\kappa\psi)\}^{(2-p)/4}}\, {}_2F_1\left(-\frac{1}{2\psi}, -\frac{1}{2\psi}+\frac{1}{2}; \frac{p}{2}; \tanh^2(\kappa\psi)\right). \end{aligned}$$

If $\psi < 0$, then it is clear that ${}_2F_1$ of (8.14) converges since $|\tanh(\kappa\psi)| < 1$ and its value is positive by the definition of ${}_2F_1$. If $\psi > 0$, then from (B.14) we have

$$\begin{aligned} &{}_2F_1\left(-\frac{1}{2\psi}, -\frac{1}{2\psi}+\frac{1}{2}; \frac{p}{2}; \tanh^2(\kappa\psi)\right) \\ &= \frac{1}{\{\cosh(\kappa\psi)\}^{p-1+2/\psi}}\, {}_2F_1\left(\frac{p}{2}+\frac{1}{2\psi}, \frac{p}{2}+\frac{1}{2\psi}-\frac{1}{2}; \frac{p}{2}; \tanh^2(\kappa\psi)\right), \end{aligned}$$

and its value is positive since $p \geq 2$. Putting

$$\frac{1}{\psi} = -\nu - \frac{p}{2}\,(<0),\ \tanh(\kappa\psi) = -\frac{2}{A^2}\kappa(A^2, \rho|r), \quad \nu > 0,\ A \neq 0,\ \rho > 0,\ r > 0,$$

since $0 < (2/A^2)\kappa(A^2, \rho|r) < 1$, where $\kappa(A^2, \rho|r) = \rho r/\{1+(r^2+\rho^2)/A^2\}$, we observe that (8.14) reduces to the pdf of the Pearson type VII distribution on the sphere given by Shimizu and Iida (2002).

If $p = 3$, on using

$${}_2F_1\left(a, \frac{1}{2}+a; \frac{3}{2}; z^2\right) = \frac{1}{2}z^{-1}(1-2a)^{-1}\left\{(1+z)^{1-2a} - (1-z)^{1-2a}\right\},$$

(8.14) reduces, for $\psi \neq 0$, to a simpler form (Jones and Pewsey, 2005)

$$f(\boldsymbol{x}) = \frac{(1+\psi)\sinh(\kappa\psi)}{\psi\sinh\{\kappa(1+\psi)\}}\left\{\cosh(\kappa\psi)+\sinh(\kappa\psi)\boldsymbol{x}'\boldsymbol{\mu}\right\}^{1/\psi}.$$

8.4.8 Laplace Distribution

8.4.8.1 Linear Laplace Distribution

Suppose that a random vector $\boldsymbol{W}$ given $U = u\,(>0)$ is distributed as a p-dimensional normal with mean vector $\boldsymbol{\eta} \neq \boldsymbol{0}$ and variance-covariance matrix I/u. The pdf of $\boldsymbol{W}|\,(U=u)$ is given by

$$f(\boldsymbol{w}|u) = \left(\frac{u}{2\pi}\right)^{p/2} \exp\left\{-\frac{u}{2}(\boldsymbol{w}-\boldsymbol{\eta})'(\boldsymbol{w}-\boldsymbol{\eta})\right\}, \quad \boldsymbol{w} \in \mathbb{R}^p.$$

If a weighting function for U is specified by an inverse gamma distribution with shape parameter α and scale parameter 2β, whose pdf is given by

$$g(u) = \frac{(2\beta)^\alpha}{\Gamma(\alpha)}u^{-(\alpha+1)}\exp\left(-\frac{2\beta}{u}\right), \quad \alpha > 0,\ \beta > 0,$$

then the unconditional distribution of $\boldsymbol{W}$ has pdf

$$\begin{aligned} f(\boldsymbol{w}) &= \int_0^\infty f(\boldsymbol{w}|u)g(u)du \\ &= \frac{2\beta^{\alpha/2+p/4}}{\pi^{p/2}\Gamma(\alpha)}\delta_{\boldsymbol{w}}^{\alpha/2-p/4}K_{p/2-\alpha}\left(2\sqrt{\beta\delta_{\boldsymbol{w}}}\right), \end{aligned} \tag{8.15}$$

where $\delta_{\boldsymbol{w}} = (\boldsymbol{w}-\boldsymbol{\eta})'(\boldsymbol{w}-\boldsymbol{\eta})$ and K_ν denotes the modified Bessel function of the third kind and order ν, given by (B.30). The distribution with pdf (8.15) is called the generalized p-dimensional Laplace distribution (Kotz et al., 2001).

Note that the modified Bessel function of the third kind has a simpler form for $\nu = \pm 1/2$, namely (B.33). As a special case, the standard Laplace distribution on the line with location parameter η is obtained when $p = 1$, $\alpha = 1$, and $\beta = 1/4$. Its pdf is given by

$$f(w) = \frac{1}{\sqrt{2\pi}}|w-\eta|^{1/2}K_{-1/2}(|w-\eta|) = \frac{1}{2}e^{-|w-\eta|}.$$

Since the standard Laplace distribution can be obtained by joining the same exponential distribution with its reflection at η, it is often referred to as the double-exponential distribution.

8.4.8.2 Hyper-Spherical Laplace Distribution

The generalized symmetric Laplace distribution with parameters α, β, ρ, and $\boldsymbol{\tau}$ ($\boldsymbol{\eta} = \rho\,\boldsymbol{e}(\boldsymbol{\tau})$, $\rho > 0$) on the $(p-1)$-dimensional sphere (Siew and Shimizu, 2008) is constructed from the generalized p-dimensional Laplace distribution in Euclidean space by the conditioning method. After the polar transformation $\boldsymbol{w} = r\,\boldsymbol{e}(\boldsymbol{t})$, the angular representation of the joint pdf is given by

$$\begin{aligned} f_{\mathrm{GL}}(\boldsymbol{t}|r) &= \frac{1}{2\pi^{p/2}S_p(\alpha,\beta,\rho)}\left(1-\frac{2\rho r\langle \boldsymbol{e}(\boldsymbol{t}),\boldsymbol{e}(\boldsymbol{\tau})\rangle}{r^2+\rho^2}\right)^{\alpha/2-p/4} \\ &\quad \times K_{p/2-\alpha}\left(2\sqrt{\beta\delta_t}\right)\prod_{j=1}^{p-2}\cos^{p-j-1}t_j, \end{aligned}$$

where $\delta_t = r^2+\rho^2-2\rho r\langle \boldsymbol{e}(\boldsymbol{t}),\boldsymbol{e}(\boldsymbol{\tau})\rangle$ and

$$S_p(\alpha,\beta,\rho) = \sum_{j=0}^\infty \frac{1}{j!\Gamma(j+p/2)}\left(\frac{\beta\rho^2r^2}{r^2+\rho^2}\right)^j K_{2j-\alpha+p/2}\left(2\sqrt{\beta(r^2+\rho^2)}\right).$$

It is clear that a random unit vector $\boldsymbol{X}$ has the generalized Laplace distribution on the sphere with respect to the uniform distribution if the conditional pdf of $\boldsymbol{X}$ on S^{p-1} given $R = r$ is

$$f_{\mathrm{GL}}(\boldsymbol{x}|r) = \frac{1}{\Gamma(p/2)S_p(\alpha,\beta,\rho)}\left(1-\frac{2\rho r}{r^2+\rho^2}\boldsymbol{\mu}'\boldsymbol{x}\right)^{\alpha/2-p/4}K_{p/2-\alpha}(2\sqrt{\beta\,\delta_{\boldsymbol{x}}}), \tag{8.16}$$

where $\boldsymbol{\mu} = \boldsymbol{e}(\boldsymbol{\tau})$ and $\delta_{\boldsymbol{x}} = r^2 + \rho^2 - 2\rho r\boldsymbol{\mu}'\boldsymbol{x}$. The distribution is rotationally symmetric about $\boldsymbol{\mu}$, the mean direction is $\boldsymbol{\mu}$, and the mean resultant length is calculated as

$$L = \rho r\sqrt{\frac{\beta}{r^2+\rho^2}\frac{S_{p+2}(\alpha,\beta,\rho)}{S_p(\alpha,\beta,\rho)}}.$$

The density can be widely dispersed or highly concentrated around the mean direction $\boldsymbol{e}(\boldsymbol{\tau})$, bell-shaped or heavy-tailed, depending on the choice of the parameters. When $\rho \to 0$, the distribution reduces to the uniform distribution.

The distribution on the sphere equivalent to the double exponential distribution on the line is derived by setting $\alpha - p/2 = 1/2$, $\beta = 1/(8r^2)$, and $\rho = r$. Its pdf is

$$f_{\mathrm{DE}}(\boldsymbol{t}|r) = C^{-1}\sqrt{\frac{\pi}{2}}\exp\left\{-\sqrt{1-\langle \boldsymbol{e}(\boldsymbol{t}), \boldsymbol{e}(\boldsymbol{\tau})\rangle}\right\}\prod_{j=1}^{p-2}\cos^{p-j-1}t_j$$

with

$$C = 2\pi^{p/2}S_p((p+1)/2, 1/(8r^2), r) = 2\pi^{p/2}\sum_{j=0}^{\infty}\frac{K_{2j-1/2}(1)}{j!\Gamma(j+p/2)2^{4j}}$$

$$= \frac{2\pi^{p/2}}{e}\sqrt{\frac{\pi}{2}}\left\{1+\sum_{j=1}^{\infty}\frac{1}{j!\Gamma(j+p/2)2^{4j}}\sum_{k=0}^{2j-1}\frac{\Gamma(2j+k)}{k!\Gamma(2j-k)2^k}\right\},$$

since the modified Bessel function of order $n+1/2$ can be expressed as

$$K_{n+1/2}(z) = \sqrt{\frac{\pi}{2z}}e^{-z}\sum_{k=0}^{n}\frac{(n+k)!}{k!(n-k)!(2z)^k}.$$

The von Mises–Fisher, cardioid and the generalized cardioid distributions arise as limiting cases of the generalized Laplace distribution. First, on using the result $K_\nu(z) \approx \sqrt{\pi/(2z)}e^{-z}$ as $z \to \infty$, (8.16) tends to the von Mises–Fisher pdf in (8.10) as $\rho \to \infty$, with $\kappa_1 = 2\sqrt{\beta}\,r$, which again approximates a cardioid pdf $\{\rho_1^{p/2-1}/\Gamma(p/2)\}(1+2\rho_1\boldsymbol{\mu}'\boldsymbol{x})$ for small β, where $\rho_1 = \kappa_1/2$. Second, on using $K_\nu(z) \approx \{\Gamma(\nu)/2\}(2/z)^\nu$ as $z \to 0^+$ together with the duplication formula for the gamma function, (8.16) tends to the pdf in (8.14) for small β, $\alpha < p/2$ and $\rho \neq r$, with $\alpha - p/2 = 1/\psi$ and $2\rho r/(r^2+\rho^2) = -\tanh(\kappa\psi)$.

8.4.9 Another Distribution Motivated by Stereographic Projection

We derive a p-dimensional generalization of the pdf (4.49) using the p-dimensional version of stereographic projection in Section 8.2.5, namely

$$g(\phi, \theta_1, \dots, \theta_{p-2})$$
$$= \{C_{\psi,p}(\kappa_1,\kappa_2)\}^{-1}\frac{\{1+\tanh(\kappa_1\psi)\cos(\theta_{p-2}-\mu)\}^{1/\psi+1-p/2}}{\{1-\tanh(\kappa_2\psi)\cos(\theta_{p-2}-\mu)\}^{1/\psi+p/2}}\prod_{j=1}^{p-2}\sin^j\theta_j, \quad (8.17)$$

where $\psi > 0,\ \kappa_1,\ \kappa_2 \geq 0,\ p \geq 3$, and $C_{\psi,p}(\kappa_1, \kappa_2)$ is the normalizing constant expressed by

$$\begin{aligned}
&C_{\psi,p}(\kappa_1, \kappa_2) \\
&= 2^{p-2}\omega_p \frac{\{1+\tanh(\kappa_1\psi)\}^{1+1/\psi-p/2}}{\{1-\tanh(\kappa_2\psi)\}^{1/\psi+p/2}} B\left(\frac{p-1}{2}, \frac{p-1}{2}\right) \\
&\times F_1\left(\frac{p-1}{2}; -\left(1+\frac{1}{\psi}-\frac{p}{2}\right), \frac{1}{\psi}+\frac{p}{2}, p-1; z_1; z_2\right) \\
&= 2^{p-2}\omega_p \frac{\{1+\tanh(\kappa_1\psi)\}^{1+1/\psi-p/2}}{\{1-\tanh(\kappa_2\psi)\}^{1/\psi+p/2}} (1-z_2)^{-(p-1)/2} B\left(\frac{p-1}{2}, \frac{p-1}{2}\right) \\
&\times {}_2F_1\left(\frac{p-1}{2}, -\left(1+\frac{1}{\psi}-\frac{p}{2}\right); p-1; \frac{z_1-z_2}{1-z_2}\right),
\end{aligned}$$

where F_1 and ${}_2F_1$ denote Appel's function and the Gauss hypergeometric function, respectively.

The normalizing constant reduces to a simpler form when $p = 3$, namely

$$\begin{aligned}
&\int_{-\pi}^{\pi}\int_0^{\pi} \frac{\{1+\tanh(\kappa_1\psi)\cos\theta\}^{-1/2+1/\psi}}{\{1-\tanh(\kappa_2\psi)\cos\theta\}^{3/2+1/\psi}} \sin\theta d\theta d\phi \\
&= 8\pi \frac{\{1+\tanh(\kappa_1\psi)\}^{-1/2+1/\psi}}{\{1-\tanh(\kappa_2\psi)\}^{3/2+1/\psi}} F_1\left(1; \frac{1}{2}-\frac{1}{\psi}, \frac{3}{2}+\frac{1}{\psi}; 2; z_1, z_2\right) \\
&= 8\pi \frac{\{1+\tanh(\kappa_1\psi)\}^{-1/2+1/\psi}}{\{1-\tanh(\kappa_2\psi)\}^{3/2+1/\psi}} \frac{1}{(1/2+1/\psi)(z_1-z_2)} \\
&\times \left\{1-\left(\frac{1-z_1}{1-z_2}\right)^{1/2+1/\psi}\right\}.
\end{aligned}$$

In $(p-1)$-dimensional Euclidean space, $\mathbb{R}^{p-1}$, a multivariate Pearson type VII distribution has pdf

$$f_{p-1}(\boldsymbol{x}) = \frac{\Gamma(\nu+(p-1)/2)}{\Gamma(\nu)} (\pi A^2)^{-(p-1)/2} \left(1+\frac{\boldsymbol{x}'\boldsymbol{x}}{A^2}\right)^{-(p-1)/2-\nu},$$

where $\boldsymbol{x} \in \mathbb{R}^{p-1}, A \neq 0, \nu > 0$. By an application of (8.5), the corresponding pdf $g(\phi, \theta_1, \ldots, \theta_{p-2})$ on S_p^v is

$$\begin{aligned}
&g(\phi, \theta_1, \ldots, \theta_{p-2}) \\
&= \frac{\Gamma(\nu+(p-1)/2)}{\Gamma(\nu)} (\pi A^2)^{-(p-1)/2} \left[1+\left\{\frac{v\sin\theta_{p-2}}{A(1+\cos\theta_{p-2})}\right\}^2\right]^{-(p-1)/2-\nu} |J|.
\end{aligned}$$

A special case of the Pearson type VII distribution, with $\nu = n/2$ and $A^2 = n$, is the multivariate t-distribution with n degrees of freedom. When transformed onto the sphere using inverse stereographic projection, the resulting distribution has pdf

$$g(\phi, \theta_1, \dots, \theta_{p-2})$$

$$= \frac{\Gamma((n+p-1)/2)}{\Gamma(n/2)} (\pi n)^{-(p-1)/2} \left\{ 1 + \frac{1}{n} \left(\frac{\upsilon \sin \theta_{p-2}}{1 + \cos \theta_{p-2}} \right)^2 \right\}^{-(n+p-1)/2} |J|. \quad (8.18)$$

Note that as $n \to \infty$, this pdf converges to a density on S_p^υ induced by the $(p-1)$-dimensional standard normal distribution, $N(0, I)$, namely

$$g(\phi, \theta_1, \dots, \theta_{p-2}) = \frac{1}{(2\pi)^{(p-1)/2}} \exp\left\{ -\frac{\upsilon^2 \sin^2 \theta_{p-2}}{2(1 + \cos \theta_{p-2})^2} \right\} |J|.$$

Formally, $p \geq 3$. However, if we set $p = 2$, we obtain the normal distribution on the circle with pdf

$$g(\theta_0) = \frac{\upsilon}{(2\pi)^{1/2}(1 + \cos \theta_0)} \exp\left\{ -\frac{\upsilon^2 \sin^2 \theta_0}{2(1 + \cos \theta_0)^2} \right\}.$$

Replacing θ_0 by θ, this pdf is equivalent to that in (4.48), which is the pdf of the circular normal induced by the modified inverse stereographic projection.

An alternative form for (8.18) is

$$g(\phi, \theta_1, \dots, \theta_{p-2})$$

$$= \frac{\Gamma((n+p-1)/2)\upsilon^{p-1}}{\Gamma(n/2)(\pi n)^{(p-1)/2}(1 + u_0^2)^{(n+p-1)/2}} \frac{(1 + \cos \theta_{p-2})^{(n-(p-1))/2}}{(1 - w \cos \theta_{p-2})^{(n+p-1)/2}} \prod_{j=1}^{p-2} \sin^j \theta_j,$$

where $u_0 = \upsilon/\sqrt{n}$ and $w = (u_0^2 - 1)/(u_0^2 + 1)$. Multiplying the two cosine components appearing in this last pdf by $\tanh(\kappa_1 \psi)$ and $\tanh(\kappa_2 \psi)$, respectively, and adjusting the normalizing constant appropriately, we obtain the pdf in (8.17). Thus, (8.17) can indeed be considered to be the p-dimensional generalization of the pdf in (4.49).

8.4.10 Exit Distribution

The exit distribution (Durrett, 1984) Exit $_p(\boldsymbol{\eta})$ on S^{p-1} has pdf of the form

$$f(\boldsymbol{x}) = \frac{1}{\omega_p} \frac{1 - \| \boldsymbol{\eta} \|^2}{\| \boldsymbol{x} - \boldsymbol{\eta} \|^p}, \quad \boldsymbol{x} \in S^{p-1}, \quad (8.19)$$

where $\boldsymbol{\eta}$ is in the unit sphere, i.e. $\boldsymbol{\eta} \in \{\boldsymbol{\zeta} \in \mathbb{R}^p \mid \| \boldsymbol{\zeta} \| < 1\}$. Here $\| \cdot \|$ denotes the Euclidean norm. The exit distribution is unimodal and rotationally symmetric about $\boldsymbol{x} = \boldsymbol{\eta}/ \| \boldsymbol{\eta} \|$. The vector parameter $\boldsymbol{\eta}$ controls the concentration of the distribution. In fact, when $\boldsymbol{\eta} = \boldsymbol{0}$, (8.19) reduces to the uniform pdf, and as $\| \boldsymbol{\eta} \|$ goes to 1, Exit $_p(\boldsymbol{\eta})$ tends to a degenerate distribution concentrated at $\boldsymbol{x} = \boldsymbol{\eta}$.

The family $H'(\eta, \nu)$ in Section 4.7 arises as an exit distribution for Brownian motion in $\mathbb{R}^p$, provided that $p = 2\nu + 2$ is a positive integer not less than 2. A particle starts from the origin at time $t = 0$. A random vector $\boldsymbol{X} = (X_1, \ldots, X_p)'$ which represents the point at which the particle first hits the unit sphere is uniformly distributed over the sphere and X_1 is distributed as $H'(0, p/2 - 1)$. If the particle starts at the point $\boldsymbol{\eta} = (\eta, 0, \ldots, 0)'$ with $-1 < \eta < 1$, the particle will eventually hit the unit sphere at the point $\boldsymbol{X}^* = (X_1^*, \ldots, X_p^*)'$, and the distribution of $\boldsymbol{X}^*$ with respect to Lebesgue measure on the unit sphere is an exit distribution with pdf

$$g(\boldsymbol{x}^*) = \frac{1}{\omega_p} \frac{1 - \|\boldsymbol{\eta}\|^2}{\|\boldsymbol{x}^* - \boldsymbol{\eta}\|^p} = \frac{1}{\omega_p} \frac{1 - \eta^2}{(1 - 2\eta x_1^* + \eta^2)^{p/2}}, \quad \boldsymbol{x}^* \in S^{p-1}.$$

When $p = 2$, the distribution represents a wrapped Cauchy distribution on the circle. The marginal distribution of X_1^* is $H'(\eta, p/2 - 1)$.

Kato and Jones (2013) present an extension of the exit distribution. The pdf is given by

$$f(\boldsymbol{x}) = \frac{1}{\omega_p} \frac{(1 + \rho_1^2\rho_2^2 - 2\rho_1\rho_2\boldsymbol{\eta}_1'\boldsymbol{\eta}_2)^{p/2}}{1 - \rho_1^2\rho_2^2} \frac{1 - \rho_1^2}{\|\boldsymbol{x} - \rho_1\boldsymbol{\eta}_1\|^p} \frac{1 - \rho_2^2}{\|\boldsymbol{x} - \rho_2\boldsymbol{\eta}_2\|^p}, \quad \boldsymbol{x} \in S^{p-1}$$

for $\boldsymbol{\eta}_1, \boldsymbol{\eta}_2 \in S^{p-1}$ and $0 \le \rho_1, \rho_2 < 1$. When $\rho_2 = 0$, the distribution reduces to the exit distribution. Now, a Cauchy family on the unit sphere S^{p-1}

$$f(\boldsymbol{x}) \propto \frac{1}{(c^2 + \|\boldsymbol{x} - \boldsymbol{a}\|^2)^p}, \quad \boldsymbol{x} \in S^{p-1}$$

for $\boldsymbol{a} \in \mathbb{R}^p$ and $c > 0$ can be seen in Dunau and Sénateur (1988). Another type of wrapped Cauchy distribution on the unit sphere S^{p-1} is obtained by using a conformal transformation of the uniform pdf (Shimizu, 2013):

$$f(\boldsymbol{x}) = \frac{1}{\omega_p} \left(\frac{1 - \alpha^2}{1 + \alpha^2 - 2\alpha\boldsymbol{\omega}'\boldsymbol{x}} \right)^{p-1}, \quad \boldsymbol{x} \in S^{p-1}$$

for a p-dimensional real vector $\boldsymbol{a} \neq \boldsymbol{0}$, $\alpha = \tanh(\|\boldsymbol{a}\|)$ and $\boldsymbol{\omega} = \boldsymbol{a}/\|\boldsymbol{a}\|$. When $p = 2$, the distribution reduces to a wrapped Cauchy distribution on the circle.

8.4.11 A Family of Spherical Gradient Distributions

Sei (2013) gives a family of distributions in terms of cost-convex functions. Denote by $d(\boldsymbol{x}, \boldsymbol{y})$ the geodesic distance between $\boldsymbol{x}$ and $\boldsymbol{y}$ in S^{p-1}, which is $\cos^{-1}(\boldsymbol{x}'\boldsymbol{y})$ if Euclidean coordinates in $\mathbb{R}^p$ are used to express S^{p-1}, where the range of $\cos^{-1}(.)$ is $[0, \pi]$. The cost function is $c(\boldsymbol{x}, \boldsymbol{y}) = (1/2)\{d(\boldsymbol{x}, \boldsymbol{y})\}^2$ and the c-transform ϕ^c of a function $\phi : S^{p-1} \to \mathbb{R}$ is defined by

$$\phi^c(\boldsymbol{y}) = \sup_{\boldsymbol{x} \in S^{p-1}} \{-c(\boldsymbol{x}, \boldsymbol{y}) - \phi(\boldsymbol{x})\}.$$

The function ϕ is called cost-convex if $(\phi^c)^c = \phi$.

In the examples below, we consider combinations of rotationally symmetric functions $f(d(\boldsymbol{x},\boldsymbol{z}))$, where $f \in C^2([0,\pi])$, whose kth derivative is denoted by $f^{(k)}$. Let $\mathcal{F}$ be the set of functions f on $[0,\pi]$ which satisfy $f^{(1)}(0) = f^{(1)}(\pi) = 0$ and $f^{(2)}(r) > -1$ for almost all $r \in [0,\pi]$. Choosing n pairs $\{(f_j,\boldsymbol{z}_j)\}_{j=1}^n$ from $\mathcal{F} \times S^{p-1}$, we can define the spherical gradient model

$$\phi_\theta(\boldsymbol{x}) = \sum_{j=1}^{n} \theta_j f_j(d(\boldsymbol{x},\boldsymbol{z}_j)), \quad \theta = (\theta_1,\dots,\theta_n)' \in \Theta,$$

where Θ is a convex subset of $\mathbb{R}^n$ such that ϕ_θ belongs to the set of all wrapping potential functions, $W(S^{p-1})$.

An explicit formula for pdf is given in Sei (2013):

$$p(\boldsymbol{x}|\theta) = \{(\sin \parallel v_\theta \parallel)/ \parallel v_\theta \parallel\}^{p-1} \det\left(\boldsymbol{x}\boldsymbol{x}' + H_\theta + \sum_{j=1}^{n}\theta_j K_j\right),$$

$$v_\theta = -\sum_{j=1}^{n}\theta_j f_j^{(1)}(\alpha_j)\boldsymbol{e}_j,\ \alpha_j = \cos^{-1}(\boldsymbol{x}'\boldsymbol{z}_j),\ \boldsymbol{e}_j = \frac{\boldsymbol{z}_j - \boldsymbol{x}\cos\alpha_j}{\sin\alpha_j},$$

$$H_\theta = \boldsymbol{e}_\theta\boldsymbol{e}_\theta' + \frac{\alpha_\theta\cos\alpha_\theta}{\sin\alpha_\theta}(I - \boldsymbol{x}\boldsymbol{x}' - \boldsymbol{e}_\theta\boldsymbol{e}_\theta'),\ \boldsymbol{e}_\theta = \frac{v_\theta}{\parallel v_\theta \parallel},\ \alpha_\theta = \parallel v_\theta \parallel,$$

$$K_j = f_j^{(2)}(\alpha_j)\boldsymbol{e}_j\boldsymbol{e}_j' + \frac{f_j^{(1)}(\alpha_j)\cos\alpha_j}{\sin\alpha_j}(I - \boldsymbol{x}\boldsymbol{x}' - \boldsymbol{e}_j\boldsymbol{e}_j'),$$

with Euclidean coordinates in $\mathbb{R}^p$ to express S^{p-1}, where $\parallel\cdot\parallel$ denotes the Euclidean norm and I the identity matrix. Note that the above formula does not involve any special functions.

Examples of pairs $(f_j,\boldsymbol{z}_j)$ are given below.

Example 8.1 ***(Linear Potential)***
Let $f_j(\xi) = \cos\xi$ for all j. Then $\phi_\theta(\boldsymbol{x}) = \sum_{j=1}^n \theta_j \cos\{d(\boldsymbol{x},\boldsymbol{z}_j)\} = \sum_{j=1}^n \theta_j \boldsymbol{x}'\boldsymbol{z}_j$ is in $W(S^{p-1})$ as long as $\sum_{j=1}^n |\theta_j| \le 1$. A potential function $\phi_\mu(\boldsymbol{x}) = \boldsymbol{\mu}'\boldsymbol{x}$ is in $W(S^{p-1})$ if $\parallel \boldsymbol{\mu} \parallel \le 1$. We call ϕ_μ the linear potential and the resulting statistical model the linear-potential model, which is rotationally symmetric.

Example 8.2 ***(Quadratic Potential)***
Let $f_j(\xi) = \cos\xi$ for $j = 1,\dots,n_1$ and $f_j(\xi) = \cos(2\xi)/4$ for $j = n_1+1,\dots,n$. Then the potential is

$$\phi_\theta(\boldsymbol{x}) = \sum_{j=1}^{n_1}\theta_j\boldsymbol{x}'\boldsymbol{z}_j + \sum_{j=n_1+1}^{n}\frac{\theta_j}{4}\{2(\boldsymbol{x}'\boldsymbol{z}_j)^2 - 1\}.$$

Suppose that $\boldsymbol{\mu} \in \mathbb{R}^p$ and A is a $p \times p$ symmetric matrix. Denote by $|A|_1$ the trace norm of A defined by the sum of absolute eigenvalues of A. Then a potential function $\phi_{\mu,A}(\boldsymbol{x}) = \boldsymbol{x}'\boldsymbol{\mu} + \boldsymbol{x}'A\boldsymbol{x}/2$ is in $W(S^{p-1})$ if $(\boldsymbol{\mu}, A)$ satisfies $\| \boldsymbol{\mu} \| + |A|_1 \leq 1$. The model is called the quadratic-potential model.

8.5 Distributions on the Complex Sphere

8.5.1 Bingham Distribution

The Bingham distribution on the $(k-1)$-dimensional complex unit sphere $\mathbb{C}S^{k-1} = \{\boldsymbol{z} = (z_1, \dots, z_k)' \in \mathbb{C}^k \mid \boldsymbol{z}^*\boldsymbol{z} = 1\}$, denoted $\mathbb{C}B_{k-1}(A)$, has pdf (Kent, 1994; Mardia and Jupp, 2000)

$$f(\boldsymbol{z}) = c(A)^{-1} \exp(\boldsymbol{z}^* A \boldsymbol{z}), \quad \boldsymbol{z} \in \mathbb{C}S^{k-1},$$

where $\boldsymbol{z}^* = \overline{\boldsymbol{z}'}$ is the complex conjugate of the transpose of $\boldsymbol{z}$, A denotes a $k \times k$ Hermitian matrix ($A^* = A$), and $c(A)$ is a normalizing constant. By splitting $\boldsymbol{z}$ into its real and imaginary parts, one can regard the complex Bingham distribution in $\mathbb{C}^k$ as a special case of the Bingham distribution in $\mathbb{R}^{2k}$. The Bingham distribution on the complex sphere is invariant under scalar rotation.

Since $\boldsymbol{z}^*\boldsymbol{z} = 1$ for $\boldsymbol{z} \in \mathbb{C}S^{k-1}$, the parameter matrices A and $A + \alpha I$ for all $\alpha \in \mathbb{C}$ define the same complex Bingham distribution $\mathbb{C}B_{k-1}(A)$ with $c(A + \alpha I) = c(A)\exp(\alpha)$. To remove this non-identifiability, we adopt the convention $\lambda_1 < \cdots < \lambda_k = 0$, where $\lambda_1, \dots, \lambda_k$ denote the eigenvalues of A, assumed distinct for simplicity. Under this convention we evaluate the normalizing constant as follows (Kent, 1994).

Since A is Hermitian, there exists a $k \times k$ unitary matrix U ($U^*U = I$) such that $A = U\Lambda U^*$, where $\Lambda = \text{diag}(\lambda_1, \dots, \lambda_k)$ is a diagonal matrix whose diagonal elements λ_j are the corresponding eigenvalues of A. Let $\boldsymbol{w} = (w_1, \dots, w_k)' = U^*\boldsymbol{z}$ ($\boldsymbol{z} = U\boldsymbol{w}$). Then, since

$$\boldsymbol{z}^* A \boldsymbol{z} = (U\boldsymbol{w})^* A (U\boldsymbol{w}) = \boldsymbol{w}^* U^* A U \boldsymbol{w} = \boldsymbol{w}^* \Lambda \boldsymbol{w}$$

and

$$\left|\left|\frac{\partial \boldsymbol{z}}{\partial \boldsymbol{w}}\right|\right| = ||U|| = |\pm 1| = 1,$$

it holds that

$$f(\boldsymbol{z})d\boldsymbol{z} = c(A)^{-1} \exp(\boldsymbol{z}^* A \boldsymbol{z}) d\boldsymbol{z} = c(\Lambda)^{-1} \exp(\boldsymbol{w}^* \Lambda \boldsymbol{w}) d\boldsymbol{w},$$

i.e. the normalizing constant $c(A) = c(\Lambda)$ depends only on the eigenvalues of A. Set $w_j = x_j + iy_j = t_j^{1/2} \exp(i\theta_j)$, where $t_j = |w_j|^2 \geq 0$ and $\theta_j = \arg(w_j)$ ($0 \leq \theta_j < 2\pi, j = 1, \dots, k$). Since

$$\frac{\partial(x_j, y_j)}{\partial(t_j, \theta_j)} = \begin{vmatrix} \frac{1}{2\sqrt{t_j}}\cos\theta_j & \frac{1}{2\sqrt{t_j}}\sin\theta_j \\ -\sqrt{t_j}\sin\theta_j & \sqrt{t_j}\cos\theta_j \end{vmatrix} = \frac{1}{2},$$

we have from $\partial(x_1, y_1, \dots, x_k, y_k)/\partial(t_1, \theta_1, \dots, t_k, \theta_k) = 1/2^k$ that $dx_1 dy_1 \cdots dx_k dy_k = 2^{-k} dt_1 d\theta_1 \cdots dt_k d\theta_k$. And setting $t_j = r^2 s_j \ (j = 1, \dots, k-1)$ and $t_k = r^2(1 - \sum_{j=1}^{k-1} s_j)$ leads to

$$\frac{\partial(t_1, \dots, t_{k-1}, t_k)}{\partial(s_1, \dots, s_{k-1}, r)} = \begin{vmatrix} r^2 & & & & -r^2 \\ & r^2 & & 0 & -r^2 \\ & & \ddots & & \vdots \\ & 0 & & r^2 & -r^2 \\ 2rs_1 & 2rs_2 & \cdots & 2rs_{k-1} & 2r(1 - \sum_{j=1}^{k-1} s_j) \end{vmatrix} = 2r^{2k-1}$$

and $dx_1 dy_1 \cdots dx_k dy_k = 2^{1-k} r^{2k-1} d\theta_1 \cdots d\theta_k ds_1 \cdots ds_{k-1} dr$. Since $\sum_{j=1}^{k} t_j = \sum_{j=1}^{k} |w_j|^2 = 1$, putting $r = 1$ yields $dx_1 dy_1 \cdots dx_k dy_k = 2^{1-k} d\theta_1 \cdots d\theta_k ds_1 \cdots ds_{k-1}$. Hence, $c(\Lambda)$ is represented from $\boldsymbol{w}^* \Lambda \boldsymbol{w} = \sum_{j=1}^{k} \lambda_j |w_j|^2 = \sum_{j=1}^{k} \lambda_j s_j$ as

$$\begin{aligned} c(\Lambda) &= \int_{\mathbb{C}S^{k-1}} \exp(\boldsymbol{w}^* \Lambda \boldsymbol{w}) d\boldsymbol{w} \\ &= \int_{S_{k-1}} \int_{[0,2\pi]^k} \exp\left(\sum_{j=1}^{k} \lambda_j s_j\right) 2^{1-k} d\theta_1 \cdots d\theta_k ds_1 \cdots ds_{k-1} \\ &= 2\pi^k \int_{S_{k-1}} \exp\left(\sum_{j=1}^{k} \lambda_j s_j\right) ds_1 \cdots ds_{k-1}, \end{aligned}$$

where $S_{k-1} = \left\{ (s_1, \dots, s_{k-1})' \mid s_j \geq 0 \text{ and } \sum_{j=1}^{k-1} s_j \leq 1 \right\}$ $(s_k = 1 - \sum_{j=1}^{k-1} s_j)$ is the $(k-1)$-dimensional unit simplex in $\mathbb{R}^{k-1}$. By induction we have

$$\int_{\alpha S_{k-1}} \exp\left(\sum_{j=1}^{k} \lambda_j s_j\right) ds_1 \cdots ds_{k-1} = \sum_{j=1}^{k} b_j \exp(\alpha \lambda_j)$$

with notation $\alpha S_{k-1} = \left\{ (s_1, \dots, s_{k-1})' \mid s_j \geq 0 \text{ and } \sum_{j=1}^{k-1} s_j \leq \alpha \right\}$, $s_k = \alpha - \sum_{j=1}^{k-1} s_j$ and $b_j^{-1} = \prod_{\ell=1, \ell \neq j}^{k} (\lambda_j - \lambda_\ell)$. Thus, the normalizing constant of $\mathbb{C}B_{k-1}(A)$ is obtained as

$$c(A) = 2\pi^k \sum_{j=1}^{k} b_j \exp(\lambda_j).$$

Simulation for the complex Bingham distribution is studied by Kent et al. (2004), and application to shape analysis is illustrated in Kent (1994).

8.5.2 *t*-Distribution

Similar to circular and spherical Pearson type VII distributions, we obtain the Pearson type VII distribution on the complex sphere (Shimizu et al., 2008) as a scale mixture of complex Bingham distributions:

$$f(z) = \tilde{c}(A)^{-1} \int_0^\infty \frac{1}{\sigma^k} \exp\left(\frac{1}{\sigma^2} z^* A z\right) dG(\sigma), \quad z \in \mathbb{C}S^{k-1}.$$

Here the weight function $G(\sigma)$ is

$$\frac{d}{d\sigma} G(\sigma) = \frac{2^{1-\nu} a^{2\nu}}{\Gamma(\nu)} \sigma^{-1-2\nu} \exp\left(-\frac{a^2}{2\sigma^2}\right)$$

for $\sigma > 0$, $a \neq 0$ and $\nu > 0$. Similar to the case of the Bingham distribution on the complex sphere, the normalizing constant $\tilde{c}(A)$ is given by

$$\tilde{c}(A) = 2\pi^k \left(\frac{a^2}{2}\right)^{k/2-1} \frac{\Gamma(\nu - k/2 + 1)}{\Gamma(\nu)} \sum_{j=1}^k b_j \left(1 - \frac{2}{a^2}\lambda_j\right)^{-(\nu - k/2 + 1)},$$
$$\nu - k/2 + 1 > 0$$

with $b_j^{-1} = \prod_{\ell=1,\ell\neq j}^k (\lambda_j - \lambda_\ell)$, and $f(z)$ is obtained as

$$\begin{aligned} f(z) &= \tilde{c}(A)^{-1} \left(\frac{a^2}{2}\right)^{-k/2} \frac{\Gamma(\nu + k/2)}{\Gamma(\nu)} \left(1 - \frac{2}{a^2} z^* A z\right)^{-(\nu+k/2)} \\ &= \left\{ 2\pi^k \left(\frac{a^2}{2}\right)^{k-1} \frac{\Gamma(\nu - k/2 + 1)}{\Gamma(\nu + k/2)} \sum_{j=1}^k b_j \left(1 - \frac{2}{a^2}\lambda_j\right)^{-(\nu - k/2 + 1)} \right\}^{-1} \\ &\quad \times \left(1 - \frac{2}{a^2} z^* A z\right)^{-(\nu+k/2)}. \end{aligned}$$

The Pearson type VII distribution on the complex sphere with $\nu = n/2$ and $a^2 = n$ is the t-distribution on the complex sphere with n degrees of freedom. When $n \to \infty$, it is shown from Stirling's formula (B.3) that the t-distribution on the complex sphere with n degrees of freedom converges to the Bingham distribution on the complex sphere $\mathbb{C}B_{k-1}(A)$.

8.6 Statistical Inference

Some useful and fairly recent results from statistical inference for spherical distributions will be presented now. For earlier and more recent results, the reader is referred to the books on statistical inference for directional data, by e.g. Fisher et al. (1993), Mardia and Jupp (2000) and Ley and Verdebout (2017).

Beran (1968, 1979) obtained optimal invariant tests of uniformity on a compact homogeneous space, which may be adapted for the sphere. Results on inference related to distributions have mostly appeared, due to its elegant form and ease in analysis, for Fisher or von Mises–Fisher distribution. The MLEs of its parameters are available in closed form. Interestingly, unlike the multivariate normal distribution (MND) on $\mathbb{R}^3$, the simultaneous MLE of the mean direction parameter vectors, $\boldsymbol{\mu}_j\,(j = 1, \ldots, k)$, of several independent Fisher distributions is admissible (SenGupta and Maitra, 1998). This result has been extended by Kanika et al. (2015) to study the best equivariant property under the similar setting. Hayakawa (1990) studied the behavior of power for several large sample tests for $\boldsymbol{\mu}$. Exact optimal tests for $\boldsymbol{\mu}$ are also available. SenGupta and Jammalamadaka (1991) derived multiparameter locally optimal tests, e.g. best first and second directional derivative and locally most mean powerful (LMMPU) (SenGupta and Vermeire, 1986) tests for $\boldsymbol{\mu}$ when κ is known, and proposed C_α-type tests for the case when κ is unknown. They (SenGupta and Jammalamadaka, 1998) have also derived predictive densities for the Fisher distribution and a generalization of it. Analogous to MANOVA with MND, tests have been proposed for testing homogeneity of mean direction vectors, analysis of spherical mean directions (ANOSMED), of several independent Fisher distributions. However, absence of orthogonality of the between and within sums of squares matrices has constrained its development under large sample approximations and that too has produced tests with small powers. This shortcoming has been overcome recently through an Integrated Likelihood ratio test (ILRT) obtained by Kulkarni and SenGupta (2022), They also established the robustness of their test against the family of FB_6 distributions. Bagchi and Kadane (2012) have studied in detail Laplace and Tierney–Kadane approximations for large sample approximations to posterior moments and marginal distributions related to distributions on the sphere (and circle and cylinder). Figueiredo (2007) has compared several tests of uniformity defined on the hyper-sphere. Bagchi and Guttman (1988) have derived theorems for the multivariate von Mises–Fisher (or the Langevin distribution) parallel to the well-known theorems of Lindley and Smith (1972).

Inference procedures for the high-dimensional case, i.e. when $p \to \infty$, are becoming increasingly important. Watson (1988) argued that in this case, to get useful results one must let $\kappa \to \infty$ at the rate $q^{1/2}$, i.e. enhance the distribution $M_p(\boldsymbol{\mu}, \kappa p^{1/2})$. In this case, MLEs of $\boldsymbol{\mu}$ and κ are obtained in very simple forms. A test for $\boldsymbol{\mu}$ can be obtained based on a simple test statistic with an asymptotic Chi-squared null distribution which is, surprisingly, free of κ. The application of this distribution to obtain approximate permutation distributions of the sample product–moment correlation and multiple correlation coefficients has also been demonstrated in Watson (1988).

8.7 Applications

The spherically projected multivariate linear (SPML) model is introduced by Presnell et al. (1998). The SPML model assumes the representation $\boldsymbol{u}_j = \boldsymbol{y}_j / \parallel \boldsymbol{y}_j \parallel$ for the unit vectors $\boldsymbol{u}_j$, where the random vectors $\boldsymbol{y}_1, \dots, \boldsymbol{y}_n$ are independently distributed as a normal with common variance-covariance matrix and mean vectors $\boldsymbol{\mu}_j = B'\boldsymbol{x}_j$ with a parameter matrix B and covariates $\boldsymbol{x}_1, \dots, \boldsymbol{x}_n$. Examples in the paper by Presnell et al. (1998) include movements of blue periwinkles cited in Fisher et al.'s (1993) book and butterfly migrations.

Spherical distributions play important roles in a variety of problems in applied research, e.g. especially in astrophysics, environmental sciences and geophysics. The axial distribution obtained as a marginal from the von Mises–Fisher distribution in Section 8.4.3 has been used to detect attraction of CD8 T cells to an infection site in the study of malaria vaccine candidates which induce large numbers of these malaria-specific cells (Zenkov et al., 2022).

Bimodal or multi-modal data on S^{p-1} may be modeled by using a mixture of von Mises–Fisher distributions. Another application of fitting the mixture distribution is to find clusters of points on the sphere. Banerjee et al. (2005) derive two clustering algorithms based on the Expectation Maximization framework for estimating the mean direction and concentration parameters of a mixture of the von Mises–Fisher distributions. Noting that the von Mises–Fisher distribution is constrained to modeling symmetric clusters, Kim and SenGupta (2018) enhanced their two generalizations of AGvM distribution (see Section 8.3.3) for modeling clusters with possibly asymmetric, bimodal within-cluster data.

As an application, Wang et al. (2013) used latitude, longitude, and magnitude data on foreshocks during 72 hours before the earthquake of the Pacific coast of Tohoku, Japan, which occurred with a magnitude 9.0R (Richter scale) (MW) at 14:46 JST (05:46 UTC) on 11 March, 2011, and having the epicenter 38.30° for latitude and 142.37° for longitude, approximately 70 km east of the coast. The method proposed by Banerjee et al. (2005) was used to find clusters of earthquakes which occurred near Japan. In the period studied, earthquakes greater than or equal to 4R in the entire world occurred 127 times, i.e. 90 times in [4R, 5R), 35 times in [5R, 7R), and twice for greater than or equal to 7R. In the area near Japan, the numbers for these categories were 37, 25, and 2 respectively, totaling to 64, which was about half of 127. Kulkarni and SenGupta (2022) conducted ILRT for ANOSMED on Vectorcardiogram and Compositional spherical data sets using Fisher distributions. The same approach was taken by Faizan et al. (2023) to study thalassemia patients over different demographical classes.

Scealy and Welsh (2011) discussed the possibility of analyzing compositional data using distributions on the hyper-sphere and used the Kent distribution after transforming compositional data to directional data by the square root transformation. The mean direction of the Kent distribution is regressed as a function of a vector of covariates. The paper applied the Kent distribution to model compositions of foraminifer, a marine micro-organism (Aitchison, 1986). See also Scealy and Welsh (2014) for compositional data with small concentration. A directional mixed effects model on the Kent distribution is proposed by Scealy and Welsh (2017). The paper models compositional data from the 2009–2010 Australian Household Expenditure Survey Confidentialized Unit Record File.

It is shown in Nakayama et al. (2011) that the normalizing constant of the Fisher–Bingham distribution is a holonomic function. This fact is exploited to get maximum likelihood estimates for a sample from the Fisher–Bingham distribution. Nakayama et al. (2011) fitted the Fisher–Bingham distribution to astronomical and magnetism data sets. Sei and Kume (2015) implemented the holonomic gradient method to compute the normalizing constant of the Bingham distribution, and recently Kume and Sei (2018) (see also references therein) studied the maximum likelihood inference of the Fisher–Bingham distribution using an adjusted holonomic gradient method with revisit to the dataset in Nakayama et al. (2011).

In the paper by Sei (2013), high-frequency potential as a generalization of the examples in Section 8.4.11 is provided, and an illustration is given for a real data set of the locations of 188 stars of magnitude brighter than or equal to 3.0. The quadratic model is selected as an optimal model in the sense of minimum AIC between the quadratic model and the null model (isotropic or uniform distribution).

Data on high-dimensional spheres with novel applications are rapidly emerging in a variety of scientific disciplines such as Gait analysis and text mining – see also Banerjee et al. (2005). Watson (1988) gives a possible application of his high-dimensional Langevin distribution for the analysis of compositional data by considering a scenario where the analysis of a rock specimen yields the proportions of p substances it is composed of.

References

Abe, T., Shimizu, K. and Pewsey, A. (2010). Symmetric unimodal models for directional data motivated by inverse stereographic projection, *Journal of the Japan Statistical Society*, 40, 45–61.

Abramowitz, M. and Stegun, I. A. Eds. (1972). *Handbook of Mathematical Functions with Formulas, Graphs, and Mathematical Tables*, Dover, New York.

Aitchison, J. (1986). *The Statistical Analysis of Compositional Data*, Chapman and Hall, London.

Bagchi, P. and Guttman, I. (1988). Spuriosity and outliers in directional data, *Journal of Applied Statistics*, 17, 341–350.

Bagchi, P. and Kadane, J. B. (1991). Laplace approximations to posterior moments and marginal distributions on circles, spheres, and cylinders, *The Canadian Journal of Statistics*, 19, 67–77. Erratum: *The Canadian Journal of Statistics*, 2012, 40, 770–771.

Banerjee, A., Dhillon, I. S., Ghosh, J. and Sra, S. (2005). Clustering on the unit hypersphere using von Mises–Fisher distributions, *Journal of Machine Learning Research*, 6, 1345–1382.

Beran, R. J. (1968). Testing for uniformity on a compact homogeneous space, *Journal of Applied Probability*, 5, 177–195.

Beran, R. J. (1979). Exponential models for directional data, *The Annals of Statistics*, 7, 1162–1178.

Chikuse, Y. (2003). *Statistics on Special Manifolds*, Lecture Notes in Statistics 174, Springer, New York.

Dunau, J. L. and Sénateur, H. (1988). Une caracterisation du type de la loi de Cauchy-conforme sur $\mathbb{R}^n$, *Probability Theory and Related Fields*, 77, 129–135.

Durrett, R. (1984). *Brownian Motion and Martingales in Analysis*, Wadsworth, Belmont, CA.

Faizan, M., Rashid, N., Hussain, S., Khan, A., Khan, J., Zeb, J. and Ahmed, S. Q. (2023). Prevalence and clinical features of thalassemia minor cases, *Pakistan Journal of Medical and Health Sciences*, 17, 456–459.

Fang, K. T., Kotz, S. and Ng, K. W. (1990). *Symmetric Multivariate and Related Distributions*, Chapman and Hall, London.

Figueiredo, A. (2007). Comparison of tests of uniformity defined on the hypersphere, *Statistics & Probability Letters*, 77, 329–334.

Fisher, N. I., Lewis, T. and Embleton, B. J. J. (1993). *Statistical Analysis of Spherical Data*, Cambridge University Press, Cambridge.

Gupta, A. K. and Song, D. (1997). L_p-norm spherical distribution, *Journal of Statistical Planning and Inference*, 60, 241–260.

Hayakawa, T. (1990). On tests for the mean direction of the Langevin distribution, *Annals of the Institute of Statistical Mathematics*, 42, 359–373.

Jones, M. C. and Pewsey, A. (2005). A family of symmetric distributions on the circle, *Journal of the American Statistical Association*, 100, 1422–1428.

Kanika, Kumar, S. and SenGupta, A. (2015). A unified approach to decision-theoretic properties of the MLEs for the mean directions of several Langevin distributions, *Journal of Multivariate Analysis*, 133, 160–172.

Kato, S. and Jones, M. C. (2013). An extended family of circular distributions related to wrapped Cauchy distributions via Brownian motion, *Bernoulli*, 19, 154–171.

Kent, J. T. (1982). The Fisher-Bingham distribution on the sphere, *Journal of the Royal Statistical Society*, B44, 71–80.

Kent, J. T. (1994). The complex Bingham distribution and shape analysis, *Journal of the Royal Statistical Society*, B56, 285–299.

Kent, J. T., Constable, P. D. L. and Er, F. (2004). Simulation for the complex Bingham distribution, *Statistics and Computing*, 14, 53–57.

Kim, S. and SenGupta, A. (2013). A three-parameter generalized von Mises distribution, *Statistical Papers*, 54, 685–693.

Kim, S. and SenGupta, A. (2018). Clustering methods for spherical data: An overview and a new generalization, Chapter 11 in *Proceedings of the Pacific Rim Statistical Conference for Production Engineering - Big Data, Production Engineering and Statistics*, ICSA Book Series in Statistics, Springer, 155–164.

Kotz, S., Kozubowski, T. J. and Podgórski, K. (2001). *The Laplace Distribution and Generalizations: A Revisit with Applications to Communications, Economics, Engineering, and Finance*, Birkhäuser, Boston, MA.

Kulkarni, H. V. and SenGupta, A. (2022). An efficient test for homogeneity of mean directions on the hyper-sphere, *International Statistical Review*, 90, 41–61.

Kume, A. and Sei, T. (2018). On the exact maximum likelihood inference of Fisher–Bingham distributions using an adjusted holonomic gradient method, *Statistics and Computing*, 28, 835–847.

Langevin, M. P. (1905). Magnétisme et théorie des électrons, *Annales de Chimie et de Physique*, Ser8.T5, 70–271.

Ley, C. and Verdebout, T. (2017). *Modern Directional Statistics*, Chapman & Hall/CRC, Boca Raton, FL.

Lindley, D. V. and Smith, A. F. M. (1972). Bayes estimates for the linear model, *Journal of the Royal Statistical Society*, B34, 1–41.

Mardia, K. V. and Jupp, P. E. (2000). *Directional Statisitcs*, Wiley, Chichester.

Muirhead, R. J. (1982). *Aspects of Multivariate Statistical Theory*, Wiley, New York.

Nakayama, H., Nishiyama, K., Noro, M., Ohara, K., Sei, T., Takayama, N. and Takemura, A. (2011). Holonomic gradient descent and its application to the Fisher–Bingham integral, *Advances in Applied Mathematics*, 47, 639–658.

Presnell, B., Morrison, S. P. and Littell, R. C. (1998). Projected multivariate linear models for directional data, *Journal of the American Statistical Association*, 93, 1068–1077.

Scealy, J. L. and Welsh, A. H. (2011). Regression for compositional data by using distributions defined on the hypersphere, *Journal of the Royal Statistical Society*, B73, 351–375.

Scealy, J. L. and Welsh, A. H. (2014). Fitting Kent models to compositional data with small concentration, *Statistics and Computing*, 24, 165–179.

Scealy, J. L. and Welsh, A. H. (2017). A directional mixed effects model for compositional expenditure data, *Journal of the American Statistical Association*, 112, 24–36.

Sei, T. (2013). A Jacobian inequality for gradient maps on the sphere and its application to directional statistics, *Communications in Statistics - Theory and Methods*, 42, 2525–2542.

Sei, T. and Kume, A. (2015). Calculating the normalising constant of the Bingham distribution on the sphere using the holonomic gradient method, *Statistics and Computing*, 25, 321–332.

SenGupta, A. (2004). On the constructions of probability distributions for directional data, *Bulletin of Calcutta Mathematical Society*, 96, 139–154.

SenGupta, A. and Jammalamadaka, S. R. (1991). On locally optimal tests for the mean direction of the Langevin distribution, *Statistics & Probability Letters*, 12, 537–544.

SenGupta, A. and Jammalamadaka, S. R. (1998). Predictive inference for directional data, *Statistics & Probability Letters*, 40, 247–257.

SenGupta, A. and Kim, S. (2016). Statistical inference for homologous gene pairs between two circular genomes: A new circular-circular regression model, *Statistical Methods & Applications*, 25, 421–432.

SenGupta, A. and Maitra, R. (1998). On best equivariance and admissibility of simultaneous MLE for mean direction vectors of several Langevin distributions, *Annals of the Institute of Statistical Mathematics*, 50, 715–727.

SenGupta, A. and Vermeire, L. (1986). Locally optimal tests for multiparameter hypotheses, *Journal of the American Statistical Association*, 81, 819–825.

Shimizu, K. (2013). Transforming the von Mises–Fisher distribution via generalized Möbius transformation, *International Conference on Statistical Distributions and Applications*, Michigan.

Shimizu, K. and Iida, K. (2002). Pearson type VII distributions on spheres, *Communications in Statistics - Theory and Methods*, 31, 513–526.

Shimizu, K., Abe, T. and Kato, S. (2008). *t*-distributions on various manifolds, In *Multivariate Statistical Methods: Theory and Applications*, SenGupta, A. ed., Macmillan, New Delhi, pp. 56–75.

Siew, H. Y. and Shimizu, K. (2008). The generalized symmetric Laplace distribution on the sphere, *Statistical Methodology*, 5, 487–501.

Szabłowski, P. J. (1998). Uniform distributions on spheres in finite dimensional L_α and their generalizations, *Journal of Multivariate Analysis*, 64, 103–117.

Wang, M. Z., Shimizu, K. and Uesu, K. (2013). An analysis of earthquakes latitude, longitude and magnitude data by use of directional statistics, *Japanese Journal of Applied Statistics*, 42, 29–44. In Japanese.

Watson, G. S. (1983). *Statistics on Spheres*, Wiley, New York.

Watson, G. S. (1988). The Langevin distribution on high dimensional spheres, *Journal of Applied Statistics*, 15, 123–130. (Also freely available in *50 Years of Journal of Applied Statistics: Directional Statistics: Advances in Theory and Emerging Applications*, Edited by Ashis SenGupta, 2024.)

Wood, A. T. A. (1982). A bimodal distribution on the sphere, *Applied Statistics*, 31, 52–58.

Wood, A. T. A. (1987). The simulation of spherical distributions in the Fisher–Bingham family, *Communications in Statistics - Simulation and Computation*, 16, 885–898.

Yfantis, E. A. and Borgman, L. E. (1982). An extension of the von Mises distribution, *Communications in Statistics - Theory and Methods*, 11, 1695–1706.

Zenkov, V. S., O'Connor, J. H., Cockburn, I. A. and Ganusov, V. V. (2022). A new method based on the von Mises-Fisher distribution shows that a minority of liver-localized CD8 T cells display hard-to-detect attraction to Plasmodium-infected hepatocytes, *Frontiers in Bioinformatics*, 1, Article 770448.

9

Distributions on the (Ring) Torus and Hyper-Torus

As an extension of circular distributions, we deal with distributions of bivariate and, more generally, multivariate circular variables in this chapter. Such distributions are conventionally called distributions on the torus and hyper-torus. The reason will be explained soon in Section 9.1.1. After presenting some preliminary notions and general methods of construction, we introduce several toroidal distributions.

9.1 Preliminary Notions

9.1.1 Definitions and Representations

A distribution on the torus or bivariate circular distribution means the joint distribution of two circular random variables Θ_1 and Θ_2 such as two wind directions measured at two different locations at the same time or measured at two points in time, for example, at 6 a.m. and 12 noon, at a monitoring site. The reason why this terminology is used is that a pair of realization $(\theta_1, \theta_2)'$ can be identified with a point on the torus surface as shown in Figure 9.1.

The pdf $f(\theta_1, \theta_2)$ of a continuous random vector $(\Theta_1, \Theta_2)'$ on the torus satisfies the following properties:

a) $f(\theta_1, \theta_2) \geq 0$ for all $(\theta_1, \theta_2)' \in [0, 2\pi)^2$,

b) $\displaystyle\int_0^{2\pi} \int_0^{2\pi} f(\theta_1, \theta_2) d\theta_1 d\theta_2 = 1$,

c) $f(\theta_1, \theta_2) = f(\theta_1 + 2k_1\pi, \theta_2 + 2k_2\pi)$ on $\mathbb{R}^2$ for any integers k_1 and k_2.

Conversely, any function f satisfying the above conditions (a), (b), and (c) simultaneously is a toroidal pdf. Sometimes the support $[0, 2\pi)^2$ may be replaced by $[-\pi, \pi)^2$. The distribution function is defined by

$$F(\theta_1, \theta_2) = \int_0^{\theta_1} \int_0^{\theta_2} f(t_1, t_2) dt_1 dt_2, \quad (\theta_1, \theta_2)' \in [0, 2\pi)^2.$$

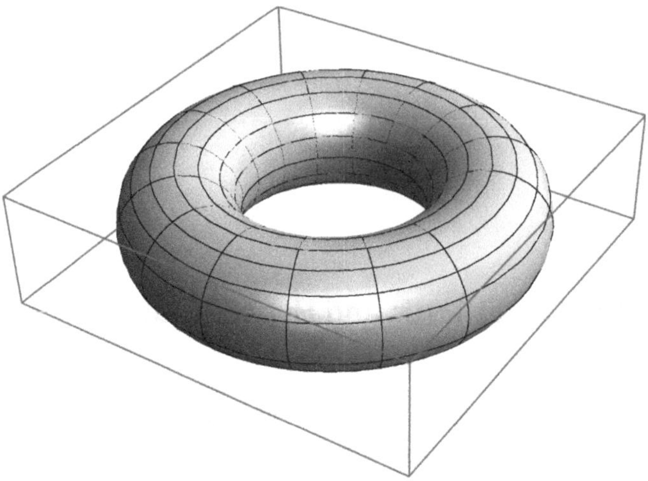

Figure 9.1 Torus surface.

The marginal distribution of Θ_1 has density

$$f_{\Theta_1}(\theta_1) = \int_0^{2\pi} f(\theta_1, \theta_2) d\theta_2$$

and the conditional density of Θ_2 given $\Theta_1 = \theta_1$ is

$$f_{\Theta_2|\Theta_1}(\theta_2|\theta_1) = \frac{f(\theta_1, \theta_2)}{f_{\Theta_1}(\theta_1)}$$

provided that $f_{\Theta_1}(\theta_1) \neq 0$.

Planar contour plots or 3D plots are often used to display toroidal pdfs. Figure 9.2 shows an example of these plots of a toroidal pdf. Note that the four corners, the

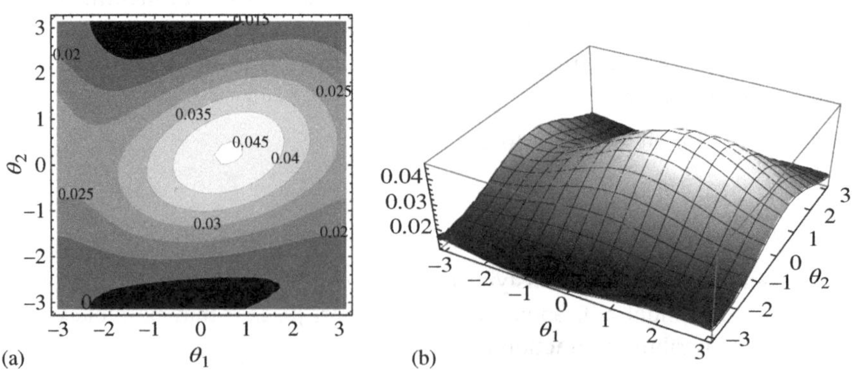

Figure 9.2 Plots of a toroidal density function $f(\theta_1, \theta_2)$: (a) contour plot and (b) 3D plot.

upper and lower left corners and the upper and lower right corners, in Figure 9.2a represent an identical point.

9.1.2 Characteristic Function and Bivariate Fourier Series

For a random vector $(\Theta_1, \Theta_2)'$ on the torus with pdf $f(\theta_1, \theta_2)$, the characteristic function (c.f.) or (p, q)th (joint) trigonometric moment is defined by

$$\phi_{p,q} = E\left\{e^{i(p,q)(\Theta_1,\Theta_2)'}\right\} = \int_0^{2\pi}\int_0^{2\pi} e^{i(p\theta_1+q\theta_2)} f(\theta_1, \theta_2) d\theta_1 d\theta_2$$

with $i = \sqrt{-1}$. The bivariate or double Fourier series of $f(\theta_1, \theta_2)$ is represented by

$$\begin{aligned} f(\theta_1, \theta_2) = \sum_{q=0}^{\infty}\sum_{p=0}^{\infty} \varepsilon_{p,q} \{ & \alpha_{p,q} \cos(p\theta_1)\cos(q\theta_2) + \beta_{p,q} \sin(p\theta_1)\sin(q\theta_2) \\ & + \gamma_{p,q} \cos(p\theta_1)\sin(q\theta_2) + \delta_{p,q} \sin(p\theta_1)\cos(q\theta_2)\}, \end{aligned}$$

where

$$\varepsilon_{0,0} = \frac{1}{4\pi^2},\ \varepsilon_{p,0} = \frac{1}{2\pi^2}\ (p > 0),\ \varepsilon_{0,q} = \frac{1}{2\pi^2}\ (q > 0),\ \varepsilon_{p,q} = \frac{1}{\pi^2}\ (p, q > 0).$$

The coefficient $\alpha_{p,q}$ is given by

$$\alpha_{p,q} = E\{\cos(p\Theta_1)\cos(q\Theta_2)\} = \int_0^{2\pi}\int_0^{2\pi} \cos(p\theta_1)\cos(q\theta_2) f(\theta_1, \theta_2) d\theta_1 d\theta_2,$$

which corresponds to the (p, q)th cosine–cosine joint moment of $(\Theta_1, \Theta_2)'$. The other (p, q)th sine–sine, cosine–sine and sine–cosine trigonometric moments are similarly defined:

$$\beta_{p,q} = E\{\sin(p\Theta_1)\sin(q\Theta_2)\} = \int_0^{2\pi}\int_0^{2\pi} \sin(p\theta_1)\sin(q\theta_2) f(\theta_1, \theta_2) d\theta_1 d\theta_2,$$

$$\gamma_{p,q} = E\{\cos(p\Theta_1)\sin(q\Theta_2)\} = \int_0^{2\pi}\int_0^{2\pi} \cos(p\theta_1)\sin(q\theta_2) f(\theta_1, \theta_2) d\theta_1 d\theta_2,$$

and

$$\delta_{p,q} = E\{\sin(p\Theta_1)\cos(q\Theta_2)\} = \int_0^{2\pi}\int_0^{2\pi} \sin(p\theta_1)\cos(q\theta_2) f(\theta_1, \theta_2) d\theta_1 d\theta_2.$$

It holds that

$$\alpha_{p,q} = \alpha_{-p,q} = \alpha_{p,-q} = \alpha_{-p,-q}, \quad \beta_{p,q} = -\beta_{-p,q} = -\beta_{p,-q} = \beta_{-p,-q},$$

$$\gamma_{p,q} = \gamma_{-p,q} = -\gamma_{p,-q} = -\gamma_{-p,-q} \quad \text{and} \quad \delta_{p,q} = -\delta_{-p,q} = \delta_{p,-q} = -\delta_{-p,-q}.$$

The c.f. $\psi(\boldsymbol{p})$ of a wrapped symmetric bivariate distribution is obtained from the c.f. $\psi(\boldsymbol{t})$ of the corresponding linear bivariate distribution by replacing $\boldsymbol{t}$ by $\boldsymbol{p}$. Now, assuming the location parameters are zeroes, we have

$$\psi(\boldsymbol{p}) = E[\exp(i\boldsymbol{p}'\boldsymbol{\theta})] = E[\cos(\boldsymbol{p}'\boldsymbol{\theta})] = E[(\cos(p_1\theta_1)\cos(p_2\theta_2) \\ - \sin(p_1\theta_1)\sin(p_2\theta_2)],$$

the second equality following by symmetry. Thus we need to only evaluate the two trigonometric product moments (the marginals follow from these as special cases) to represent the corresponding toroidal density explicitly in its Fourier series form as given above.

9.1.3 Distribution of the Sum

If a random vector $(\Theta_1, \Theta_2)'$ on the torus has joint pdf $f(\theta_1, \theta_2)$, then the pdf $h(\theta)$ of the sum $\Theta_1 + \Theta_2$ in the sense of mod 2π can be calculated by the convolution

$$h(\theta) = \int_0^{2\pi} f(\xi, \theta - \xi)d\xi.$$

If two circular random variables Θ_1 and Θ_2 are independent and have characteristic functions or trigonometric moments $\phi_p^{(1)}$ and $\phi_p^{(2)}$, respectively, then the c.f. ϕ_p of $\Theta = \Theta_1 + \Theta_2$ is

$$\phi_p = E(e^{ip\Theta}) = E(e^{ip\Theta_1})E(e^{ip\Theta_2}) = \phi_p^{(1)}\phi_p^{(2)}$$

and, from the one-to-one correspondence between the distribution and the characteristic function, a distribution with ϕ_p is the distribution of Θ.

9.1.4 Circular–Circular Correlation

The embedding approach leads to a circular–circular correlation (Jupp and Mardia, 1980; Mardia and Jupp, 2000, p. 248). Let $(\Theta, \Psi)'$ be a random vector on the torus. We consider Pearson's product moment correlation coefficient between two linear combinations $\boldsymbol{a}'\boldsymbol{U}$ and $\boldsymbol{b}'\boldsymbol{V}$, where $\boldsymbol{a}$ and $\boldsymbol{b}$ are real constant vectors, $\boldsymbol{U} = (\cos\Theta, \sin\Theta)'$ and $\boldsymbol{V} = (\cos\Psi, \sin\Psi)'$. Then

$$\mathrm{Corr}(\boldsymbol{a}'\boldsymbol{U}, \boldsymbol{b}'\boldsymbol{V}) = \frac{\mathrm{Cov}(\boldsymbol{a}'\boldsymbol{U}, \boldsymbol{b}'\boldsymbol{V})}{\sqrt{\mathrm{Var}(\boldsymbol{a}'\boldsymbol{U})\mathrm{Var}(\boldsymbol{b}'\boldsymbol{V})}} = \frac{\boldsymbol{a}'\Sigma_{12}\boldsymbol{b}}{\sqrt{(\boldsymbol{a}'\Sigma_{11}\boldsymbol{a})(\boldsymbol{b}'\Sigma_{22}\boldsymbol{b})}},$$

where

$$\Sigma_{11} = \begin{pmatrix} \mathrm{Var}(\cos\Theta) & \mathrm{Cov}(\cos\Theta, \sin\Theta) \\ \mathrm{Cov}(\sin\Theta, \cos\Theta) & \mathrm{Var}(\sin\Theta) \end{pmatrix},$$

$$\Sigma_{22} = \begin{pmatrix} \mathrm{Var}(\cos\Psi) & \mathrm{Cov}(\cos\Psi, \sin\Psi) \\ \mathrm{Cov}(\sin\Psi, \cos\Psi) & \mathrm{Var}(\sin\Psi) \end{pmatrix},$$

and

$$\Sigma_{12} = \begin{pmatrix} \mathrm{Cov}(\cos\Theta, \cos\Psi) & \mathrm{Cov}(\cos\Theta, \sin\Psi) \\ \mathrm{Cov}(\sin\Theta, \cos\Psi) & \mathrm{Cov}(\sin\Theta, \sin\Psi) \end{pmatrix} (= \Sigma'_{21}).$$

The canonical correlation coefficient between $\boldsymbol{U}$ and $\boldsymbol{V}$ can be obtained by maximizing the correlation coefficient $\mathrm{Corr}(\boldsymbol{a}'\boldsymbol{U}, \boldsymbol{b}'\boldsymbol{V})$ with respect to $\boldsymbol{a}$ and $\boldsymbol{b}$ subject to $\boldsymbol{a}'\Sigma_{11}\boldsymbol{a} = 1$ and $\boldsymbol{b}'\Sigma_{22}\boldsymbol{b} = 1$. Differentiating the Lagrangian function

$$L(\boldsymbol{a}, \boldsymbol{b}) = \boldsymbol{a}'\Sigma_{12}\boldsymbol{b} - \frac{\lambda}{2}(\boldsymbol{a}'\Sigma_{11}\boldsymbol{a} - 1) - \frac{\mu}{2}(\boldsymbol{b}'\Sigma_{22}\boldsymbol{b} - 1)$$

with respect to $\boldsymbol{a}$ and $\boldsymbol{b}$ and equating the derivatives to zero, we conclude that the Lagrange multipliers are $\lambda = \boldsymbol{a}'\Sigma_{12}\boldsymbol{b} = \boldsymbol{b}'\Sigma_{21}\boldsymbol{a} = \mu$ and λ^2 satisfies the determinantal equation

$$|\Sigma_{11}^{-1}\Sigma_{12}\Sigma_{22}^{-1}\Sigma_{21} - \lambda^2 I| = 0,$$

where I denotes the 2×2 identity matrix. Direct calculation leads to the fact that the sum of the squared canonical correlation coefficients is given by

$$\begin{aligned} r^2 &= \mathrm{tr}(\Sigma_{11}^{-1}\Sigma_{12}\Sigma_{22}^{-1}\Sigma_{21}) \\ &= \frac{1}{(1 - r_1^2)(1 - r_2^2)} [r_{cc}^2 + r_{cs}^2 + r_{sc}^2 + r_{ss}^2 \\ &\quad +2\{(r_{cc}r_{ss} + r_{cs}r_{sc})r_1 r_2 - (r_{cc}r_{cs} + r_{sc}r_{ss})r_2 - (r_{cc}r_{sc} + r_{cs}r_{ss})r_1\}], \end{aligned} \tag{9.1}$$

where $r_{cc} = \mathrm{Corr}(\cos\Theta, \cos\Psi)$, $r_{cs} = \mathrm{Corr}(\cos\Theta, \sin\Psi)$, $r_{sc} = \mathrm{Corr}(\sin\Theta, \cos\Psi)$, $r_{ss} = \mathrm{Corr}(\sin\Theta, \sin\Psi)$, $r_1 = \mathrm{Corr}(\cos\Theta, \sin\Theta)$, and $r_2 = \mathrm{Corr}(\cos\Psi, \sin\Psi)$ are correlation coefficients. The value r^2 can be used as a circular–circular correlation coefficient. Note that $0 \le r^2 \le 2$ since $0 \le \lambda^2 \le 1$, and one can use a correlation coefficient $r^2/2$, which is in the interval $[0, 1]$. The largest eigenvalue of $\Sigma_{11}^{-1}\Sigma_{12}\Sigma_{22}^{-1}\Sigma_{21}$, as a correlation coefficient (Johnson and Wehrly, 1977), is obviously in $[0, 1]$.

Fisher and Lee (1983) define a different circular–circular correlation. Suppose that $(\Theta_1, \Psi_1)'$ and $(\Theta_2, \Psi_2)'$ denote independent and identically distributed random vectors as $(\Theta, \Psi)'$. Then the circular–circular correlation between Θ and Ψ is defined by

$$\rho = \frac{E\{\sin(\Theta_1 - \Theta_2)\sin(\Psi_1 - \Psi_2)\}}{\sqrt{E\{\sin^2(\Theta_1 - \Theta_2)\}E\{\sin^2(\Psi_1 - \Psi_2)\}}}.$$

To estimate ρ from a random sample $(\theta_j, \psi_j)'$ for $j = 1, \ldots, n\ (n \ge 2)$, we consider a pair of two discrete circular random variables Θ and Ψ, $(\Theta, \Psi)'$, whose joint pmf is given by

$$P(\Theta = \theta_j, \Psi = \psi_j) = \frac{1}{n}, \quad j = 1, \ldots, n.$$

Then the circular–circular correlation is represented by

$$\hat{\rho} = \frac{\sum_{1\le k<j\le n} \sin(\theta_k - \theta_j)\sin(\psi_k - \psi_j)}{\left\{\sum_{1\le k<j\le n}\sin^2(\theta_k - \theta_j)\sum_{1\le k<j\le n}\sin^2(\psi_k - \psi_j)\right\}^{1/2}}. \tag{9.2}$$

This means that if we have a bivariate circular sample $(\theta_j, \psi_j)'$ of size n, $\hat{\rho}$ gives an estimate of ρ. It is obvious that $-1 \le \hat{\rho} \le 1$. A different expression (Fisher, 1993, p. 151) of (9.2) is

$$\hat{\rho} = \frac{4(AB - CD)}{\{(n^2 - E^2 - F^2)(n^2 - G^2 - H^2)\}^{1/2}},$$

where

$$A = \sum_{j=1}^{n} \cos\theta_j \cos\psi_j,\ B = \sum_{j=1}^{n} \sin\theta_j \sin\psi_j,\ C = \sum_{j=1}^{n} \cos\theta_j \sin\psi_j,$$

$$D = \sum_{j=1}^{n} \sin\theta_j \cos\psi_j,\ E = \sum_{j=1}^{n} \cos(2\theta_j),\ F = \sum_{j=1}^{n} \sin(2\theta_j),$$

$$G = \sum_{j=1}^{n} \cos(2\psi_j) \quad \text{and} \quad H = \sum_{j=1}^{n} \sin(2\psi_j).$$

Note that it is possible to write the numerator of (9.2) as

$$\sum_{1\le k<j\le n} \sin(\theta_k - \theta_j)\sin(\psi_k - \psi_j) = \sum_{1\le k<j\le n} \begin{vmatrix} \sin\theta_k & \sin\theta_j \\ \cos\theta_k & \cos\theta_j \end{vmatrix} \times \begin{vmatrix} \sin\psi_k & \sin\psi_j \\ \cos\psi_k & \cos\psi_j \end{vmatrix}.$$

It is easy to have a similar expression for the denominator of (9.2). Masuyama (1939) defines a correlation between "tensor quantities" in 3D Euclidean space:

$$R(\boldsymbol{A}; \boldsymbol{B}) = \frac{|[a_i b_j]|}{\sqrt{|[a_i a_j]| \times |[b_i b_j]|}}, \tag{9.3}$$

where $\boldsymbol{A} = (\boldsymbol{A}_1, \boldsymbol{A}_2, \dots, \boldsymbol{A}_n)'$, $\boldsymbol{B} = (\boldsymbol{B}_1, \boldsymbol{B}_2, \dots, \boldsymbol{B}_n)'$, $\boldsymbol{A}_k = (a_{1k}, a_{2k}, a_{3k})'$, and $\boldsymbol{B}_k = (b_{1k}, b_{2k}, b_{3k})'$ for $k = 1, 2, \dots, n$ and

$$|[a_i b_j]| = \sum_{1\le\alpha<\beta<\gamma\le n} \begin{vmatrix} a_{1\alpha} & a_{1\beta} & a_{1\gamma} \\ a_{2\alpha} & a_{2\beta} & a_{2\gamma} \\ a_{3\alpha} & a_{3\beta} & a_{3\gamma} \end{vmatrix} \times \begin{vmatrix} b_{1\alpha} & b_{1\beta} & b_{1\gamma} \\ b_{2\alpha} & b_{2\beta} & b_{2\gamma} \\ b_{3\alpha} & b_{3\beta} & b_{3\gamma} \end{vmatrix}.$$

The denominator of (9.3) is similarly defined.

Baba (1981) define another measure of circular–circular correlation. The same measure is also defined in Jammalamadaka and Sarma (1988). Let $(\Theta, \Psi)'$ be a random vector on the torus, and let μ and ν be the mean directions of Θ and Ψ. Then their measure is defined by

$$\rho_{\mathrm{BJS}} = \frac{E\{\sin(\Theta - \mu)\sin(\Psi - \nu)\}}{\sqrt{E\{\sin^2(\Theta - \mu)\}E\{\sin^2(\Psi - \nu)\}}}. \tag{9.4}$$

Observe that $\text{Var}\{\sin(\Theta - \mu)\} = E\{\sin^2(\Theta - \mu)\}$ and $\text{Var}\{\sin(\Psi - \nu)\} = E\{\sin^2(\Psi - \nu)\}$ since $E\{\sin(\Theta - \mu)\} = E\{\sin(\Psi - \nu)\} = 0$. It is possible to rewrite (9.4) as

$$\rho_{\text{BJS}} = \frac{E[\cos\{\Theta - \Psi - (\mu - \nu)\} - \cos\{\Theta + \Psi - (\mu + \nu)\}]}{2\sqrt{E\{\sin^2(\Theta - \mu)\}E\{\sin^2(\Psi - \nu)\}}}. \tag{9.5}$$

The circular–circular correlation $\rho_{\text{BJS}} = \rho_{\text{BJS}}(\Theta, \Psi)$ in (9.4) or (9.5) satisfies the following properties (Jammalamadaka and Sarma, 1988):

a) ρ_{BJS} does not depend on the zero direction used for either variable.
b) $\rho_{\text{BJS}}(\Theta, \Psi) = \rho_{\text{BJS}}(\Psi, \Theta)$.
c) $-1 \leq \rho_{\text{BJS}} \leq 1$.
d) $\rho_{\text{BJS}} = 0$ if Θ and Ψ are independent, although the converse need not be true.
e) If Θ and Ψ have full support, $\rho_{\text{BJS}} = 1$ if and only if (iff) $\Theta = \Psi+$ constant (mod 2π) and $\rho_{\text{BJS}} = -1$ iff $\Theta + \Psi =$ constant (mod 2π).

A list of correlations can be seen in Mardia and Jupp (2000), Table 11.3, p. 256. Nonparametric circular–circular correlations have been studied in Fisher and Lee (1982, 1983), Shieh et al. (1994, 2011), and Zhan et al. (2019).

9.1.5 Circulas

The circular analogue of copulas is called circulas in Jones et al. (2015). In this section, we deal with the bivariate case for simplicity.

A general bivariate circular density or density on the unit torus can be written as

$$f(\theta_1, \theta_2) = 4\pi^2 f_1(\theta_1) f_2(\theta_2) c(2\pi F_1(\theta_1), 2\pi F_2(\theta_2)), \quad 0 \leq \theta_1, \theta_2 < 2\pi$$

in terms of its circula $c(\cdot, \cdot)$, its marginal circular density functions f_1 and f_2, and their distribution functions F_1 and F_2. The circula needs to satisfy periodicity

$$c(\theta_1 + 2j\pi, \theta_2 + 2k\pi) = c(\theta_1, \theta_2), \quad 0 \leq \theta_1, \theta_2 < 2\pi$$

for all integers j and k.

Suppose that Θ_1 is distributed as the circular uniform, and consider a circular random variable Ω with density g, distributed independently of Θ_1. Then the sum or the difference $\Theta_2 \equiv \Omega + q\Theta_1$ (mod 2π) for $q \in \{1, -1\}$ follows the circular uniform distribution (see Section 4.2). Thus, the vector $(\Theta_1, \Theta_2)'$ gives a bivariate distribution with circular uniform distributions as marginals. Since the conditional density of Θ_2 given $\Theta_1 = \theta_1$ is $g(\theta_2 - q\theta_1)$, the circula density is

$$c_q(\theta_1, \theta_2) = \frac{1}{2\pi} g(\theta_2 - q\theta_1),$$

and the joint density of Θ_1 and Θ_2 is

$$f(\theta_1, \theta_2) = 2\pi f_1(\theta_1) f_2(\theta_2) g(2\pi(F_2(\theta_2) - qF_1(\theta_1))).$$

As seen in Section 9.2.4, such distributions can be found in Wehrly and Johnson (1980). In addition, it is noted that Rivest (1982) has given the construction of a bivariate circular distribution by using $x = \theta + \psi + \phi \,(\text{mod}\, 2\pi)$ and $y = \theta - \psi \,(\text{mod}\, 2\pi)$ for independent random variables $\theta,\ \psi$, and ϕ to construct $c_1(\theta_1, \theta_2)$ above.

If g is the circular uniform density, then it is obvious that the joint circula density and distribution functions are, respectively,

$$c_I(\theta_1, \theta_2) = \frac{1}{4\pi^2} \quad \text{and} \quad C_I(\theta_1, \theta_2) = \frac{\theta_1\theta_2}{4\pi^2}, \quad 0 \le \theta_1, \theta_2 < 2\pi.$$

In general, the circula distribution function is given by Jones et al. (2015) as

$$C_q(\theta_1, \theta_2) = \frac{q}{2\pi} \left\{ W(\theta_2) + W(-q\theta_1) - W(\theta_2 - q\theta_1) \right\}, \quad 0 \le \theta_1, \theta_2 < 2\pi,$$

where

$$W(\omega) = \int_0^{\omega} \int_0^{b} g(a) da db = \int_0^{\omega} (\omega - a) g(a) da$$

for $-2\pi \le \omega \le 4\pi$. If g is the cardioid density $g(\omega) = (2\pi)^{-1}(1 + 2\rho \cos\omega)$ for $0 \le \rho \le 1/2$, then $W(\omega) = (2\pi)^{-1} \left\{ \omega^2/2 + 2\rho(1 - \cos\omega) \right\}$ and

$$C_q(\theta_1, \theta_2) = \frac{1}{4\pi^2} \left[\theta_1\theta_2 + 2q\rho \left\{ 1 - \cos\theta_1 - \cos\theta_2 + \cos(\theta_1 - q\theta_2) \right\} \right].$$

If $\rho = 0$ in the cardioid distribution, then the distribution reduces to the circular uniform and the circula distribution function C_q reduces to C_I.

9.2 General Methods of Construction

9.2.1 Generalizing Univariate Characteristic Function and Properties

The n-variate circular or hyper-torus pdf $f(\theta_1, \dots, \theta_n)$ with random vector $\boldsymbol{\Theta} = (\Theta_1, \dots, \Theta_n)'$ is similarly defined to the torus case. The characteristic function or $(p_1, \dots, p_n)$th trigonometric moment is defined by

$$\phi_{p_1,\dots,p_n} = E\left(e^{i\mathbf{p}'\boldsymbol{\Theta}}\right) = \int_0^{2\pi} \cdots \int_0^{2\pi} e^{i(p_1\theta_1 + \dots + p_n\theta_n)} f(\theta_1, \dots, \theta_n) \prod_{j=1}^{n} d\theta_j,$$

where $\mathbf{p} = (p_1, \dots, p_n)'$. The joint pdf $f(\theta_1, \dots, \theta_n)$ is expressed by the inversion formula

$$f(\theta_1, \dots, \theta_n) = \left(\frac{1}{2\pi}\right)^n \sum_{p_1=-\infty}^{\infty} \cdots \sum_{p_n=-\infty}^{\infty} e^{-i(p_1\theta_1 + \dots + p_n\theta_n)} \phi_{p_1,\dots,p_n}.$$

It is obvious that if the random variables $\Theta_1, \dots, \Theta_n$ are mutually independent and each Θ_j has characteristic function ϕ_{p_j}, then the $(p_1, \dots, p_n)$th trigonometric moment is $\phi_{p_1,\dots,p_n} = \prod_{j=1}^{n} \phi_{p_j}$.

9.2.2 Maximum Entropy Characterizations

Suppose that we are interested in distributions defined over a space S and that these distributions are to be represented by densities relative to some familiar measure such as Lebesgue and Harr. Let $t_1, \dots, t_q$ represent specified q real valued measurable functions over S such that no linear combination of $t_1, \dots, t_q$ is constant. The following statement (Mardia, 1975b) is applicable to generate bivariate circular distributions, even distributions on the cylinder.

If for a (joint) pdf $f(\boldsymbol{x})$

i) S_1 is the support of $f(\boldsymbol{x})$ where $\boldsymbol{x} \in S_1$, $S_1 \subset S$,
ii) $E\{t_j(\boldsymbol{x})\} = a_j$ (fixed) for $j = 1, \dots, q$,
iii) the entropy is maximized,

then $f(\boldsymbol{x})$ should be of the form

$$f(\boldsymbol{x}) = \exp\left\{ b_0 + \sum_{j=1}^{q} b_j t_j(\boldsymbol{x}) \right\}, \quad \boldsymbol{x} \in S_1, \tag{9.6}$$

provided there exist $b_0, b_1, \dots, b_q$, such that (9.6) satisfies (i) and (ii). Further, if there exists such a density, then it is unique.

9.2.3 Specified Conditionals

Arnold and Strauss (1991) characterize the class of bivariate distributions such that the conditional distributions belong to any specified exponential families. Let $f_1(x; \boldsymbol{\eta})$ and $f_2(x; \boldsymbol{\tau})$ denote members of two ℓ_1- and ℓ_2-parameter exponential families:

$$f_1(x; \boldsymbol{\eta}) = r_1(x)\beta_1(\boldsymbol{\eta}) \exp\left\{ \sum_{j=1}^{\ell_1} \eta_j p_j(x) \right\}, \quad \boldsymbol{\eta} = (\eta_1, \dots, \eta_{\ell_1})'$$

and

$$f_2(y; \boldsymbol{\tau}) = r_2(y)\beta_2(\boldsymbol{\tau}) \exp\left\{ \sum_{j=1}^{\ell_2} \tau_j q_j(y) \right\}, \quad \boldsymbol{\tau} = (\tau_1, \dots, \tau_{\ell_2})',$$

where we assume that the $\{p_j\}$ are linearly independent, and similarly for $\{q_j\}$. Let $f(x, y)$ be a bivariate density whose conditional densities satisfy

$$f(x|y) = f_1(x; \eta(y)) \quad \text{and} \quad f(y|x) = f_2(y; \tau(x))$$

for some functions $\eta(y)$ and $\tau(x)$, where f_1 and f_2 belong to the exponential families. Then $f(x, y)$ must be of the form

$$f(x, y) = r_1(x)r_2(y) \exp\left[\{\boldsymbol{p}(x)\}'\boldsymbol{M}\boldsymbol{q}(y)\right]$$

for suitable choice of M, a $(\ell_1 + 1) \times (\ell_2 + 1)$ matrix of constant parameters subject to the requirement that $\int_{\mathbb{R}^2} f(x, y)dxdy = 1$. Here

$$\boldsymbol{p}(x) = (1, p_1(x), \ldots, p_{\ell_1}(x))' \quad \text{and} \quad \boldsymbol{q}(y) = (1, q_1(y), \ldots, q_{\ell_2}(y))'.$$

Even though the above theorem states the bivariate case in Euclidean space, the result can be extended to the case of distributions on the torus, cylinder and their multivariate generalizations.

9.2.4 Specified Marginals

9.2.4.1 A Bivariate Circular Model

Wehrly and Johnson (1980) constructed a bivariate circular model or a distribution on the torus. Let $f_1(\theta_1)$ and $f_2(\theta_2)$ be pdfs on the circle, $F_1(\theta_1)$ and $F_2(\theta_2)$ be their distribution functions, and g be pdf of a circular distribution. Then, a bivariate function

$$f(\theta_1, \theta_2) = 2\pi\, g[2\pi\{F_1(\theta_1) \pm F_2(\theta_2)\}]f_1(\theta_1)f_2(\theta_2) \tag{9.7}$$

for $0 \le \theta_1, \theta_2 < 2\pi$ or $-\pi \le \theta_1, \theta_2 < \pi$ provides a bivariate circular model whose marginal pdfs are $f_1(\theta)$ and $f_2(\theta)$. Fernández-Durán (2007) studied the model whose marginals are trigonometric sums distributions (Section 5.2.3) which possibly represent multimodality and skewness. See Section 9.11.2 for von Mises and asymmetric von Mises marginals.

It is obvious that the conditional distribution of Θ_1 given $\Theta_2 = \theta_2$ has density

$$f_{\Theta_1|\Theta_2}(\theta_1|\theta_2) = \frac{f(\theta_1, \theta_2)}{f_2(\theta_2)} = 2\pi\, g[2\pi\{F_1(\theta_1) \pm F_2(\theta_2)\}]f_1(\theta_1).$$

If a transformation $U_1 = F_1(\Theta_1)$, $U_2 = F_2(\Theta_2)$ is applied, then the marginal distributions of U_1 and U_2 are uniform and the joint pdf of $(U_1, U_2)'$ with

$$c(u_1, u_2) = 2\pi\, g(2\pi(u_1 \pm u_2)), \quad 0 \le u_1, u_2 < 1$$

gives a circular copula.

Equation (9.7) leads to a family of distributions for a Markov process on the circle. Let $f(\cdot)$ and $g(\cdot)$ be pdfs on the circle and $F(\cdot)$ be the distribution function of $f(\cdot)$. Further, let $\Theta_0, \Theta_1, \ldots$ be random variables taking values on the circle such that

$$p(\theta_0) = f(\theta_0),\ p(\theta_n|\theta_0, \ldots, \theta_{n-1}) = p(\theta_n|\theta_{n-1}) = 2\pi g[2\pi\{F(\theta_n) - F(\theta_{n-1})\}]f(\theta_n).$$

Then $p(\theta_0)$ is the initial pdf, and $p(\theta_n|\theta_{n-1})$ is the stationary transition pdf. An alternative type of Markov process by using the Möbius circle transformation is studied by Kato (2010).

9.2.4.2 A Four-Variate Circular Model

In a similar manner to the construction of Wehrly and Johnson's (1980) model, Kato and Shimizu (2008) obtained a four-variate circular pdf whose bivariate marginals are specified. Indeed, the model is applicable to the construction of distributions on two tori, cylinders or discs. Let $(X_1, X_2)'$ have a specified pdf $f_1(x_1,x_2)$ with distribution function $F_1(x_1,x_2)$ on the support $M_1 \subset \mathbb{R}^2$ and $(Y_1, Y_2)'$ a specified pdf $f_2(y_1,y_2)$ with distribution function $F_2(y_1,y_2)$ on the support $M_2 \subset \mathbb{R}^2$. Suppose that $f_{11}(x_1)$ $(f_{21}(y_1))$ is the marginal pdf of X_1 (Y_1) and $F_{11}(x_1)$ $(F_{21}(y_1))$ its distribution function. Let $g(\cdot)$ be a pdf on the circle and $h(\cdot,\cdot)$ a pdf on the torus. Then

$$\begin{aligned} p_{1\pm}(x_1,x_2,y_1,y_2) &= 2\pi f_1(x_1,x_2) f_2(y_1,y_2) \\ &\quad \times g\left[2\pi\left\{\frac{1}{f_{11}(x_1)}\frac{\partial F_1(x_1,x_2)}{\partial x_1} \pm \frac{1}{f_{21}(y_1)}\frac{\partial F_2(y_1,y_2)}{\partial y_1}\right\}\right] \end{aligned} \tag{9.8}$$

and

$$\begin{aligned} p_{2\pm}(x_1,x_2,y_1,y_2) &= 4\pi^2 f_1(x_1,x_2) f_2(y_1,y_2) \\ &\quad \times h\left[2\pi\left\{F_{11}(x_1) \pm F_{21}(y_1)\right\}, 2\pi\left\{\frac{1}{f_{11}(x_1)}\frac{\partial F_1(x_1,x_2)}{\partial x_1} \pm \frac{1}{f_{21}(y_1)}\frac{\partial F_2(y_1,y_2)}{\partial y_1}\right\}\right] \end{aligned} \tag{9.9}$$

are pdfs on $M_1 \times M_2$, where $(x_1,x_2)' \in M_1$ and $(y_1,y_2)' \in M_2$. Both have the marginal pdfs $f_1(x_1,x_2)$ and $f_2(y_1,y_2)$ of $(X_1,X_2)'$ and $(Y_1,Y_2)'$, respectively. Here the right side of (9.9) permits the four combinations of signs, that is, $(+,+), (+,-), (-,+)$ and $(-,-)$.

A family of distributions for a Markov process on a manifold $M \subset \mathbb{R}^2$ can be constructed through (9.8). Let $f(\cdot,\cdot)$ be a pdf on M, $F(\cdot,\cdot)$ be its distribution function, $f_1(x_{n1})$ be its marginal pdf of X_{n1} and $g(\cdot)$ be a pdf on the circle. Further, let $\boldsymbol{X}_j = (X_{j1}, X_{j2})'$, $j = 0, 1, 2, \ldots$ be random vectors taking values on M such that

$$\begin{aligned} p(\boldsymbol{x}_0) &= f(x_{01},x_{02}), p(\boldsymbol{x}_n|\boldsymbol{x}_0,\boldsymbol{x}_1,\ldots,\boldsymbol{x}_{n-1}) = p(\boldsymbol{x}_n|\boldsymbol{x}_{n-1}) = 2\pi f(x_{n1},x_{n2}) \\ &\quad \times g\left[2\pi\left\{\frac{1}{f_1(x_{n1})}\frac{\partial F(x_{n1},x_{n2})}{\partial x_{n1}} \pm \frac{1}{f_1(x_{n-1,1})}\frac{\partial F(x_{n-1,1},x_{n-1,2})}{\partial x_{n-1,1}}\right\}\right]. \end{aligned}$$

Then $p(\boldsymbol{x}_0)$ is the initial pdf, and $p(\boldsymbol{x}_n|\boldsymbol{x}_{n-1})$ is the stationary transition pdf. Similarly another Markov process on M can be constructed by using (9.9).

It is remarked that the model by Kato and Shimizu is reduced to the following. Let $(X_1, X_2)'$ have a specified pdf $f_1(x_1,x_2)$ with distribution function $F_1(x_1,x_2)$ on the support $M_1 \subset \mathbb{R}^2$, $f_{11}(x_1)$ be the marginal pdf of X_1, and Y have a specified pdf

$f_2(y)$ with distribution function $F(y)$ on the support $M_2 \subset \mathbb{R}$. Further, let $g(\cdot)$ be a pdf on the circle. Then

$$p(x_1, x_2, y) = 2\pi f_1(x_1, x_2) f_2(y) g\left[2\pi\left\{\frac{1}{f_{11}(x_1)}\frac{\partial F_1(x_1, x_2)}{\partial x_1} \pm F_2(y)\right\}\right] \tag{9.10}$$

is a joint pdf on $M_1 \times M_2$, where $(x_1, x_2)' \in M_1$ and $y \in M_2$, and it has the marginal pdf $f_1(x_1, x_2)$ ($f_2(y)$) of $(X_1, X_2)'$ (Y). This may be viewed as an incomplete model of (9.8) in the sense that, for example, (9.10) gives a model of incomplete pairs of wind directions at two locations at two points in time.

9.2.4.3 Models on the Hyper-Torus

An extension of the bivariate circular model by Wehrly and Johnson (1980) to the hyper-torus has been studied by Kim et al. (2016). This topic will be treated in Section 9.11.1.

9.2.5 Parametric Mixtures

This Section is owing to Ong and SenGupta (2012). Let $(\Theta|x, \zeta)$ and $(\Phi|x, \zeta)$ be two independent and identically distributed random variables on the circle with pdfs $f_\Theta(\theta|x, \zeta)$ and $f_\Phi(\phi|x, \zeta)$, respectively, where x is a parameter and ζ may be a vector of parameters. If x is a random variable X with pdf $g(x; \eta)$ having a (vector) parameter η, then Θ and Φ are jointly distributed on the torus with pdf

$$\text{Model I}: \ f(\theta, \phi; \zeta, \eta) = \int f_\Theta(\theta|x, \zeta) f_\Phi(\phi|x, \zeta) g(x; \eta) dx.$$

The marginal pdf of Θ on the circle is given by $f_\Theta(\theta; \zeta, \eta) = \int f_\Theta(\theta|x, \zeta) g(x; \eta) dx$. The assumption of identical distribution for $(\Theta|x, \zeta)$ and $(\Phi|x, \zeta)$ may be relaxed. Consider $f_\Theta(\theta|x_1, \zeta)$ and $f_\Phi(\phi|x_2, \zeta)$ such that x_1 and x_2 are jointly distributed with pdf $g(x_1, x_2; \eta)$. Then $(\Theta, \Phi)'$ has joint pdf

$$\text{Model II}: \ f(\theta, \phi; \zeta, \eta) = \iint f_\Theta(\theta|x_1, \zeta) f_\Phi(\phi|x_2, \zeta) g(x_1, x_2; \eta) dx_1 dx_2.$$

The marginal pdf of Θ is given by $f_\Theta(\theta; \zeta, \eta) = \iint f_\Theta(\theta|x_1, \zeta) g(x_1, x_2; \eta) dx_1 dx_2$.

Fourier series expansion and trigonometric moments are obtainable as follows. For Model I, with

$$f_\Theta(\theta|\lambda) = \frac{1}{2\pi}\left[1 + 2\sum_{p=1}^{\infty}\{\alpha_p(\lambda)\cos(p\theta) + \beta_p(\lambda)\sin(p\theta)\}\right],$$

$$\alpha_p(\lambda) = \int_0^{2\pi} \cos(p\theta_1) f_\Theta(\theta_1|\lambda) d\theta_1, \ \beta_p(\lambda) = \int_0^{2\pi} \sin(p\theta_1) f_\Theta(\theta_1|\lambda) d\theta_1,$$

$$f_\Phi(\phi|\lambda) = \frac{1}{2\pi}\left[1 + 2\sum_{q=1}^{\infty}\{\alpha'_q(\lambda)\cos(q\phi) + \beta'_q(\lambda)\sin(q\phi)\}\right],$$

$$\alpha_q'(\lambda) = \int_0^{2\pi} \cos(q\theta_2) f_\Phi(\theta_2|\lambda)d\theta_2,\ \beta_q'(\lambda) = \int_0^{2\pi} \sin(q\theta_2) f_\Phi(\theta_2|\lambda)d\theta_2,$$

and the mixing pdf $g(\lambda)$, we have

$$\begin{aligned} f_{\Theta,\Phi}(\theta,\phi) = \left(\frac{1}{2\pi}\right)^2 \Bigg(1 + 2\sum_{p=1}^{\infty}\{\alpha_p \cos(p\theta) + \beta_p \sin(p\theta) + \alpha_p' \cos(p\phi) \\ + \beta_p' \sin(p\phi)\} + 4\sum_{p=1}^{\infty}\sum_{q=1}^{\infty}\Big[E_\Lambda\{\cos(p\Theta)\cos(q\Phi)\}\cos(p\theta)\cos(q\phi) \\ + E_\Lambda\{\cos(p\Theta)\sin(q\Phi)\}\cos(p\theta)\sin(q\phi) + E_\Lambda\{\sin(p\Theta)\cos(q\Phi)\} \\ \times \sin(p\theta)\cos(q\phi) + E_\Lambda\{\sin(p\Theta)\sin(q\Phi)\}\sin(p\theta)\sin(q\phi)\Big]\Bigg). \end{aligned}$$

Here

$$\alpha_p = \int_0^{2\pi} \cos(p\theta)\left\{\int f_\Theta(\theta|\lambda)g(\lambda)d\lambda\right\}d\theta$$

and so on, and

$$\begin{aligned} E_\Lambda\{\cos(p\Theta)\cos(q\Phi)\} = \int_0^{2\pi}\int_0^{2\pi} \cos(p\theta_1)\cos(q\theta_2) \\ \times\left\{\int f_\Theta(\theta_1|\lambda)f_\Phi(\theta_2|\lambda)g(\lambda)d\lambda\right\}d\theta_1 d\theta_2 \end{aligned}$$

and so on.

9.2.6 Sine-Skewing

As in Section 5.11, sine-skewing is a method for constructing skew or asymmetric circular distributions from symmetric circular ones. Likewise, the method is the basis of obtaining skew and/or multimodal distributions for toroidal settings.

9.2.6.1 Sine-Correlated Toroidal Distributions

Imoto and Abe (2021) have given a class of "sine-correlated toroidal distributions" as an extension of sine-skewed circular distributions in Section 5.11:

$$f(\theta_1,\theta_2) = \{1 + \lambda\sin(\theta_1 - \mu_1)\sin(\theta_2 - \mu_2)\}g_1(\theta_1 - \mu_1)g_2(\theta_2 - \mu_2) \tag{9.11}$$

with $-1 \le \lambda \le 1$ which controls the dependence between two variables Θ_1 and Θ_2. It is obvious that Θ_1 and Θ_2 are independent if and only if $\lambda = 0$. The marginal density of Θ_1 (Θ_2) is g_1 (g_2) and the conditional distribution of Θ_1 given $\Theta_2 = \theta_2$ is a sine-skewed circular distribution. The proposed toroidal distribution can be viewed as a special case of circulas.

An extended method can be seen in Abe et al. (2021). Indeed, for a distribution function $G(\cdot)$ whose pdf is symmetric about zero and defined on the domain including $[-1,1]$, $f_T(\theta_1, \theta_2) = 2G(\lambda \sin\theta_1 \sin\theta_2)g_1(\theta_1)g_2(\theta_2)$ with a parameter λ is a toroidal pdf. The domain of λ depends on the choice of G. For $G(x) = (1+x)/2$ as a special case, $f_T(\theta_1, \theta_2)$ reduces to (9.11).

9.2.6.2 Generalized Bivariate Sine-Skewed Distributions

Bekker et al.'s (2021) "generalized bivariate sine-skewed distributions" are defined by

$$f(\theta_1, \theta_2) = C^{-1}\left\{\frac{1+\lambda_1 \sin(\theta_1-\mu_1)+\lambda_2 \sin(\theta_2-\mu_2)}{2}\right\}^m g(\theta_1-\mu_1, \theta_2-\mu_2) \tag{9.12}$$

with skewness parameters $-1 \le \lambda_1, \lambda_2 \le 1$, which satisfy $|\lambda_1| + |\lambda_2| \le 1$, a toroidal pdf symmetric about $-\pi \le \mu_1, \mu_2 < \pi$, a positive integer m, and a normalizing constant

$$C = \int_{-\pi}^{\pi}\int_{-\pi}^{\pi}\left(\frac{1+\lambda_1 \sin t_1 + \lambda_2 \sin t_2}{2}\right)^m g(t_1, t_2)dt_1 dt_2.$$

The case $m = 1$ in (9.12) is a special case of the d-dimensional toroidal distribution defined by Eq. (2.1) in Ameijeiras-Alonso and Ley (2022) with $d = 2$.

9.3 Uniform Distribution

Since the uniform pdf $f(\theta_1, \theta_2)$ takes a constant c on $[0, 2\pi)^2$ or $[-\pi, \pi)^2$, it follows from the integral value $\int_0^{2\pi}\int_0^{2\pi} f(\theta_1, \theta_2)d\theta_1 d\theta_2 = \int_0^{2\pi}\int_0^{2\pi} c\, d\theta_1 d\theta_2 = 4\pi^2 c = 1$ that $c = 1/(4\pi^2)$. If a random vector $(\Theta_1, \Theta_2)'$ on the torus is uniformly distributed, Θ_1 and Θ_2 are independent and uniformly distributed. The characteristic function or (p, q)th trigonometric moment of the uniform distribution is

$$\phi_{p,q} = E\{e^{i(p\Theta_1+q\Theta_2)}\} = \begin{cases} 1, & p = q = 0, \\ 0, & \text{others.} \end{cases}$$

A distribution whose (1,0), (0,1), and (1,1) trigonometric moments are zero does not imply the uniform distribution as in the following example:

$$f(\theta_1, \theta_2) = \frac{1}{4\pi^2}\left\{1 + 4d\cos(2\theta_1)\cos(2\theta_2)\right\}, \quad |d| \le \frac{1}{4},$$

whose characteristic function is given by

$$\int_0^{2\pi}\int_0^{2\pi} e^{i(p\theta_1+q\theta_2)} f(\theta_1, \theta_2)d\theta_1 d\theta_2 = \begin{cases} 1, & p = q = 0, \\ d, & p = q = 2, \\ 0, & \text{others.} \end{cases}$$

9.4 Families of Bivariate von Mises Distributions

As an example of specified conditionals in Section 9.2.3, the joint pdfs on the torus with von Mises conditionals (e.g. SenGupta, 2004), whose pdfs are of the form $f(\xi) \propto \exp(a \cos \xi + b \sin \xi)$ with constants a and b, are

$$f(\theta, \psi) = \exp\left[\{\boldsymbol{p}(\theta)\}' M \boldsymbol{q}(\psi)\right], \quad (\theta, \psi) \in [0, 2\pi)^2, \tag{9.13}$$

where

$$\boldsymbol{p}(\theta) = \begin{pmatrix} 1 \\ \cos\theta \\ \sin\theta \end{pmatrix}, \boldsymbol{q}(\psi) = \begin{pmatrix} 1 \\ \cos\psi \\ \sin\psi \end{pmatrix} \quad \text{and} \quad M = \begin{pmatrix} m_{00} & m_{01} & m_{02} \\ m_{10} & m_{11} & m_{12} \\ m_{20} & m_{21} & m_{22} \end{pmatrix}.$$

Here M is a matrix of parameters with m_{00}, a function of the other m_{jk}'s, being the normalizing constant. Thus, the model has eight parameters. The form (9.13) appears in Mardia (1975a) and in Mardia and Jupp (2000, Section 3.7.1, p. 55) as

$$\begin{aligned} f(\theta_1, \theta_2) \propto \exp\{ &\kappa_1 \cos(\theta_1 - \mu_1) + \kappa_2 \cos(\theta_2 - \mu_2) \\ &+ (\cos\theta_1, \sin\theta_1) B (\cos\theta_2, \sin\theta_2)' \} \end{aligned} \tag{9.14}$$

with 2×2 real matrix B or

$$\begin{aligned} f(\theta_1, \theta_2) = c(\kappa_1, \kappa_2, A) \exp\{ &\kappa_1 \cos(\theta_1 - \mu_1) + \kappa_2 \cos(\theta_2 - \mu_2) \\ &+ (\cos(\theta_1 - \nu_1), \sin(\theta_1 - \nu_1)) A (\cos(\theta_2 - \nu_2), \sin(\theta_2 - \nu_2))' \} \end{aligned} \tag{9.15}$$

with 2×2 real matrix A. The constant $c(\kappa_1, \kappa_2, A)$ with $\nu_1 = \mu_1$ and $\nu_2 = \mu_2$ is given by Mardia (2010) as Theorem 3.1. The marginal distributions are not von Mises unless independence holds. Note as in SenGupta (2004) that the model with joint pdf (9.14) or (9.15) corresponds to the maximum entropy distribution obtained on specifying the first marginal and first joint harmonics of Θ and Ψ, that is, on specifying $E(\cos\Theta)$, $E(\sin\Theta)$, $E(\cos\Psi)$, $E(\sin\Psi)$, $E(\cos\Theta\cos\Psi)$, $E(\cos\Theta\sin\Psi)$, $E(\sin\Theta\cos\Psi)$, and $E(\sin\Theta\sin\Psi)$. Bayesian analysis in (9.14) can be seen in Mardia (2010). Submodels of (9.14) or (9.15) will be treated in Section 9.9.

As another example of specified conditionals, SenGupta and Arnold (2025) have constructed the highly flexible family of bivariate generalized (power) cardioid continuous and discrete conditionals distributions. A flexible model of this type for the continuous case is given by

$$f(\theta_1, \theta_2) = [(1, \cos\theta_1, \sin\theta_1) A (1, \cos\theta_2, \sin\theta_2)']^{\gamma}$$

where $\gamma \in R$ and A is a matrix of real parameters i.e.

$$A = \begin{pmatrix} a_{00} & a_{01} & a_{02} \\ a_{10} & a_{11} & a_{12} \\ a_{20} & a_{21} & a_{22} \end{pmatrix}$$

Note that this family is not a member of the regular multi-parameter exponential family. The entries in the parameter matrix A must be constrained to ensure that the function $f(\theta_1, \theta_2)$ is non-negative and integrates to 1 over the region $(0, 2\pi)^2$. Except for that, they may be quite arbitrary. The case $\gamma = -1$ corresponds to wrapped Cauchy conditionals, while $\gamma = 1$ yields cardioid conditionals. Not only is this a convenient flexible family of distributions with generalized (power) cardioid conditionals but it also represents the complete class of such distributions with a given fixed value of the parameter γ.

9.5 Bivariate Gamma-Mixed von Mises Distributions

Based on the results in Section 9.2.5, we can construct bivariate mixed distributions. Examples are given below.

Ong and SenGupta (2012) give a lemma to derive two types of bivariate gamma-mixed von Mises distributions: The reciprocal of a power series $\sum_{n=0}^{\infty} a_n x^n$ ($a_0 \neq 0$) is expressed by a power series itself as

$$\frac{1}{\sum_{n=0}^{\infty} a_n x^n} = \sum_{j=0}^{\infty} c_j x^j, \tag{9.16}$$

where

$$c_0 = \frac{1}{a_0} \quad \text{and} \quad c_j = -\frac{1}{a_0} \sum_{n=1}^{j} c_{j-n} a_n \quad (j \geq 1).$$

By applying this result, we have a series representation of the von Mises pdf

$$\frac{1}{2\pi I_0(\kappa)} e^{\kappa \cos(\theta-\mu)} = \frac{1}{2\pi} \sum_{j=0}^{\infty} d_j \kappa^{2j} e^{\kappa \cos(\theta-\mu)}$$

with $d_0 = 1$ and $d_j = -\sum_{n=1}^{j} \{1/(2^{2n} n! n!)\} d_{j-n}$. Consider the conditional von Mises pdf on the parameter $\kappa = x\delta$. If x follows a gamma distribution $\Gamma(\alpha, \beta)$, whose random variable X has pdf

$$g(x) = \frac{\beta^\alpha}{\Gamma(\alpha)} x^{\alpha-1} e^{-\beta x}, \quad x > 0,\ \alpha > 0,\ \beta > 0, \tag{9.17}$$

then the unconditional von Mises pdf is given by

$$f(\theta) = \frac{1}{2\pi} \sum_{j=0}^{\infty} d_j \delta^{2j} \beta^\alpha (\alpha)_{2j} \frac{1}{\{\beta - \delta \cos(\theta - \mu)\}^{2j+\alpha}},$$

where $(\alpha)_{2j}$ denotes Pochhammer's symbol.

9.5.1 Model I

Let $\Theta_k\,(k = 1, 2)$ be independent and distributed as von Mises distributions $\text{VM}(\mu_k, \kappa_k)$ conditional on the parameters $\kappa_k = x\delta_k$, respectively. If x is a gamma random variable with pdf (9.17), then we have the unconditional bivariate mixed von Mises distribution of $(\Theta_1, \Theta_2)'$ with pdf

$$f(\theta_1, \theta_2) = \left(\frac{1}{2\pi}\right)^2 \sum_{k=0}^{\infty}\sum_{j=0}^{\infty} d_k d_j \delta_1^{2k} \delta_2^{2j} \beta^{\alpha} (\alpha)_{2k+2j} \times \frac{1}{\{\beta - \delta_1 \cos(\theta_1 - \mu_1) - \delta_2 \cos(\theta_2 - \mu_2)\}^{2k+2j+\alpha}},$$

where $d_0 = 1$ and $d_j = -\sum_{n=1}^{j}\{1/(2^{2n} n!n!)\}d_{j-n}$. The marginal pdfs are

$$f(\theta_k) = \frac{1}{2\pi}\sum_{j=0}^{\infty} d_j \delta_k^{2j} \beta^{\alpha}(\alpha)_{2j} \frac{1}{\{\beta - \delta_k \cos(\theta_k - \mu_k)\}^{2j+\alpha}} \tag{9.18}$$

for $k = 1, 2$.

The joint trigonometric moments are

$$\Psi(p, q) = E\left\{e^{i(p\Theta_1 + q\Theta_2)}\right\} = e^{i(p\mu_1 + q\mu_2)} \times \sum_{j,k,m,n=0}^{\infty} \frac{d_j d_k \delta_1^{2(j+m)+p} \delta_2^{2(k+n)+q} (\alpha)_{2(j+k+m+n)+p+q}}{\Gamma(p + m + 1)m!\Gamma(q + n + 1)n! 2^{2(m+n)+p+q} \beta^{2(j+k+m+n)+p+q}}.$$

9.5.2 Model II

Let $\Theta_k\,(k = 1, 2)$ be independent and distributed as $\text{VM}(\mu_k, \kappa_k)$ conditional on the parameters $\kappa_k = x_k\delta_k$ respectively. As a mixing distribution, consider the Wicksell–Kibble bivariate gamma distribution (e.g. Mardia, 1970, Section 7.4, p. 53, Section 10.10, p. 88; Ong and SenGupta, 2012) with joint pdf

$$f(x_1, x_2) = \frac{(\beta_1\beta_2)^{\alpha}}{\Gamma(\alpha)(1 - \rho)} \left(\frac{x_1 x_2}{\rho\beta_1\beta_2}\right)^{(\alpha-1)/2} \times \exp\left(-\frac{\beta_1 x_1 + \beta_2 x_2}{1 - \rho}\right) I_{\alpha-1}\left(\frac{2\sqrt{\rho\beta_1\beta_2 x_1 x_2}}{1 - \rho}\right), \quad x_1 > 0, x_2 > 0,$$

where $\beta_1 > 0, \beta_2 > 0, \alpha > 0, 0 \le \rho < 1$, and $I_r(\cdot)$ denotes the modified Bessel function of the first kind and order r. Then the unconditional joint mixed distribution of $(\Theta_1, \Theta_2)'$ has pdf

$$f(\theta_1, \theta_2) = \left(\frac{1}{2\pi}\right)^2 \{a_1 a_2 (1 - \rho)\}^{\alpha} \sum_{m,j=0}^{\infty} d_m d_j \delta_1^{2m} \delta_2^{2j} \times \frac{1}{\{a_1 - \delta_1 \cos(\theta_1 - \mu_1)\}^{2m+\alpha} \{a_2 - \delta_2 \cos(\theta_2 - \mu_2)\}^{2j+\alpha}}$$

$$\times \sum_{k=0}^{\infty} \frac{(\alpha+k)_{2m}(\alpha+k)_{2j}(\alpha)_k}{k!}$$

$$\times \left[\frac{a_1 a_2 \rho}{\{a_1 - \delta_1 \cos(\theta_1 - \mu_1)\}\{a_2 - \delta_2 \cos(\theta_2 - \mu_2)\}} \right]^k,$$

where $a_j = \beta_j/(1-\rho)$, $j = 1, 2$. The marginal distributions of Θ_1 and Θ_2 have the same form of the pdfs (9.18) with $\beta = \beta_j$ $(j = 1, 2)$, since the marginals of the Wicksell–Kibble bivariate gamma distributions are $\Gamma(\alpha, \beta_j)$, whose pdfs are given by (9.17) with $\beta = \beta_j$ respectively.

9.6 Wrapped Bivariate Normal Distribution

The wrapped bivariate normal distribution, briefly described by Thompson (1975) and Johnson and Wehrly (1977), and studied by Baba (1981) and Jammalamadaka and Sarma (1988), has pdf

$$f(\theta, \phi) = \frac{1}{2\pi\sigma_1\sigma_2\sqrt{1-\rho^2}} \sum_{k=-\infty}^{\infty} \sum_{m=-\infty}^{\infty} \exp(-Q/2), \quad 0 \le \theta < 2\pi, \quad 0 \le \phi < 2\pi, \tag{9.19}$$

for $-\infty < \mu_1, \mu_2 < \infty$, $\sigma_1, \sigma_2 > 0$, and $-1 < \rho < 1$, where

$$Q = \frac{1}{1-\rho^2} \left\{ \frac{(\theta - 2\pi k - \mu_1)^2}{\sigma_1^2} - \frac{2\rho(\theta - 2\pi k - \mu_1)(\phi - 2\pi m - \mu_2)}{\sigma_1\sigma_2} + \frac{(\phi - 2\pi m - \mu_2)^2}{\sigma_2^2} \right\}.$$

The marginal pdf is given by

$$\int_0^{2\pi} f(\theta, \phi) d\phi = \frac{1}{\sqrt{2\pi}\sigma_1} \sum_{k=-\infty}^{\infty} \exp\left\{ -\frac{1}{2\sigma_1^2} (\theta - 2\pi k - \mu_1)^2 \right\},$$

which is the pdf of a circular wrapped normal distribution.

The (p, q)th joint trigonometric moment for a random vector $(\Theta, \Phi)'$, which follows the wrapped bivariate normal distribution is

$$E\{e^{i(p\Theta + q\Phi)}\} = \exp\left\{ i(p\mu_1 + q\mu_2) - \left(\frac{1}{2} p^2\sigma_1^2 + \frac{1}{2} q^2\sigma_2^2 + pq\sigma_1\sigma_2\rho \right) \right\},$$

and (9.19) is represented by a Fourier series

$$f(\theta, \phi) = \frac{1}{(2\pi)^2} \left[1 + 2\sum_{p=1}^{\infty} e^{-p^2\sigma_1^2/2} \cos(p\theta^*) + 2\sum_{q=1}^{\infty} e^{-q^2\sigma_2^2/2} \cos(q\phi^*) + 4\sum_{p=1}^{\infty}\sum_{q=1}^{\infty} e^{-(p^2\sigma_1^2 + q^2\sigma_2^2)/2} \right.$$

$$\left. \times \left\{ e^{pq\sigma_1\sigma_2\rho} \cos(p\theta^* - q\phi^*) + e^{-pq\sigma_1\sigma_2\rho} \cos(p\theta^* + q\phi^*) \right\} \right] ,$$

where $\theta^* = \theta - \mu_1$ and $\phi^* = \phi - \mu_2$. As σ_1 and σ_2 go to zero, the wrapped bivariate normal distribution tends to a degenerate distribution, while for large σ_1 and σ_2, it is approximated by the uniform distribution on the torus. The wrapped multivariate normal distribution is defined by a similar way to the bivariate case.

To evaluate the circular–circular correlation ρ_{BJS}, we use (9.5). Let $(X, Y)'$ follow the bivariate normal distribution $N_2(\mu_1, \mu_2, \sigma_1^2, \sigma_2^2, \rho)$. Since $\Theta + \Phi$ (mod 2π) $= X + Y$ (mod 2π) and $\Theta - \Phi$ (mod 2π) $= X - Y$ (mod 2π), that is, $\Theta + \Phi$ and $\Theta - \Phi$ follow wrapped normal distributions $\text{WN}(\mu_1 + \mu_2, \sigma_1^2 + \sigma_2^2 + 2\sigma_1\sigma_2\rho)$ and $\text{WN}(\mu_1 - \mu_2, \sigma_1^2 + \sigma_2^2 - 2\sigma_1\sigma_2\rho)$ respectively, we directly have

$$E[\cos\{\Theta + \Phi - (\mu_1 + \mu_2)\}] = \exp\left\{ -\frac{1}{2}(\sigma_1^2 + \sigma_2^2 + 2\sigma_1\sigma_2\rho) \right\}$$

and

$$E[\cos\{\Theta - \Phi - (\mu_1 - \mu_2)\}] = \exp\left\{ -\frac{1}{2}(\sigma_1^2 + \sigma_2^2 - 2\sigma_1\sigma_2\rho) \right\},$$

so that

$$\begin{aligned} &E[\cos\{\Theta - \Phi - (\mu_1 - \mu_2)\} - \cos\{\Theta + \Phi - (\mu_1 + \mu_2)\}] \\ &\qquad = \exp\left\{ -\frac{1}{2}(\sigma_1^2 + \sigma_2^2) \right\} \sinh(\sigma_1\sigma_2\rho). \end{aligned}$$

For the denominator of the correlation coefficient, from the fact that Θ is distributed as $\text{WN}(\mu_1, \sigma_1^2)$ and Φ is distributed as $\text{WN}(\mu_2, \sigma_2^2)$, we have

$$E\{\sin^2(\Theta - \mu_1)\} = \frac{1}{2}\left(1 - e^{-2\sigma_1^2}\right) = e^{-\sigma_1^2}\sinh(\sigma_1^2)$$

and

$$E\{\sin^2(\Phi - \mu_2)\} = \frac{1}{2}\left(1 - e^{-2\sigma_2^2}\right) = e^{-\sigma_2^2}\sinh(\sigma_2^2).$$

Thus, in the case of the wrapped bivariate normal distribution, ρ_{BJS} has a very simple form:

$$\rho_{\text{BJS}} = \frac{\sinh(\sigma_1\sigma_2\rho)}{\sqrt{\sinh(\sigma_1^2)\sinh(\sigma_2^2)}}.$$

It is obvious that $\rho_{\text{BJS}} = 0$ if X and Y are independent. However, even if X and Y are perfectly correlated in the linear sense, it does not follow that $\rho_{\text{BJS}} = 1$ unless $\sigma_1 = \sigma_2$. In this case, from the property (e) of the correlation coefficient, $\Theta = \Phi +$ constant (mod 2π). Similarly, when Pearson's product moment correlation coefficient between X and Y is -1 and $\sigma_1 = \sigma_2$, then $\rho_{\text{BJS}} = -1$ and $\Theta + \Phi =$ constant (mod 2π).

9.7 Wrapped Bivariate Heavy Tail Distributions

High volatility distributions on $\mathbb{R}$ or $\mathbb{R}^q$ often are not available in analytical forms, and hence their pdfs cannot be wrapped directly. This is true for the most widely popular heavy-tail families, the stable and Linnik families. However, they admit of amenable forms in terms of their characteristic functions. Wrapped circular families were obtained from the c.f.s of their corresponding linear families as was seen in Section 4.5. Here, we generalize that approach to yield their bivariate pdfs.

9.7.1 Wrapped Bivariate α-Stable Family

The c.f. of the bivariate symmetric α-stable *family* of distributions (see e.g. SenGupta and Roy, 2026a), with location parameters taken as zeroes without loss of generality, is given by

$$\psi(\boldsymbol{t}) = \exp\left[-\left\{\frac{1}{2}(\boldsymbol{t}'\Sigma\boldsymbol{t})\right\}^{\alpha/2}\right] = \exp\left[-\left\{\frac{1}{2}\left(t_1^2\sigma_1^2 + 2t_1t_2\sigma_{12} + t_2^2\sigma_2^2\right)\right\}^{\alpha/2}\right].$$

Thus the c.f. of the corresponding wrapped bivariate symmetric α-stable family of distributions is given by

$$\psi_{\text{BWS}}(p,q) = E\{e^{i(p\Theta+q\Phi)}\} = \exp\left\{-\left(\frac{1}{2}p^2\sigma_1^2 + \frac{1}{2}q^2\sigma_2^2 + pq\sigma_{12}\right)^{\alpha/2}\right\},$$

and its pdf, with location parameters μ_1 and μ_2, is represented by the Fourier series

$$f(\theta,\phi) = \frac{1}{(2\pi)^2}\left[1 + 2\sum_{p=1}^{\infty} e^{-(p^2\sigma_1^2/2)^{\alpha/2}}\cos(p\theta^*) + 2\sum_{q=1}^{\infty} e^{-(q^2\sigma_2^2/2)^{\alpha/2}}\cos(q\phi^*)\right.$$
$$\left. + 4\sum_{p=1}^{\infty}\sum_{q=1}^{\infty}\{\alpha_{p,q}\cos(p\theta^*)\cos(q(\phi^*) + \beta_{p,q}\sin(p\theta^*)\sin(q\phi^*)\}\right],$$

where $\theta^* = \theta - \mu_1$, $\phi^* = \phi - \mu_2$, and

$$\alpha_{p,q} = (1/2)(\psi_{\text{BWS}}(p,q) + \psi_{\text{BWS}}(p,-q)),$$
$$\beta_{p,q} = (1/2)(-\psi_{\text{BWS}}(p,q) + \psi_{\text{BWS}}(p,-q)).$$

The above representation is quite useful for statistical inference of the parameters of general as well as the individual distributions in this family since it displays all the trigonometric moments explicitly. This fact has been exploited by SenGupta and Roy (2026a) to develop optimal inference procedures for its most important index parameter α.

Taking $\alpha = 1$ and 2, we get the corresponding wrapped bivariate Cauchy and normal distributions respectively. Further details on the wrapped bivariate normal distribution have been given above.

9.7.2 Wrapped Bivariate Linnik Family

The c.f. of the bivariate symmetric α-Linnik *family* of distributions (see e.g., SenGupta and Roy, 2026b), with location parameters taken as zeroes without loss of generality, is given by

$$\psi(\boldsymbol{t}) = [1 + (\boldsymbol{t}'\Sigma\boldsymbol{t})^{\alpha/2}]^{-1} = \left[1 + \left(t_1^2\sigma_1^2 + 2t_1t_2\sigma_{12} + t_2^2\sigma_2^2\right)^{\alpha/2}\right]^{-1}.$$

Thus the c.f. of the corresponding wrapped bivariate symmetric α-Linnik family of distributions is given by

$$\psi_{\mathrm{BWL}}(p, q) = E\{e^{i(p\Theta+q\Phi)}\} = \left[1 + \left(p^2\sigma_1^2 + 2p_1p_2\sigma_{12} + q^2\sigma_2^2\right)^{\alpha/2}\right]^{-1}.$$

and its pdf, with location parameters μ_1 and μ_2, is represented by the Fourier series

$$f(\theta, \phi) = \frac{1}{(2\pi)^2}\left[1 + 2\sum_{p=1}^{\infty}(1 + (p\sigma_1)^{\alpha})^{-1}\cos(p\theta^*) + 2\sum_{q=1}^{\infty}(1 + (q\sigma_2)^{\alpha})^{-1}\cos(q\phi^*)\right.$$
$$\left. + 4\sum_{p=1}^{\infty}\sum_{q=1}^{\infty}\{\alpha_{p,q}\cos(p\theta^*)\cos(q(\phi^*) + \beta_{p,q}\sin(p\theta^*)\sin(q\phi^*)\}\right],$$

where $\theta^* = \theta - \mu_1$, $\phi^* = \phi - \mu_2$, and

$$\alpha_{p,q} = (1/2)(\psi_{\mathrm{BWL}}(p, q) + \psi_{\mathrm{BWL}}(p, -q)),$$
$$\beta_{p,q} = (1/2)(-\psi_{\mathrm{BWL}}(p, q) + \psi_{\mathrm{BWL}}(p, -q)).$$

As for the wrapped bivariate symmetric α-stable family of distributions, here also the above representation is quite useful for statistical inference of the parameters of general as well as the individual distributions in this family since it displays all the trigonometric moments explicitly. This fact has been exploited by SenGupta and Roy (2026b) to develop optimal inference procedures for its most important index parameter α.

Taking $\alpha = 2$, we get the corresponding wrapped bivariate Laplace distribution.

9.8 Bivariate Cardioid Distributions

It is possible to define a bivariate cardioid distribution using (9.7), that is, circular cardioid densities $f_j(\theta_j)$, $j = 1, 2$, and a certain circular density $g(\theta)$. However, we use other methods to obtain bivariate cardioid distributions.

9.8.1 Using a Structural Model

Wang and Shimizu (2012) propose a circular–circular structural model or model using a method of trivariate reduction. As a result, a bivariate cardioid distribution may be defined.

We consider a circular–circular structural model $\Theta = \xi + \delta$ and $\Phi = \eta + \varepsilon$, where ξ is a random variable which follows a cardioid distribution $C(\mu, \rho)$ and η is linked to ξ as $\eta = M(\xi)$ with the Möbius transformation

$$M(\xi) = \arg\left(\beta\,\frac{e^{i\xi} + \alpha}{1 + \overline{\alpha}e^{i\xi}}\right) = 2\tan^{-1}\left[\left(\frac{1-\rho_\alpha}{1+\rho_\alpha}\right)\tan\left\{\frac{1}{2}(\xi - \mu_\alpha)\right\}\right] + (\mu_\alpha + \mu_\beta),$$

where $\alpha = \rho_\alpha e^{i\mu_\alpha}\,(\rho_\alpha \geq 0,\ \rho_\alpha \neq 1, -\pi \leq \mu_\alpha < \pi)$, $\beta = e^{i\mu_\beta}\,(-\pi \leq \mu_\beta < \pi)$, and $\overline{\alpha}$ denotes the complex conjugate of α. Here δ and ε are independently distributed as cardioid distributions $C(0, \rho_1)$ and $C(0, \rho_2)$, respectively, and are independent of ξ. The joint pdf of Θ and Φ, whose distribution is denoted by $\mathrm{MC}_2(\mu, \rho, \mu_\alpha, \rho_\alpha, \mu_\beta, \rho_1, \rho_2)$ or MC_2 for short, is calculated as

$$\begin{aligned} f_1(\theta, \phi) = \left(\frac{1}{2\pi}\right)^2 & [1 + 2\rho_2\rho_\alpha\{1 + 2\rho\rho_1\cos(\theta - \mu)\}\cos(\mu_\alpha + \mu_\beta - \phi) \\ & + 2\rho\rho_1\cos(\theta - \mu) + 2\rho\rho_2(1 - \rho_\alpha^2)\cos(\phi - \mu - \mu_\beta) \\ & + 2\rho_1\rho_2(1 - \rho_\alpha^2)\cos(\phi - \theta - \mu_\beta) \\ & - 2\rho\rho_1\rho_2\rho_\alpha(1 - \rho_\alpha^2)\cos(\theta - \phi + \mu - \mu_\alpha + \mu_\beta)] \end{aligned}$$

for $0 \leq \rho_\alpha < 1$, and for $\rho_\alpha > 1$,

$$\begin{aligned} f_2(\theta, \phi) = \left(\frac{1}{2\pi}\right)^2 & \left[1 + 2\rho_2\{1 + 2\rho\rho_1\cos(\theta - \mu)\}\frac{1}{\rho_\alpha}\cos(\mu_\alpha + \mu_\beta - \phi)\right. \\ & + 2\rho\rho_1\cos(\theta - \mu) + 2\rho\rho_2\frac{\rho_\alpha^2 - 1}{\rho_\alpha^2}\cos(\phi + \mu - 2\mu_\alpha - \mu_\beta) \\ & + 2\rho_1\rho_2\frac{\rho_\alpha^2 - 1}{\rho_\alpha^2}\cos(\theta + \phi - 2\mu_\alpha - \mu_\beta) \\ & \left. - 2\rho\rho_1\rho_2\frac{\rho_\alpha^2 - 1}{\rho_\alpha^3}\cos(\theta + \phi + \mu - 3\mu_\alpha - \mu_\beta)\right]. \end{aligned}$$

The marginal pdfs of Θ and Φ are $C(\mu, \rho\rho_1)$ and $C(\mu_\phi, \rho_\phi)$ from the reproductive property of cardioid distributions, where

$$\mu_\phi = \gamma_2 + \tan^{-1}\left\{\frac{\rho\,(1 - \rho_\alpha^2)\sin\gamma_1}{\rho_\alpha + \rho\,|1 - \rho_\alpha^2|\cos\gamma_1}\right\}$$

and

$$\rho_\phi = \rho_2\sqrt{\rho^2(1 - \rho_\alpha^2)^2 + \rho_\alpha^2 + 2\rho\rho_\alpha(1 - \rho_\alpha^2)\cos\gamma_1}$$

with $\gamma_1 = \mu - \mu_\alpha$ and $\gamma_2 = \mu_\alpha + \mu_\beta$. The conditional pdfs are easily obtainable from the joint and marginal pdfs. If $\rho_1 = 0$ or $\rho_2 = 0$ in MC_2, then Θ is independent of Φ. If $\mu = \mu_\alpha$ or $\rho_\alpha = 0$, the joint pdf has a mode at $(\mu, \mu + \mu_\beta)$, but we do not know whether the distribution is unimodal or not.

The joint trigonometric moments $\Psi(p,q) = E\{e^{i(p\Theta+q\Phi)}\}$, $p, q = 0, \pm 1, \pm 2, \dots$, and the circular–circular correlation coefficient r^2 in (9.1) for MC_2 are given in Wang and Shimizu (2012).

9.8.2 Using Mixtures

The method of parametric mixtures in Section 9.2.5 will be used to construct bivariate cardioid distributions as in SenGupta and Ong (2014) and Ong and SenGupta (2021).

9.8.2.1 Model I

As an example (SenGupta and Ong, 2014) of Model I in Section 9.2.5, consider the joint pdf based on cardioid pdfs

$$f(\theta_1, \theta_2) = \left(\frac{1}{2\pi}\right)^2 \int_0^1 \{1 + 2x\delta_1 \cos(\theta_1 - \mu_1)\}\{1 + 2x\delta_2 \cos(\theta_2 - \mu_2)\} g(x) dx \tag{9.20}$$

for $-1/2 \leq \delta_1, \delta_2 \leq 1/2$, when the mixing distribution is the truncated exponential distribution with pdf

$$g(x) = \frac{\tau e^{-\tau x}}{1 - e^{-\tau}}, \quad 0 < x < 1, \ \tau > 0.$$

Then we have a bivariate cardioid pdf

$$f(\theta_1, \theta_2) = \left(\frac{1}{2\pi}\right)^2 \left[1 + 2\gamma_1\{\delta_1 \cos(\theta_1 - \mu_1) + \delta_2 \cos(\theta_2 - \mu_2)\} \right. \\ \left. + 4\gamma_2\delta_1\delta_2 \cos(\theta_1 - \mu_1)\cos(\theta_2 - \mu_2)\right],$$

where

$$\gamma_1 = \frac{1}{\tau(1 - e^{-\tau})}\{1 - (\tau + 1)e^{-\tau}\}, \ \gamma_2 = \frac{1}{\tau^2(1 - e^{-\tau})}\left\{2 - (\tau^2 + 2\tau + 2)e^{-\tau}\right\}.$$

It suffices to choose τ such that $0 < \gamma_1, \gamma_2 < 1$. The marginal distributions of Θ_j ($j = 1, 2$) are cardioid distributions $C(\mu_j, \gamma_1\delta_j)$.

Another choice of mixing distribution (Ong and SenGupta, 2021) in (9.20) is the beta distribution on the interval (0, 1) with pdf

$$g(x) = \frac{1}{B(\alpha, \beta)} x^{\alpha-1}(1 - x)^{\beta-1}$$

for $\alpha, \beta > 0$, with mean $E(X) = \alpha/(\alpha + \beta)$ and second moment $E(X^2) = \alpha(\alpha + 1)/\{(\alpha + \beta)(\alpha + \beta + 1)\}$. The resulting bivariate cardioid pdf is given by

$$f(\theta_1, \theta_2) = \left(\frac{1}{2\pi}\right)^2 \left[1 + 2E(X)\left\{\delta_1 \cos(\theta_1 - \mu_1) + \delta_2 \cos(\theta_2 - \mu_2)\right\} \right. \\ \left. + 4E(X^2)\delta_1\delta_2 \cos(\theta_1 - \mu_1)\cos(\theta_2 - \mu_2)\right]$$

with $-1/2 \le \delta_1, \delta_2 \le 1/2$. The marginal distributions of Θ_j $(j = 1, 2)$ are cardioid distributions $C(\mu_j, \delta_j E(X))$.

9.8.2.2 Model II(a)

Ong and SenGupta (2021) applied the bivariate beta pdf

$$g(x,y) = \frac{\Gamma(\alpha+\beta+\gamma)}{\Gamma(\alpha)\Gamma(\beta)\Gamma(\gamma)} x^{\alpha-1}y^{\beta-1}(1-x-y)^{\gamma-1}, \quad 0 < x, y < 1, x+y < 1$$

with $\alpha, \beta, \gamma > 0$ as a mixing distribution to

$$f(\theta_1, \theta_2) = \left(\frac{1}{2\pi}\right)^2 \int_0^1 \int_0^{1-y} \{1 + 2x\delta_1 \cos(\theta_1 - \mu_1)\} \times \{1 + 2y\delta_2 \cos(\theta_2 - \mu_2)\} g(x,y) dxdy. \tag{9.21}$$

The resulting bivariate cardioid pdf has the form

$$f(\theta_1, \theta_2) = \left(\frac{1}{2\pi}\right)^2 \left[1 + 2\delta_1 E(X) \cos(\theta_1 - \mu_1) + 2\delta_2 E(Y) \cos(\theta_2 - \mu_2) + 4\delta_1\delta_2 E(XY) \cos(\theta_1 - \mu_1) \cos(\theta_2 - \mu_2)\right]$$

with $E(X) = \alpha/(\alpha+\beta+\gamma)$, $E(Y) = \beta/(\alpha+\beta+\gamma)$ and $E(XY) = \alpha\beta/\{(\alpha+\beta+\gamma)(\alpha+\beta+\gamma+1)\}$.

On appealing to the general Fourier series representation of toroidal distributions given above, we find

$$E\{\cos(\theta_1 - \mu_1)\} = \delta_1 E(X), E\{\cos(\theta_2 - \mu_2)\} = \delta_2 E(Y),$$

$$E\{\cos(\theta_1 - \mu_1) \cos(\theta_2 - \mu_2)\} = \delta_1\delta_2 E(XY)$$

and all other trigonometric moments are zeroes.

Ong and SenGupta (2021) applied different beta-type mixing distributions in (9.21) to obtain Model II(b), (c), and (d).

9.9 Bivariate Asymmetric Generalized von Mises Distribution

A submodel of (9.14) or (9.15) is considered by Rivest (1988) as

$$f(\theta_1, \theta_2) = C^{-1} \exp\{\kappa_1 \cos(\theta_1 - \mu_1) + \kappa_2 \cos(\theta_2 - \mu_2) + \alpha \cos(\theta_1 - \mu_1) \cos(\theta_2 - \mu_2) + \beta \sin(\theta_1 - \mu_1) \sin(\theta_2 - \mu_2)\}, \tag{9.22}$$

which is a six-parameter model, with normalizing constant

$$C = 4\pi^2 \sum_{\substack{j,\ell=-\infty \\ j+\ell:\text{ even}}}^{\infty} I_j(\kappa_1) I_\ell(\kappa_2) I_{(j+\ell)/2}\left(\frac{\alpha+\beta}{2}\right) I_{(j-\ell)/2}\left(\frac{\alpha-\beta}{2}\right)$$

in terms of the modified Bessel function $I_j(\cdot)$ of the first kind and order j. As a special case when $\kappa_2 = 0$ and $\alpha = \beta$ in (9.22), the marginal of Θ_1 is $\text{VM}(\mu_1, \kappa_1)$ and the conditional of Θ_2 given $\Theta_1 = \theta_1$ has $\text{VM}(\theta_2 - \mu_2 - (\theta_1 - \mu_1), \alpha)$, which is the location mixture model.

Several submodels having five parameters, like a bivariate normal distribution with means, variances and correlation coefficient, have been considered. Under the constraints on the trigonometric moments $E(e^{i\Theta})$, $E(e^{i\Psi})$, and $E\{e^{i(\Theta\pm\Psi)}\}$, the maximum entropy distribution of $(\Theta, \Psi)'$ in Section 9.2.2 is given by (Mardia, 1975b)

$$f(\theta, \psi) = C^{-1} \exp\{\kappa_1 \cos(\theta - \mu) + \kappa_2 \cos(\psi - \nu) + \rho\sqrt{\kappa_1\kappa_2} \cos(\theta \pm \psi - \tau)\},$$

where $\rho \leq 1, \kappa_1, \kappa_2 > 0$ and $0 \leq \mu, \nu, \tau < 2\pi$, with normalizing constant

$$C = 4\pi^2 \left\{ I_0(\kappa_1)I_0(\kappa_2)I_0(\rho\sqrt{\kappa_1\kappa_2}) + 2\sum_{p=1}^{\infty} I_p(\kappa_1)I_p(\kappa_2)I_p(\rho\sqrt{\kappa_1\kappa_2}) \cos(p\tau) \right\}$$

in terms of the modified Bessel function $I_p(\cdot)$ of the first kind and order p.

Another example of maximum entropy distributions is, under the constraints on the trigonometric moments $E(e^{i\Theta})$ and $E(e^{i\Psi})$ and the first joint sine moment $E\{\sin(\Theta - \mu)\sin(\Phi - \nu)\}$,

$$f(\theta, \psi) \propto \exp\{\kappa_1 \cos(\theta - \mu) + \kappa_2 \cos(\psi - \nu) + \kappa_3 \sin(\theta - \mu)\sin(\psi - \nu)\}, \tag{9.23}$$

which appears in Jammalamadaka and SenGupta (2001, pp. 52–53).

A submodel with five parameters is given by Singh et al. (2002) as

$$\begin{aligned} f(\theta_1, \theta_2) = C^{-1} \exp\{&\kappa_1 \cos(\theta_1 - \mu_1) + \kappa_2 \cos(\theta_2 - \mu_2) \\ &+\lambda \sin(\theta_1 - \mu_1)\sin(\theta_2 - \mu_2)\}. \end{aligned} \tag{9.24}$$

This appears as a maximum entropy distribution (9.23). The normalizing constant has an expression (Singh et al., 2002)

$$C = 4\pi^2 \sum_{m=0}^{\infty} \binom{2m}{m} \left(\frac{\lambda^2}{4\kappa_1\kappa_2}\right)^m I_m(\kappa_1)I_m(\kappa_2)$$

in terms of a singly infinite series. Singh et al. (2002) provide the "Sine" model, while Mardia (2007) propose the "Cosine" model with also five parameters and joint pdf

$$\begin{aligned} f(\theta_1, \theta_2) = C^{-1} \exp\{&\kappa_1 \cos(\theta_1 - \mu_1) + \kappa_2 \cos(\theta_2 - \mu_2) \\ &-\kappa_3 \cos(\theta_1 - \mu_1 - \theta_2 + \mu_2)\}, \end{aligned} \tag{9.25}$$

whose normalizing constant is given by

$$C = 4\pi^2 \left\{ I_0(\kappa_1)I_0(\kappa_2)I_0(\kappa_3) + 2\sum_{p=1}^{\infty} I_p(\kappa_1)I_p(\kappa_2)I_p(\kappa_3) \right\}.$$

The hybrid model (Kent et al., 2008) has density

$$f(\theta_1,\theta_2) \propto \exp\left[\kappa_1\cos(\theta_1-\mu_1)+\kappa_2\cos(\theta_2-\mu_2)+\beta\{(\cosh\lambda-1)\right.$$
$$\left.\times\cos(\theta_1-\mu_1)\cos(\theta_2-\mu_2)+\sinh\lambda\sin(\theta_1-\mu_1)\sin(\theta_2-\mu_2)\}\right], \quad (9.26)$$

which is a reparametrization of (9.22) when β and λ are free parameters, but when $\beta = 1$, (9.26) gives another five-parameter model.

9.10 A Model Generated from Brownian Motion

Exit distributions on the sphere for Brownian motion in $\mathbb{R}^p$ were treated in Section 8.4.10. We take up Brownian motion again in this section to show generation of toroidal distributions. In fact, Kato (2009) has proposed a dependent model for a pair of p-dimensional ($p \geq 2$) unit vector. Moreover, Kato and Pewsey (2015) give bivariate wrapped Cauchy distributions.

9.10.1 Definition and Properties

Let $\{B_t; t \geq 0\}$ be $\mathbb{R}^p$-valued Brownian motion. A particle starts at $B_0 = \xi$ whose Euclidean norm $\|\xi\|$ is less than ρ in $(0,1)$ and will eventually hit a $(p-1)$-dimensional sphere with radius ρ and center at the origin in $\mathbb{R}^p$. We denote the time of the first attaining the sphere by τ_1, that is, $\tau_1 = \inf\{t; \|B_t\| = \rho\}$. The particle will hit the unit sphere S^{p-1} at the time $\tau_2 = \inf\{t; \|B_t\| = 1\}$.

A pair of random vectors defined by $(\boldsymbol{U},\boldsymbol{V})$, where $\boldsymbol{U} = QB_{\tau_1}/\|B_{\tau_1}\|$, $\boldsymbol{V} = B_{\tau_2}$ and Q is a member of the group of orthogonal transformations on $\mathbb{R}^p$, has joint pdf

$$f(\boldsymbol{u},\boldsymbol{v}) = \frac{1}{\omega_{p-1}^2}\frac{1-\rho^2}{(1-2\rho\boldsymbol{u}'Q\boldsymbol{v}+\rho^2)^{p/2}}\frac{\rho^{p-2}(\rho^2-\|\xi\|^2)}{(\rho^2-2\rho\boldsymbol{u}'Q\xi+\|\xi\|^2)^{p/2}}, \quad \boldsymbol{u},\boldsymbol{v}\in S^{p-1}. \quad (9.27)$$

Here ω_p denotes the area of the unit sphere, that is, $\omega_p = 2\pi^{p/2}/\Gamma(p/2)$. The joint pdf is derived from the fact that the distribution of $\boldsymbol{U}$ is the exit distribution $\mathrm{Exit}_p(\rho^{-1}Q\xi)$ in Section 8.4.10 and the conditional distribution of $\boldsymbol{V}$ given $\boldsymbol{U} = \boldsymbol{u}$ is $\mathrm{Exit}_p(\rho Q'\boldsymbol{u})$. The conditional distribution of $\boldsymbol{U}$ given $\boldsymbol{V} = \boldsymbol{v}$ is not of the standard form. The marginal distribution of $\boldsymbol{V}$ is $\mathrm{Exit}_p(\xi)$.

If the particle starts at $B_0 = \mathbf{0}$, i.e. $\xi = \mathbf{0}$ in (9.27), the joint pdf reduces to

$$c(\boldsymbol{u},\boldsymbol{v}) = \frac{1}{\omega_{p-1}^2}\frac{1-\rho^2}{(1-2\rho\boldsymbol{u}'Q\boldsymbol{v}+\rho^2)^{p/2}}, \quad \boldsymbol{u},\boldsymbol{v}\in S^{p-1}. \quad (9.28)$$

The domain of ρ is extended to include $\rho = 0$, whose case corresponds to the uniform distribution on $S^{p-1} \times S^{p-1}$. The distribution with pdf (9.28) is denoted by $\mathrm{BS}_p(\rho Q)$ in Kato (2009). The marginals of $\boldsymbol{U}$ and $\boldsymbol{V}$ are uniform distributions on S^{p-1}. Thus, the model with pdf (9.28) may be viewed as a copula on $S^{p-1} \times S^{p-1}$. Note that this copula has the property of periodicity for the variables. The conditional distribution of $\boldsymbol{U}$ given $\boldsymbol{V} = \boldsymbol{v}$ is $\mathrm{Exit}_p(\rho Q\boldsymbol{v})$. The conditional distribution of $\boldsymbol{V}$ given $\boldsymbol{U} = \boldsymbol{u}$ follows $\mathrm{Exit}_p(\rho Q'\boldsymbol{u})$ and $\boldsymbol{V}'Q'\boldsymbol{U}$ given $\boldsymbol{V} = \boldsymbol{v}$ follows $H'(\rho, p/2 - 1)$, whose notation appears in Section 8.4.10.

The distribution $\mathrm{BS}_p(\rho Q)$ is closed under orthogonal transformations. This means that if $(\boldsymbol{U}, \boldsymbol{V})$ is distributed as $\mathrm{BS}_p(\rho Q)$, and Q_1 and Q_2 are orthogonal transformations on $\mathbb{R}^p$, then $(Q_1\boldsymbol{U}, Q_2\boldsymbol{V})$ is distributed as $\mathrm{BS}_p(\rho Q_1 Q Q_2')$. As an application of a more general result given in Kato (2009), the moments and the Johnson and Wehrly (1977) circular–circular correlation coefficient ρ_{JW} are

$$E(\boldsymbol{U}) = E(\boldsymbol{V}) = \boldsymbol{0}, \quad E(\boldsymbol{U}\boldsymbol{U}') = E(\boldsymbol{V}\boldsymbol{V}') = \rho^{-1}I, \quad E(\boldsymbol{U}\boldsymbol{V}') = p^{-1}\rho Q$$

and

$$\rho_{\mathrm{JW}} \equiv \lambda^{1/2} = \rho,$$

where λ is the largest eigen-value of $\Sigma_{UU}^{-1}\Sigma_{UV}\Sigma_{VV}^{-1}\Sigma_{UV}'$ with variance–covariance matrices Σ_{UU} of $\boldsymbol{U}$ and Σ_{VV} of $\boldsymbol{V}$ and covariance matrix Σ_{UV} of $\boldsymbol{U}$ and $\boldsymbol{V}$.

9.10.2 The Case of the Torus

In the case of $p = 2$, that is, $\mathrm{BS}_2(\rho Q)$, the pdf is given by

$$c(\boldsymbol{u}, \boldsymbol{v}) = \frac{1 - \rho^2}{4\pi^2(1 - 2\rho\boldsymbol{u}'Q\boldsymbol{v} + \rho^2)}, \qquad \boldsymbol{u}, \boldsymbol{v} \in S^1.$$

When $(\boldsymbol{U}, \boldsymbol{V})$ is distributed as $\mathrm{BS}_2(\rho Q)$, we use transformation $(Z_U, Z_V) = (U_1 + iU_2, V_1 + iV_2)$, $i = \sqrt{-1}$, and $\psi = \rho e^{i\theta}$ for $\boldsymbol{U} = (U_1, U_2)'$, $\boldsymbol{V} = (V_1, V_2)'$ and θ which satisfies

$$Q = \begin{pmatrix} \cos\theta & -\det Q \sin\theta \\ \sin\theta & \det Q \cos\theta \end{pmatrix}, \qquad 0 \le \theta < 2\pi.$$

Then the joint pdf for $(Z_U, Z_V)'$ is given by

$$c(z_u, z_v) = \frac{1}{4\pi^2}\frac{1 - |\psi|^2}{|1 - \psi z_v z_u^{-\det Q}|^2}, \qquad z_u, z_v \in \{z \in \mathbb{C}; |z| = 1\}.$$

The marginals of Z_U and Z_V are circular uniform distributions, while the conditional distributions of Z_U given $Z_V = z_v$ and Z_V given $Z_U = z_{\boldsymbol{u}}$ are exit distributions for the circle, that is, wrapped Cauchy distributions.

9.10.3 Bivariate Wrapped Cauchy Distribution

Kato and Pewsey (2015) proposed a five-parameter bivariate wrapped Cauchy distribution with pdf

$$f(\theta_1, \theta_2) = c\left\{c_0 - c_1\cos(\theta_1 - \mu_1) - c_2\cos(\theta_2 - \mu_2) - c_3\cos(\theta_1 - \mu_1)\right.$$
$$\left.\times\cos(\theta_2 - \mu_2) - c_4\sin(\theta_1 - \mu_1)\sin(\theta_2 - \mu_2)\right\}^{-1}, \quad \theta_1, \theta_2 \in (-\pi, \pi)^2, \quad (9.29)$$

where $-\pi \le \mu_1, \mu_2 < \pi$ and $c, c_1, \dots, c_4$ are functions of $0 \le \xi_1, \xi_2 < 1$ and $-1 < \rho < 1$. The above distribution is generated from the toroidal distribution by Kato (2009) whose pdf is

$$g(\phi_1, \phi_2) = \frac{1 - \rho^2}{4\pi^2\{1 + \rho^2 - 2|\rho|\cos(q\phi_1 - \phi_2 - \nu)\}}$$

with $q = \mathrm{sgn}(\rho)$, $-\pi \le \nu < \pi$ and $-1 < \rho < 1$ through a Möbius circle transformation of each component. The marginals and conditionals of the proposed distribution are wrapped Cauchy distributions. The distribution with pdf (9.29) can also be derived as a maximum entropy distribution or a special case of (9.7). A stationary Markov process using (9.29) is constructed in the paper by Kato and Pewsey (2015).

9.11 Distributions on the Hyper-Torus

9.11.1 A General Method of Construction

Kim et al. (2016) proposed a family of multivariate extensions of Wehrly and Johnson's (1980) model as follows. Suppose that f_ℓs are circular pdfs with their distribution functions $F_\ell(\theta_\ell) = \int_{-\pi}^{\theta_\ell} f_\ell(\eta_\ell)d\eta_\ell$ for $\ell = 1, \dots, k$ and g_js are circular pdfs for $j = 1, \dots, m$. Then a function

$$f(\theta) = (2\pi)^m \prod_{j=1}^{m}\left\{g_j\left(2\pi\sum_{\ell=j}^{k}F_\ell(\theta_\ell)\right)\right\}\prod_{\ell=1}^{k}f_\ell(\theta_\ell) \quad (9.30)$$

provides a k-variate pdf, where $\theta \in (-\pi, \pi]^k$ and $1 \le m \le k - 1$. By convention, $2\pi\sum_{\ell=j}^{k}F_\ell(\theta_\ell)$ in the argument of each $g_j(\cdot)$ is interpreted as modulo 2π. Then Equation (9.30) has the following properties.

a) The random variables $\Theta_{m+1}, \dots, \Theta_k$ are jointly independent.
b) For $j = 1, \dots, m$, the conditional pdf of Θ_j given $\Theta_{j+1} = \theta_{j+1}, \dots, \Theta_k = \theta_k$ is

$$f(\theta_j|\theta_{j+1}, \dots, \theta_k) = 2\pi\, g_j\left(2\pi\sum_{\ell=j}^{k}F_\ell(\theta_\ell)\right)f_j(\theta_j).$$

c) For $j = 1, \ldots, m$, the conditional pdf of Θ_j given $\Theta_1 = \theta_1, \ldots, \Theta_{j-1} = \theta_{j-1}, \Theta_{j+1} = \theta_{j+1}, \ldots, \Theta_k = \theta_k$ is

$$f(\theta_j|\theta_1, \ldots, \theta_{j-1}, \theta_{j+1}, \ldots, \theta_k) = (2\pi)^j \prod_{s=1}^{j} g_s\left(2\pi \sum_{\ell=s}^{k} F_\ell(\theta_\ell)\right) f_j(\theta_j).$$

d) All marginal pdfs are given by $f_\ell(\theta_\ell)$ for $\ell = 1, \ldots, k$.
e) Circular variables $\Theta_1, \ldots, \Theta_k$ are jointly independent in the case when the g_js are all uniformly distributed.

As a special case, the pdf of three angular variables is

$$f(\theta_1, \theta_2, \theta_3) = 2\pi\, g\left(2\pi \sum_{\ell=1}^{3} F_\ell(\theta_\ell)\right) \prod_{\ell=1}^{3} f_\ell(\theta_\ell), \tag{9.31}$$

which gives a pairwise independent but not three-way independent distribution.

In (9.30), $(U_1, \ldots, U_k)'$ with a transformation $U_\ell = F_\ell(\Theta_\ell)$ for $\ell = 1, \ldots, k$ follows a k-variate circular distribution with circular uniform marginals and gives a circula

$$c(\theta_1, \theta_2, \ldots, \theta_k) = \frac{1}{(2\pi)^{k-m}} \prod_{j=1}^{m} g_j\left(\sum_{\ell=j}^{k} \theta_\ell\right).$$

A related approach to the above argument is the following. As a multivariate extension of the bivariate circula in Section 9.1.5, Jones et al. (2015) considered the joint density of a circular uniform random variable Φ and circular random variables $\Theta_k \equiv (\Omega_k + q_k\Phi) (\mathrm{mod}\, 2\pi)$ for $k = 1, \ldots, d\, (d \geq 3)$, where Ω_k has density g_k and independently of each other and of Φ, and $q_k = \{1, -1\}$. The density is

$$c_{d+1}(\phi, \theta_1, \ldots, \theta_d) = \frac{1}{2\pi} \prod_{k=1}^{d} g_k(\theta_k - q_k\phi).$$

The multivariate circula density is

$$c_d(\theta_1, \ldots, \theta_d) = \frac{1}{2\pi} \int_0^{2\pi} \prod_{k=1}^{d} g_k(\theta_k - q_k\phi) d\phi$$

and thus the circula density of the (j, k)th bivariate marginal of c_d is of the same form as the bivariate circula density:

$$\frac{1}{2\pi} h_{jk}(\theta_k - q_{jk}\theta_j), \quad q_{jk} = q_j q_k \in \{1, -1\},$$

where

$$h_{jk}(\omega) = \int_0^{2\pi} g_j(\psi) g_k(\omega + q_{jk}\psi) d\psi,$$

which is the density of $(\Omega_k - q_{jk}\Omega_j) \pmod{2\pi}$. From the above relation, if gs are all of the same form and closed under convolution, hs are of the same form as the gs form. If g_k is symmetric about 0 and its mean resultant length is ρ_k $(k = 1, \dots, d)$, then $(\Omega_k - q_{jk}\Omega_j)$ has the mean resultant length $\rho_{jk} = E\{\cos(\Omega_k - q_{jk}\Omega_j)\} = E(\cos\Omega_j)E(\cos\Omega_k) = \rho_j\rho_k$, which gives the dependence structure of the circula.

9.11.2 von Mises and Asymmetric von Mises Marginals

For the bivariate model (9.7), Shieh and Johnson (2005) utilized von Mises marginals, while generalized von Mises marginals which are possibly asymmetric were treated in Shieh et al. (2011). As the marginals of the model (9.31), Kim et al. (2016) used 3-parameter possibly asymmetric bimodal generalized von Mises densities by Kim and SenGupta (2013) given in Section 5.3.3.

Mardia et al. (2008) proposed a multivariate extension of the bivariate Singh et al. (2002) model (9.24). The pdf of a p-dimensional random vector $\boldsymbol{\Theta} = (\Theta_1, \dots, \Theta_p)'$ is given by

$$f(\boldsymbol{\theta}) = \{T(\boldsymbol{\kappa}, \boldsymbol{\Lambda})\}^{-1} \exp\{\boldsymbol{\kappa}' c(\boldsymbol{\theta}, \boldsymbol{\mu}) + s(\boldsymbol{\theta}, \boldsymbol{\mu})' \boldsymbol{\Lambda} s(\boldsymbol{\theta}, \boldsymbol{\mu})/2\}, \tag{9.32}$$

where $\boldsymbol{\theta} = (\theta_1, \dots, \theta_p)'$, $-\pi \le \theta_j < \pi$, $\boldsymbol{\mu} = (\mu_1, \dots, \mu_p)'$, $-\pi \le \mu_j < \pi$, $\boldsymbol{\kappa} = (\kappa_1, \dots, \kappa_p)'$, $\kappa_j \ge 0$, $\boldsymbol{\Lambda} = (\lambda_{jk})$, $-\infty < \lambda_{jk} = \lambda_{kj} < \infty$ $(j \ne k)$, $\lambda_{jj} = 0$,

$$c(\boldsymbol{\theta}, \boldsymbol{\mu}) = (\cos(\theta_1 - \mu_1), \dots, \cos(\theta_p - \mu_p))'$$

and

$$s(\boldsymbol{\theta}, \boldsymbol{\mu}) = (\sin(\theta_1 - \mu_1), \dots, \sin(\theta_p - \mu_p))'$$

for $j, k = 1, \dots, p$ with $T(\boldsymbol{\kappa}, \boldsymbol{\Lambda})$ a normalizing constant. The distribution of $\boldsymbol{\Theta} - \boldsymbol{\mu}$ is the same as that of $-(\boldsymbol{\Theta} - \boldsymbol{\mu})$. Thus, the marginals are symmetric though their densities do not have familiar forms. For $p = 1$ the distribution is a von Mises, and for $p = 2$ it reduces to the bivariate Singh et al. (2002) distribution (9.24). For $p > 2$ the normalizing constant is unknown in any explicit form. The model (9.32) is unimodal if $\mathrm{diag}(\kappa_1, \dots, \kappa_p) - \Lambda$ is positive definite (Mardia and Voss, 2014). Concentrated multivariate sine models are studied in Mardia et al. (2012).

The conditional distribution of $\Theta_1, \dots, \Theta_r$ given $\Theta_{r+1} = \theta_{r+1}, \dots, \Theta_p = \theta_p$ for (9.32) is an extended r-variate von Mises distribution with density

$$f(\theta_1, \dots, \theta_r | \theta_{r+1}, \dots, \theta_p) \propto \exp\left(\sum_{j=1}^{r} \kappa_j \cos\theta_j + \sum_{j=1}^{r-1} \sum_{\ell=j+1}^{r} \lambda_{j\ell} \sin\theta_j \sin\theta_\ell + \sum_{j=1}^{r} c_j \sin\theta_j \right),$$

where c_j $(j = 1, \dots, r)$ is constant with respect to $\theta_1, \dots, \theta_r$.

9.11.3 Product of Cardioid Densities

Shimizu and Imoto (2021) presented a d-dimensional cardioid-type product pdf on the hypertorus defined by

$$f(\theta_1, \dots, \theta_d) = C_d^{-1} \prod_{j=1}^{d} \left(\{1 + 2\xi_j \cos(\theta_j - \mu_j)\} \left[1 + 2\delta_j \cos\{2(\theta_j - \mu_j)\}\right] \right) \times \left[1 + \lambda \sin \left\{ \sum_{j=1}^{d} q_j(\theta_j - \mu_j) - \eta \right\} \right], \quad 0 \le \theta_1, \dots, \theta_d < 2\pi \tag{9.33}$$

with $q_1 = 1$, $q_j = \pm 1\,(j = 2, \dots, d)$, $-1 \le \lambda \le 1$, $0 \le \eta < \pi$, $0 \le \mu_j < 2\pi$, $0 \le \xi_j \le 1/2$, and $-1/2 \le \delta_j \le 1/2\,(j = 1, \dots, d)$. Here the normalizing constant is simply expressed as

$$C_d = (2\pi)^d \left[1 - \lambda \prod_{j=1}^{d} \{\xi_j(1 + \delta_j) \sin \eta\} \right].$$

When $d = 2$ as a particular case of (9.33), the toroidal distribution is denoted TCTP_2. The role of the parameters is: μ_1 and μ_2 are location parameters, η is a skewing parameter and controls location, $\xi_1, |\xi_2|, \delta_1, |\delta_2|$ control concentrations, and λ is a dependence and skewing parameter. It is obvious that Θ_1 is independent of Θ_2 if and only if $\lambda = 0$. The model is a member of double trigonometric sums, a special case of multiple nonnegative trigonometric sums distributions proposed by Fernández-Durán and Gregorio-Domínguez (2014). However, the normalizing constant C_2 is quite simple. The marginal and conditional distributions of TCTP_2 belong to the CTP_2 family in Shimizu and Imoto (2021). It seems difficult to express the pdf of TCTP_2 as the form of (9.7).

Discretization for a circular distribution has been studied in Section 6.2.2. A method for constructing a bivariate circular distribution on the prefixed equidistant points $2\pi j/n_1$ $(j = 0, 1, 2, \dots, n_1 - 1)$ and $2\pi k/n_2\,(k = 0, 1, 2, \dots, n_2 - 1)$ and its application to TCTP_2 can be seen in Shimizu and Imoto (2021). It seems possible to extend the method to the multivariate case.

9.12 Statistical Inference

Hypothesis test of joint independence in the model (9.31) with von Mises g whose density is given by

$$f(\phi, \psi, \chi) = \frac{1}{2\pi I_0(\kappa)} f_1(\phi) f_2(\psi) f_3(\chi) \exp\{\cos(\gamma - \delta)\},$$

where $\gamma = 2\pi\{F_1(\phi) + F_2(\psi) + F_3(\chi)\}$ modulo 2π was studied in Kim et al. (2016). The uniformly most powerful invariant (UMPI) test for the null hypothesis rejects when $V = \sum_{m=1}^{n} \cos(\gamma_m - \hat{\delta}) > v_0$ for some $v_0 > 0$, where $\hat{\delta}$ denotes the maximum-likelihood estimate of δ, which is the sample mean direction of γ_m for $m = 1, \ldots, n$. The UMPI test for independence between (ϕ, ψ) and χ, the UMP test for independence between ϕ and ψ, and the UMPI test for serial correlation of ϕ are also dealt with in Kim et al. (2016).

For inference of the model (9.32), Mardia et al. (2008) considered method of moments, maximum-likelihood estimation, maximum pseudo-likelihood method, and hypothesis tests. For $p = 3$, likelihood ratio tests of $\mu_1 = \mu_3$, $\mu_1 = \mu_3 = \mu_2 - \pi$ and independence of ϕ_1, ϕ_2, and ϕ_3 are illustrated.

For both the bivariate wrapped α-stable and Linnik families of distributions, SenGupta and Roy (2026a,b) have obtained universal (i.e. for the entire support of the parameter) estimators of the index parameter α, which is the crucial parameter determining the tail behaviors of the distributions. They also establish the CAN properties of these estimators and their asymptotic distributions. These yield large-sample consistent tests for the parameters.

9.13 Applications

Recently studies related to problems in the area of bioinformatics can be seen in the literature. Singh et al. (2002) fitted the model (9.24) in Section 9.9 to paired observations of the dihedral angles of the proline residue. Mardia et al. (2012) study modality of the "Sine" (9.24) and "Cosine" (9.25) models and fit mixtures of the "Cosine" model to Malate dehydrogenase and Myoglobin conformational angles. Fernández-Durán (2007) used the Wehrly–Johnson (1980) model (9.7) with nonnegative trigonometric sums for g, f_1 and f_2 to apply it to dihedral angles. Mardia et al. (2008) analyzed the gamma-turns data consisting of Glycine-Phenylalanine-Threonine sequences in proteins. Kim et al. (2016) illustrated their methods using the gamma-turns and compared their model to Mardia's model and the multivariate nonnegative trigonometric sums model by Fernández-Durán and Gregorio-Domínguez (2014). See also Fernández-Durán and Gregorio-Domínguez (2016) concerning an R package for the statistical analysis using nonnegative trigonometric sums. An illustration using the bivariate wrapped Cauchy distribution was made by Kato and Pewsey (2015) for a subset of the dihedral angles of Tyrosine. Bayesian procedures using a Dirichlet process mixture of bivariate von Mises sine distributions were developed by Lennox et al. (2009a,b) and an extension to angle pairs at multiple sequence positions can be seen in Lennox et al. (2010). Di Marzio et al. (2011) studied a toroidal density kernel estimator and illustrated an estimate for dihedral angles.

Shieh et al. (2011) used a family of bivariate distributions whose joint pdf has the same form as (9.7) but marginals follow generalized von Mises pdfs to study the structure organizations for each of the pairs of bacterial genomes: (*Thermotoga*, *Sulfolobus*), (*Clostridium*, *Thermotoga*), (*Lactobacillus*, *Clostridium*), (*Staphylococcus*, *Lactobacillus*), (*Oceanobacillus*, *Staphylococcus*), (*Bacillus*, *Oceanobacillus*), (*Bacillus anthracis*, *Bacillus subtitlis*), and (*Bacillus anthracis Ames*, *Bacillus anthracis Sterne*). It is reported that for asymmetric and bimodal marginal of shared orthologs, it is reasonable to fit generalized von Mises distributions.

Environmental and meteorological sciences have also many kinds of applications. Shieh and Johnson (2005) studied the maximum-likelihood estimation of parameters when $f_1(\theta_1)$ and $f_2(\theta_2)$ are von Mises $\mathrm{VM}(\mu_1, \kappa_1)$ and $\mathrm{VM}(\mu_2, \kappa_2)$ pdfs and g is also a von Mises $\mathrm{VM}(\mu_{12}, \kappa_{12})$ pdf in (9.7). They gave an illustrative example of the wind direction data (Johnson and Wehrly, 1977) measured at 6 in the morning and at noon for 21 consecutive days at a weather station in Milwaukee and showed a good fit of the model to the dataset. Kato and Jones (2010) used a circular–circular regression model and Di Marzio et al. (2013) a nonparametric regression model to analyze wind directions at 6:00 a.m. and 12:00 a.m. in Texas. Polsen and Taylor (2015) reviewed circular–circular regression models with discussion about parameter estimation, inference and diagnostic analysis, and gave analysis of wind directions with comparison between models. Lagona (2016) used the multivariate von Mises density by Mardia et al. (2008) with the GLM setting for animal orientation data. Wang and Shimizu (2012) fitted MC_2 derived from a structural model with cardioid measurement errors in Section 9.8.1 to the wind direction dataset given in Johnson and Wehrly (1977). Wrapped Cauchy or von Mises measurement errors have not yet been treated at the present moment.

SenGupta and Roy (2026a,b) provide statistical inference for the most important index parameter α of both the wrapped bivariate α-stable and α-Linnik family of distributions and apply those to real-life financial datasets.

References

Abe, T., Imoto, T., Shiohama, T. and Miyata, Y. (2021). On some flexible models for circular, toroidal, and cylindrical data, *Directional Statistics for Innovative Applications: A Bicentennial Tribute to Florence Nightingale*, SenGupta, A. and Arnold, B. C. (Eds.), Springer, Singapore, 229–243.

Ameijeiras-Alonso, J. and Ley, C. (2022). Sine-skewed toroidal distributions and their application in protein bioinformatics, *Biostatistics*, 23, 685–704.

Arnold, B. C. and Strauss, D. J. (1991). Bivariate distributions with conditionals in prescribed exponential families, *Journal of the Royal Statistical Society*, B53, 365–375.

Baba, Y. (1981). Statistics of angular data: Wrapped normal distribution model, *Proceedings of the Institute of Statistical Mathematics*, 28, 41–54 (in Japanese).

Bekker, A., Rad, N. N., Arashi, M. and Ley, C. (2021). Generalized skew-symmetric circular and toroidal distributions, *Directional Statistics for Innovative Applications: A Bicentennial Tribute to Florence Nightingale*, SenGupta, A. and Arnold, B. C. (Eds.), Springer, Singapore, 161–186.

Di Marzio, M., Panzera, A. and Taylor, C. C. (2011). Kernel density estimation on the torus, *Journal of Statistical Planning and Inference*, 141, 2156–2173.

Di Marzio, M., Panzera, A. and Taylor, C. C. (2013). Non-parametric regression for circular responses, *Scandinavian Journal of Statistics*, 40, 238–255.

Fernández-Durán, J. J. (2007). Models for circular-linear and circular-circular data constructed from circular distributions based on nonnegative trigonometric sums, *Biometrics*, 63, 579–585.

Fernández-Durán, J. J. and Gregorio-Domínguez, M. M. (2014). Modeling angles in proteins and circular genomes using multivariate angular distributions based on multiple nonnegative trigonometric sums, *Statistical Applications in Genetics and Molecular Biology*, 13, 1–18.

Fernández-Durán, J. J. and Gregorio-Domínguez, M. M. (2016). CircNNTSR: An R package for the statistical analysis of circular, multivariate circular, and spherical data using nonnegative trigonometric sums, *Journal of Statistical Software*, 70 (6), 1–19. https://www.jstatsoft.org/article/view/v070i06.

Fisher, N. I. (1993). *Statistical Analysis of Circular Data*, Cambridge University Press, Cambridge.

Fisher, N. I. and Lee, A. J. (1982). Nonparametric measures of angular-angular association, *Biometrika*, 69, 315–321.

Fisher, N. I. and Lee, A. J. (1983). A correlation coefficient for circular data, *Biometrika*, 70, 327–332.

Imoto, T. and Abe, T. (2021). Simple construction of a toroidal distribution from independent circular distributions, *Journal of Multivariate Analysis*, 186, 104799.

Jammalamadaka, S. R. and Sarma, Y. R. (1988). A correlation coefficient for angular variables, *Statistical Theory and Data Analysis II*, Matusita, K. (Ed.), North Holland, Amsterdam, 349–364.

Jammalamadaka, S. R. and SenGupta, A. (2001). *Topics in Circular Statistics*, World Scientific, Singapore.

Johnson, R. A. and Wehrly, T. E. (1977). Measures and models for angular correlation and angular-linear correlation, *Journal of the Royal Statistical Society*, B39, 222–229.

Jones, M. C., Pewsey, A. and Kato, S. (2015). On a class of circulas: Copulas for circular distributions, *Annals of the Institute of Statistical Mathematics*, 67, 843–862.

Jupp, P. E. and Mardia, K. V. (1980). A general correlation coefficient for directional data and related regression problems, *Biometrika*, 67, 163–173.

Kato, S. (2009). A distribution for a pair of unit vectors generated by Brownian motion, *Bernoulli*, 15, 898–921.

Kato, S. (2010). A Markov process for circular data, *Journal of the Royal Statistical Society*, B72, 655–672.

Kato, S. and Jones, M. C. (2010). A family of distributions on the circle with links to, and applications arising from, Möbius transformation, *Journal of the American Statistical Association*, 105, 249–262.

Kato, S. and Pewsey, A. (2015). A Möbius transformation-induced distribution on the torus, *Biometrika*, 102, 359–370.

Kato, S. and Shimizu, K. (2008). Dependent models for observations which include angular ones, *Journal of Statistical Planning and Inference*, 138, 3538–3549.

Kent, J. T., Mardia, K. V. and Taylor, C. C. (2008). Modelling strategies for bivariate circular data, *Proc. LASR 2008: Art Sci. Statist. Bioinformatics*, Barber, S., Baxter, P. D., Gusnantoand, A. and Mardia, K. V. (Eds.), University of Leeds, Leeds, 70–73.

Kim, S. and SenGupta, A. (2013). A three-parameter generalized von Mises distribution, *Statistical Papers*, 54, 685–693.

Kim, S., SenGupta, A. and Arnold, B. C. (2016). A multivariate circular distribution with applications to the protein structure prediction problem, *Journal of Multivariate Analysis*, 143, 374–382.

Lagona, F. (2016). Regression analysis of correlated circular data based on the multivariate von Mises distribution, *Environmental and Ecological Statistics*, 23, 89–113.

Lennox, K. P., Dahl, D. B., Vannucci, M. and Tsai, J. W. (2009a). Density estimation for protein conformation angles using a bivariate von Mises distribution and Bayesian nonparametrics, *Journal of the American Statistical Association*, 104, 586–596.

Lennox, K. P., Dahl, D. B., Vannucci, M. and Tsai, J. W. (2009b). Correction: Density estimation for protein conformation angles using a bivariate von Mises distribution and Bayesian nonparametrics, *Journal of the American Statistical Association*, 104, 1728.

Lennox, K. P., Dahl, D. B., Vannucci, M., Day, R. and Tsai, J. W. (2010). A Dirichlet process mixture of hidden Markov models for protein structure prediction, *The Annals of Applied Statistics*, 4, 916–942.

Mardia, K. V. (1970). *Families of Bivariate Distributions*, Griffin, London.

Mardia, K. V. (1975a). Statistics of directional data (with discussion), *Journal of the Royal Statistical Society*, B37, 349–393.

Mardia, K. V. (1975b). Characterizations of directional distributions, *A Modern Course on Statistical Distributions in Scientific Work*, Patil, G. P., Kotz, S. and Ord, J.K. (Eds.), 3 edn, Reidel, Dordrecht, 365–385.

Mardia, K. V. (2007). On some recent advancements in applied shape analysis and directional statistics, Leeds Annual Statistical Research Workshop 2007,

Department of Statistics, University of Leeds, *Systems Biology & Statistical Bioinformatics*, Barber, S., Baxter, P. D. and Mardia, K. V. (Eds.), Leeds University Press, Leeds, 9–17.

Mardia, K. V. (2010). Bayesian analysis for bivariate von Mises distributions, *Journal of Applied Statistics*, 37, 515–528.

Mardia, K. V. and Jupp, P. E. (2000). *Directional Statistics*, Wiley, Chichester.

Mardia, K. V. and Voss, J. (2014). Some fundamental properties of a multivariate von Mises distribution, *Communications in Statistics–Theory and Methods*, 43, 1132–1144.

Mardia, K. V., Hughes, G., Taylor, C. C. and Singh, H. (2008). A multivariate von Mises distribution with applications to bioinformatics, *The Canadian Journal of Statistics*, 36, 99–109.

Mardia, K. V., Kent, J. T., Zhang, Z., Taylor, C. C. and Hamelryck, T. (2012). Mixtures of concentrated multivariate sine distributions with applications to bioinformatics, *Journal of Applied Statistics*, 39, 2475–2492.

Masuyama, M. (1939). Correlation between tensor quantities, *Proceedings of the Physico-mathematical Society of Japan, 3rd Series*, 21, 638–647.

Ong, S. H. and SenGupta, A. (2012). Bivariate and multivariate circular distributions by mixtures, *Journal of the Indian Statistical Association*, 50, 193–204.

Ong, S. H. and SenGupta, A. (2021). Bivariate cardioid distributions, *Directional Statistics for Innovative Applications: A Bicentennial Tribute to Florence Nightingale*, SenGupta, A. and Arnold, B. C. (Eds.), Springer, Singapore, 245–262.

Polsen, O. and Taylor, C. C. (2015). Parametric circular-circular regression and diagnostic analysis, *Geometry Driven Statistics*, Chapter 5, Dryden, I. L. and Kent, J. T. (Eds.), Wiley, Hoboken, 115–128.

Rivest, L. P. (1982). Some statistical methods for bivariate circular data, *Journal of the Royal Statistical Society*, B44, 81–90.

Rivest, L. P. (1988). A distribution for dependent unit vectors, *Communications in Statistics–Theory and Methods*, 17, 461–483.

SenGupta, A. (2004). On the constructions of probability distributions for directional data, *Bulletin of the Calcutta Mathematical Society*, 96, 139–154.

SenGupta, A. and Arnold, B. C. (2025). Circular and toroidal families of discrete distributions. *Sankhya A*, https://doi.org/10.1007/s13171-025-00424-7.

SenGupta, A. and Ong, S. H. (2014). A unified approach for construction of probability models for bivariate linear and directional data, *Communications in Statistics–Theory and Methods*, 43, 2563–2569.

SenGupta, A. and Roy, M. (2026a). On universal and efficient estimation of the index parameter of the wrapped bivariate stable family of distributions, Submitted.

SenGupta, A. and Roy, M. (2026b). On universal and efficient estimation of the index parameter of the wrapped bivariate Linnik family of distributions, Submitted.

Shieh, G. S. and Johnson, R. A. (2005). Inferences based on a bivariate distribution with von Mises marginals, *Annals of the Institute of Statistical Mathematics*, 57, 789–802.

Shieh, G. S., Johnson, R. A. and Frees, E. W. (1994). Testing independence of bivariate circular data and weighted degenerate U-statistics, *Statistica Sinica*, 4, 729–747.

Shieh, G. S., Zheng, S., Johnson, R. A., Chang, Y. F., Shimizu, K., Wang, C. C. and Tang, S. L. (2011). Modeling and comparing the organization of circular genomes, *Bioinformatics*, 27, 912–918.

Shimizu, K. and Imoto, T. (2021). A circular distribution constructed from the product of cardioid-type densities with (hyper-) toroidal extension, *Directional Statistics for Innovative Applications: A Bicentennial Tribute to Florence Nightingale*, SenGupta, A. and Arnold, B. C. (Eds.), Springer, Singapore, 211–227.

Singh, H., Hnizdo, V. and Demchuk, E. (2002). Probabilistic model for two dependent circular variables, *Biometrika*, 89, 719–723.

Thompson, J. W. (1975). Discussion of Professor Mardia's Paper, *Journal of the Royal Statistical Society*, B37, 379.

Wang, M. Z. and Shimizu, K. (2012). On applying Möbius transformation to cardioid random variables, *Statistical Methodology*, 9, 604–614.

Wehrly, T. E. and Johnson, R. A. (1980). Bivariate models for dependence of angular observations and a related Markov process, *Biometrika*, 67, 255–256.

Zhan, X., Ma, T., Liu, S. and Shimizu, K. (2019). On circular correlation for data on the torus, *Statistical Papers*, 60, 1827–1847.

10

Distributions on the (Vertical) Cylinder and Hyper-Cylinder

In the present chapter, we consider bivariate distributions of circular and linear variables, called distributions on the cylinder or cylindrical distributions, as an extension of circular distributions. Discrete or mixed cylindrical distributions, i.e. where one of the variables—linear or circular, is discrete are also derived. These distributions are also distributions for bivariate directional data but in contrast to that of toroidal distributions in Chapter 9, one of the variables here is linear. Multivariate distributions or distributions on the hyper-cylinder where some of the variables are circular and some are linear are presented. Finally, some results on statistical inference and emerging applications of these distributions are briefly discussed.

10.1 Preliminary Notions

10.1.1 Definitions and Properties

A distribution on the cylinder or cylindrical distribution means the joint distribution of a circular random variable Θ and a linear random variable X such as the distribution of wind direction and wind speed, or that of wave direction and wave height. The terminology, distribution on the cylinder, is used because a pair of realization (θ, x) can be identified with a point on the cylinder as shown in Figure 10.1.

A continuous distribution on the cylinder with random vector $(\Theta, X)'$ has joint pdf $f(\theta, x)$, where $0 \leq \theta < 2\pi$ (or $-\pi \leq \theta < \pi$) and $-\infty < x < \infty$, which satisfies

(a) $f(\theta, x) \geq 0$ for all $(\theta, x)' \in [0, 2\pi) \times \mathbb{R}$,

(b) $\displaystyle\int_{-\infty}^{\infty} \int_{0}^{2\pi} f(\theta, x) d\theta dx = 1$,

(c) $f(\theta, x) = f(\theta + 2k\pi, x)$ on $\mathbb{R}^2$ for any integer k.

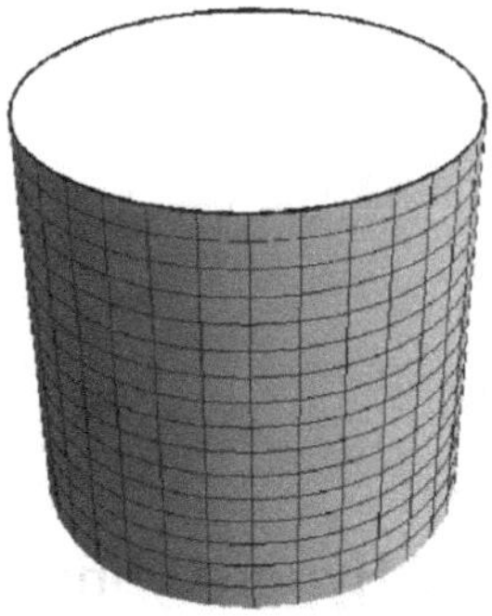

Figure 10.1 Cylinder surface.

The marginal pdf of Θ is given by

$$f_{\Theta}(\theta) = \int_{-\infty}^{\infty} f(\theta, x)dx$$

and that of X is

$$f_X(x) = \int_0^{2\pi} f(\theta, x)d\theta.$$

The conditional pdf of Θ given $X = x$ is $f_{\Theta|X}(\theta|x) = f(\theta, x)/f_X(x)$ provided that the denominator is positive, and the conditional pdf of X given $\Theta = \theta$ is similarly given by $f_{X|\Theta}(x|\theta) = f(\theta, x)/f_{\Theta}(\theta)$ provided that the denominator is positive. A distribution on the hyper-cylinder whose random vector is expressed by $(\Theta, X_1, X_2)'$ with circular Θ and linear $(X_1, X_2)'$, for example, has pdf $f(\theta, x_1, x_2)$ similarly defined to the aforementioned cylindrical pdf. The marginal pdf of Θ is $f_{\Theta}(\theta) = \int_{-\infty}^{\infty}\int_{-\infty}^{\infty} f(\theta, x_1, x_2)dx_1dx_2$, the marginal pdf of $(\Theta, X_1)'$ is $f_{\Theta,X_1}(\theta, x_1) = \int_{-\infty}^{\infty} f(\theta, x_1, x_2)dx_2$, the conditional pdf of Θ given $X_1 = x_1$ and $X_2 = x_2$ is $f_{\Theta|X_1,X_2}(\theta|x_1, x_2) = f(\theta, x_1, x_2)/f_{X_1,X_2}(x_1, x_2)$, and so on.

10.1.2 Circular–Linear Correlations

A circular–linear correlation is defined through the multiple correlation coefficient between a linear variable and a vector whose components correspond to cosine and sine of an angular variable (cf. Mardia, 1976; Johnson and Wehrly, 1977; Mardia and Jupp, 2000, p. 245), Jammalamadaka and SenGupta, 2001, p. 187).

Let $(\Theta, X)'$ be a vector of circular and linear random variables Θ and X, i.e. a random vector on the cylinder.

The squared multiple correlation coefficient of X and $\boldsymbol{U} = (\cos\Theta, \sin\Theta)'$ is

$$R_{x\theta}^2 = \frac{r_{xc}^2 + r_{xs}^2 - 2r_{cs}r_{xc}r_{xs}}{1 - r_{cs}^2},$$

where $r_{cs} = \mathrm{Corr}(\cos\Theta, \sin\Theta)$, $r_{xc} = \mathrm{Corr}(X, \cos\Theta)$ and $r_{xs} = \mathrm{Corr}(X, \sin\Theta)$. $R_{x\theta}^2$ can be used as a circular-linear correlation coefficient.

10.2 Uniform Distribution

The uniform distribution on $[0, 2\pi)^n \times (0, a)^m$ with fixed constant a (> 0) has pdf

$$f(\theta_1, \ldots, \theta_n, x_1, \ldots, x_m) = c,$$

where c is a constant which satisfies

$$\int_0^{2\pi} \cdots \int_0^{2\pi} \int_0^a \cdots \int_0^a f(\theta_1, \ldots, \theta_n, x_1, \ldots, x_m) d\theta_1 \cdots d\theta_n dx_1 \cdots dx_m = 1.$$

Thus, we have $f(\theta_1, \ldots, \theta_n, x_1, \ldots, x_m) = 1/\{(2\pi)^n a^m\}$.

If a random vector $(\Theta_1, \ldots, \Theta_n, X_1, \ldots, X_m)'$ on the hyper-cylinder $[0, 2\pi)^n \times (0, a)^m$ is uniformly distributed, $\Theta_1, \ldots, \Theta_n$ are mutually independent and each Θ_j ($j = 1, \ldots, n$) follows a uniform distribution on $[0, 2\pi)$, and $X_1, \ldots, X_m$ independently distributed of Θ_js are mutually independent and each X_k ($k = 1, \ldots, m$) follows a uniform distribution on $(0, a)$. The characteristic function of the uniform distribution is

$$\begin{aligned}\phi_{p_1,\ldots,p_n,q_1,\ldots,q_m} &= E\{e^{i(p_1\Theta_1+\cdots+p_n\Theta_n+q_1X_1+\cdots+q_mX_m)}\} \\ &= \begin{cases} \displaystyle\prod_{k=1}^m \frac{1}{iq_k a}(e^{iq_k a} - 1), & p_1 = \cdots = p_n = 0, \\ 0, & \text{others.} \end{cases}\end{aligned}$$

10.3 General Methods for Obtaining Distributions on the Cylinder

We introduced the methods for specified marginals, specified conditionals and maximum entropy characterization in Chapter 9 as some of the general methods for constructing probability distributions on the torus. Recall that it was noted there that the methods are applicable to the construction of distributions on the cylinder also. Several more general methods specifically enhanced for the construction of cylindrical distributions are also presented next.

10.3.1 Distributions with Specified Marginals

We consider now some general approaches and extensions, with examples, of deriving cylindrical distributions when both its circular and linear distributions are specified.

Johnson and Wehrly's Family

Johnson and Wehrly's (1978) original family of distributions on the cylinder with specified marginals has pdf of the form

$$f(\theta,x) = 2\pi f_1(\theta)f_2(x)g[2\pi\{F_1(\theta) \pm F_2(x)\}],$$

where $F_1(\theta)$ ($F_2(x)$) and $f_1(\theta)$ ($f_2(x)$) are the marginal distribution function and pdf of Θ (X). Examples (Johnson and Wehrly, 1978) are as follows.

Example 1

Set $f_1(\theta) = (2\pi)^{-1}$ (circular uniform), $f_2(x) = \phi(x) = (1/\sqrt{2\pi})\exp(-x^2/2)$ (standard normal) and $g(\xi) = \{2\pi I_0(\kappa)\}^{-1}\exp\{\kappa\cos(\xi-\mu)\}$ (von Mises). Then,

$$f(\theta,x) = \frac{1}{2\pi I_0(\kappa)}\exp[\kappa\cos\{\theta \pm 2\pi\Phi(x) - \mu\}]\phi(x),$$

where $\Phi(x)$ denotes the standard normal distribution function. This joint distribution has the circular uniform and the standard normal distributions as marginals.

Example 2

Set $f_1(\theta) = \{2\pi I_0(\kappa_1)\}^{-1}\exp\{\kappa_1\cos(\theta-\mu_1)\}$, and let $f_2(x)$ and $g(\xi)$ be defined in Example 1. Then,

$$f(\theta,x) = \frac{1}{I_0(\kappa)}\exp[\kappa\cos\{2\pi(F_1(\theta) \pm \Phi(x)) - \mu\}]f_1(\theta)\phi(x),$$

where $F_1(\theta) = \int_0^\theta f_1(\xi)d\xi$. This joint distribution has the von Mises and the standard normal distributions as marginals.

As an extension of Johnson and Wehrly's (1978) original family, we give a Johnson–Wehrly type four-variate distribution with two marginal distributions on the cylinder (Kato and Shimizu, 2008) under obvious notation:

$$p_\pm(\theta_1,x_1,\theta_2,x_2) = 2\pi f_1(\theta_1,x_1)f_2(\theta_2,x_2)$$
$$\times\, g\left[2\pi\left\{\frac{1}{f_{11}(\theta_1)}\frac{\partial F_1(\theta_1,x_1)}{\partial\theta_1} \pm \frac{1}{f_{21}(\theta_2)}\frac{\partial F_2(\theta_2,x_2)}{\partial\theta_2}\right\}\right].$$

Here $f_1(\theta_1,x_1)$ and $f_2(\theta_2,x_2)$ are specified pdfs on the cylinder with distribution functions $F_1(\theta_1,x_1)$ and $F_2(\theta_2,x_2)$, $f_{11}(\theta_1)$ and $f_{21}(\theta_2)$ their marginal pdfs on the circle and g a circular pdf. Its incomplete models with respect to Θ_2 are given by

$$h_{1\pm}(\theta_1,x_1,x_2) = 2\pi f_1(\theta_1,x_1)f_{22}(x_2)g\left[2\pi\left\{\frac{1}{f_{11}(\theta_1)}\frac{\partial F_1(\theta_1,x_1)}{\partial\theta_1} \pm F_{22}(x_2)\right\}\right]$$

and

$$h_{2\pm}(\theta_1, x_1, x_2) = 2\pi f_1(\theta_1, x_1) f_{22}(x_2) g\left[2\pi\left\{\frac{1}{f_{12}(x_1)}\frac{\partial F_1(\theta_1, x_1)}{\partial x_1} \pm F_{22}(x_2)\right\}\right],$$

where $F_{22}(x_2)$ and $f_{22}(x_2)$ are the marginal distribution function and pdf of X_2 and $f_{12}(x_1)$ the marginal pdf of X_1.

Abe et al. (2022) gave a bi-cylindrical sine-skewed distribution using the method for constructing toroidal distributions in Chapter 9 as follows. Let $G(x)$ be the distribution function of a density function symmetric about zero. Its simple example is given by $G(x) = (1 + x)/2$, the distribution function of a uniform distribution on the interval $(-1, 1)$. Then, for $f_j(x_j, \theta_j)$, $j = 1, 2$, pdfs on the cylinder such that $f(x_j, \theta_j) = f(x_j, -\theta_j)$ for arbitrary x_j on the support, a function

$$f(x_1, \theta_1, x_2, \theta_2) = 2G(\lambda \sin\theta_1 \sin\theta_2) f_1(x_1, \theta_1) f_2(x_2, \theta_2) \tag{10.1}$$

for $-1 \le \lambda \le 1$ provides a bi-cylindrical density function. If a random variable with (10.1) is denoted $(X_1, \Theta_1, X_2, \Theta_2)'$, then the marginal distribution of $(X_j, \Theta_j)'$ has the cylindrical density $f_j(x_j, \theta_j)$ for $j = 1, 2$, and the conditional distribution of $(X_1, \Theta_1)'$ given $(X_2, \Theta_2)' = (x_2, \theta_2)'$ has a cylindrical density $2G(\lambda \sin\theta_1 \sin\theta_2) f_1(x_1, \theta_1)$.

10.3.2 Distributions with Specified Conditionals

Following Section 3.2 of SenGupta (2004), recall the derivation discussed in Chapter 9 for toroidal distributions with specified conditionals. Consider now the case where both the (circular and linear) conditional distributions of $\Theta|(X = x)$ and of $X|(\Theta = \theta)$ are specified to be members of m and n dimensional regular exponential families, respectively. Then, as pointed out in Section 9.2.3 also, it follows that

Result 10.1. The general form of the joint pdf of Θ and X defined on the cylinder with their m and n dimensional conditionals, respectively, specified in regular exponential families is given by

$$f(\theta, x) = \exp\{p(\theta)' M q(x)\}, \quad (\theta, x) \in [0, 2\pi) \times \mathbb{R},$$

where $p(\theta)$ and $q(x)$ are $(m + 1)$ and $(n + 1)$ dimensional vectors respectively, and M is a $(m + 1) \times (n + 1)$ matrix of constant parameters. Θ and X are independent when $\tilde{M}$, the $m \times n$ lower subdiagonal matrix of M, is null.

The aforementioned result provides the conditional approach for constructing bivariate distributions on the cylinder, similar to that for toroidal distributions.

Examples

We present now some examples based on conditional specifications.

Let the conditional distribution of $\Theta|(X = x)$ be specified to be a von Mises distribution and that of $X|(\Theta = \theta)$ be a normal distribution. Then Result 10.1 gives the general form of the von Mises-normal conditionals distribution of Θ and X on the cylinder with $p(\theta) = (1 \;\; \cos\theta \;\; \sin\theta)'$ and $q(x) = (1 \; x \; x^2)'$.

Similarly, von Mises-exponential, von Mises-beta, generalized von Mises-normal, etc. conditionals distributions on the cylinder may be easily derived.

10.3.3 Maximum Entropy Based Distributions

Mardia (1975) used his Theorem 3.1 on the generalization of maximum entropy characterization for distributions on the circle, sphere, and torus. However, that result may be exploited to construct cylindrical distributions also (see, e.g. p. 144 of SenGupta, 2004). In this section we give some cylindrical models derived from the maximum-entropy method, which includes circular–linear generalizations of the von Mises and generalized von Mises distributions.

Model 1

The distribution (Johnson and Wehrly, 1978) whose joint pdf is given by

$$f(\theta, x) = \frac{\sqrt{\lambda^2 - \kappa^2}}{2\pi} \exp\{-\lambda x + \kappa x \cos(\theta - \mu)\}, \quad 0 \le \theta < 2\pi, \; x > 0, \tag{10.2}$$

with $0 \le \kappa < \lambda$ and $0 \le \mu < 2\pi$ is the maximum entropy distribution subject to $E(X)$, $E(X\cos\Theta)$, and $E(X\sin\Theta)$ taking specified values consistent with the expectations with respect to the distribution (10.2).

Properties

(a) The marginal distribution of Θ is a wrapped Cauchy distribution whose pdf is given by

$$f_\Theta(\theta) = \frac{1-\rho^2}{2\pi\{1+\rho^2 - 2\rho\cos(\theta-\mu)\}}, \qquad 0 \le \theta < 2\pi$$

with $\rho = \kappa(\lambda + \sqrt{\lambda^2 - \kappa^2})^{-1}$.

(b) The marginal pdf of X is

$$f_X(x) = \sqrt{\lambda^2 - \kappa^2}\, I_0(\kappa x) e^{-\lambda x}, \quad x > 0,$$

where $I_0(\cdot)$ denotes the modified Bessel function of the first kind and order zero.

(c) The conditional distribution of Θ given $X = x$ is a von Mises distribution $\mathrm{VM}(\mu, \kappa x)$ with pdf

$$g_{\Theta|X}(\theta|x) = \frac{1}{2\pi I_0(\kappa x)} \exp\{\kappa x \cos(\theta - \mu)\}, \quad 0 \le \theta < 2\pi.$$

(d) The conditional distribution of X given $\Theta = \theta$ is an exponential distribution with pdf

$$g_{X|\Theta}(x|\theta) = \{\lambda - \kappa\cos(\theta - \mu)\}\ \exp[-\{\lambda - \kappa\cos(\theta - \mu)\}x], \quad x > 0.$$

Concerning (b), according to Johnson and Wehrly (1978), "the marginal distribution of X does not follow a familiar distribution." However, the distribution may be interpreted as a mixture distribution. Note that (see, e.g. Jammalamadaka and SenGupta, 2001, p. 288),

$$I_0(x) = \sum_{j=0}^{\infty}(x/2)^{2j}/\Gamma^2(j+1), \quad x > 0.$$

In fact, consider an Erlang distribution with pdf

$$g_j(x) = \frac{\lambda^{2j+1}}{\Gamma(2j+1)}\, x^{2j}e^{-\lambda x}, \quad x > 0$$

and a negative binomial distribution with pmf

$$\begin{aligned} p_{\mathrm{NB}}(j) &= \binom{-1/2}{j}\sqrt{1-\frac{\kappa^2}{\lambda^2}}\left(-\frac{\kappa^2}{\lambda^2}\right)^j \\ &= \binom{j-1/2}{j}\sqrt{1-\frac{\kappa^2}{\lambda^2}}\left(\frac{\kappa}{\lambda}\right)^{2j}, \quad j = 0, 1, 2, \ldots . \end{aligned}$$

Then the marginal pdf of X is represented by

$$f_X(x) = \sum_{j=0}^{\infty} g_j(x)p_{\mathrm{NB}}(j), \quad x > 0.$$

As an extension, consider a transformation $\lambda X = (\beta Y)^\alpha$ with a linear shape $\alpha > 0$ and a linear scale $\beta > 0$, and replace κ/λ with $\tanh(\kappa^*)$. Then (10.2) is transformed into

$$f(\theta, y) = \frac{\alpha\beta^\alpha}{2\pi\ \cosh(\kappa^*)}\ y^{\alpha-1}\exp[-(\beta y)^\alpha\{1 - \tanh(\kappa^*)\cos(\theta - \mu)\}].$$

If we further think of sine-skewing, then we finally reach the WeiSSVM (a cylindrical distribution obtained by the sine-skewed von Mises distribution (circular part) with the Weibull distribution (linear part)) by Abe and Ley (2017):

$$\begin{aligned} f(\theta, y) =& \frac{\alpha\beta^\alpha}{2\pi\ \cosh(\kappa^*)}\ \{1 + \lambda\sin(\theta - \mu)\}y^{\alpha-1} \\ &\times \exp\left[-(\beta y)^\alpha\{1 - \tanh(\kappa^*)\cos(\theta - \mu)\}\right] \end{aligned}$$

with a circular skewing parameter $-1 \le \lambda \le 1$. The other motivation of Abe and Ley (2017) (in addition to the asymmetry of the circular distributions) to add flexibility to the model of Johnson and Wehrly (1978) was to extend the linear variation of the concentration parameter with x to non-linearity. This is also achieved through the component x^{α} in Weibull distribution. The marginal and conditional distributions of Θ and $\Theta|(X = x)$ are sine-skewed (with $\lambda \neq 0$) wrapped Cauchy, $\text{SSWC}(\mu, \tanh(\kappa/2)/(1+\tanh^2(\kappa/2)), \lambda)$, and sine-skewed von Mises, $\text{SSvM}(\mu, \tanh(\kappa)\beta^{\alpha}x^{\alpha}, \lambda)$, respectively. The conditional distribution of X is Weibull(β, α). The marginal distribution of X, however, is not in any known standard form.

A further extension was studied by Imoto et al. (2019) as follows. A Weibull distribution can be obtained by transforming an exponential distribution, and a generalized Pareto distribution, heavier-tailed than the Weibull, can be provided through a generalized gamma mixture. In such a way, a generalized Pareto-type cylindrical distribution (symmetric about $\theta = \mu$) with pdf

$$f_{\text{GPar}}(\theta, x) = \frac{\sqrt{1-\kappa^2}}{2\pi\sigma\delta}\left(\frac{x}{\sigma}\right)^{1/\delta-1}\left[1+\frac{\tau}{\delta}\left(\frac{x}{\sigma}\right)^{1/\delta}\{1-\kappa\cos(\theta-\mu\}\right]^{-(\delta/\tau+1)}$$

for $0 \leq \theta < 2\pi$ and $x > 0$ with parameters $\sigma > 0,\ \delta > 0,\ \tau > 0,\ 0 \leq \mu < 2\pi$, and $0 \leq \kappa < 1$ has been proposed. See Imoto et al. (2019) for more information.

Model 2

The distribution (Johnson and Wehrly, 1978) whose joint pdf is given by

$$f(\theta, x) = C^{-1}\ \exp\left\{-\frac{(x-\lambda)^2}{2\sigma^2} + \frac{\kappa x}{\sigma^2}\cos(\theta-\mu)\right\}, \quad 0 \leq \theta < 2\pi, -\infty < x < \infty, \tag{10.3}$$

with $-\infty < \lambda < \infty,\ \kappa > 0$ and $0 \leq \mu < 2\pi$, is the maximum entropy distribution subject to $E(X), E(X^2), E(X\cos\Theta)$ and $E(X\sin\Theta)$ taking specified values consistent with expectation with respect to the distribution (10.3). The normalizing constant is expressible (Kato and Shimizu, 2008) as

$$C = \sqrt{2\pi}\sigma \exp\left(\frac{\kappa^2}{4\sigma^2}\right)\left[2\pi\left\{I_0\left(\frac{\kappa\lambda}{\sigma^2}\right)I_0\left(\frac{\kappa^2}{4\sigma^2}\right) + 2\sum_{j=1}^{\infty} I_j\left(\frac{\kappa^2}{4\sigma^2}\right)I_{2j}\left(\frac{\kappa\lambda}{\sigma^2}\right)\right\}\right]$$

in terms of the modified Bessel functions of the first kind.

The conditional distribution of Θ given $X = x$ is a von Mises distribution $\text{VM}(\mu, \kappa x/\sigma^2)$ and that of X given $\Theta = \theta$ is a normal distribution $N(\lambda + \kappa(\cos(\theta-\mu)), \sigma^2)$. It is possible to have a lognormal-type distribution on the cylinder if we use a transformation $X = \log Y$.

Both the distributions in Models 1 and 2 include the case of independence, but unfortunately that forces Θ to be isotropic, i.e. have the circular uniform distribution only. We give in the following text two distributions where this shortcoming has been overcome.

Model 3

Kato and Shimizu (2008) give the following model whose pdf is given by

$$
\begin{aligned}
f(\theta,x) = C^{-1}\exp[-\lambda x + \kappa x\cos(\theta-\mu_1) + \nu_1\cos(\theta-\mu_2) \\
+\nu_2\cos\{2(\theta-\mu_3)\}],\quad 0\le\theta<2\pi,\ x>0,
\end{aligned}
\tag{10.4}
$$

with $\lambda > \kappa \ge 0,\ \nu_1,\nu_2 \ge 0,\ 0 \le \mu_1,\mu_2 < 2\pi$ and $0 \le \mu_3 < \pi$. The normalizing constant is expressible as

$$
\begin{aligned}
C = \frac{2\pi}{\sqrt{\lambda^2-\kappa^2}}\Bigg(& I_0(\nu_1)I_0(\nu_2) + 2\sum_{j=1}^{\infty} I_0(\nu_2)I_j(\nu_1)\rho^j\cos\{j(\nu_1-\nu_2)\} \\
& +2\sum_{j=1}^{\infty} I_0(\nu_1)I_j(\nu_2)\rho^j\cos\{2j(\mu_1-\mu_3)\} \\
& +2\sum_{j,n=1}^{\infty} I_j(\nu_1)I_n(\nu_2)\left[\rho^{j+2n}\cos\{j(\mu_1-\mu_2)+2n(\mu_1-\mu_3)\}\right. \\
& \left. +\rho^{|j-2n|}\cos\{j(\mu_1-\mu_2)-2n(\mu_1-\mu_3)\}\right]\Bigg),
\end{aligned}
$$

where $\rho = (\lambda - \sqrt{\lambda^2-\kappa^2})/\kappa$. The distribution is the maximum entropy distribution subject to $E(\cos\Theta)$, $E(\sin\Theta)$, $E(X)$, $E(X\cos\Theta)$, $E(X\sin\Theta)$, $E\{X\cos(2\Theta)\}$, and $E\{X\sin(2\Theta)\}$ taking specified values consistent with expectation with respect to the distribution (10.4). The case when $\nu_2 = 0$ has already been proposed by Johnson and Wehrly (1978).

The conditional distribution of Θ given $X = x$ is a generalized von Mises distribution (Maksimov, 1967; Rukhin, 1972) denoted by $\mathrm{GvM}(\mu^*, \mu_3, \nu^*, \nu_2)$ in Section 5.3.1.1, whose pdf is given by

$$
f_{\Theta|X}(\theta|x) = C_1^{-1}\exp[\nu^*\cos(\theta-\mu^*) + \nu_2\cos\{2(\theta-\mu_3)\}],\quad 0\le\theta<2\pi.
$$

Here ν^* (≥ 0) and μ^* $(0 \le \mu^* < 2\pi)$ satisfy

$$
\begin{aligned}
\nu^*\cos\mu^* &= \kappa x\cos\mu_1 + \nu_1\cos\mu_2, \\
\nu^*\sin\mu^* &= \kappa x\sin\mu_1 + \nu_1\sin\mu_2
\end{aligned}
$$

and the normalizing constant has an expression

$$
C_1 = 2\pi\left[I_0(\nu^*)I_0(\nu_2) + 2\sum_{j=1}^{\infty} I_j(\nu_2)I_{2j}(\nu^*)\cos\{2j(\mu^*-\mu_3)\}\right].
$$

The conditional distribution of X given $\Theta = \theta$ is an exponential distribution with pdf

$$
f_{X|\Theta}(x|\theta) = \{\lambda - \kappa\cos(\theta-\mu_1)\}\ \exp[-\{\lambda-\kappa\cos(\theta-\mu_1)\}x],\quad x>0.
$$

Model 4

Another maximum entropy model (Kato and Shimizu, 2008) is

$$f(\theta,x) = C^{-1}\ \exp\left[-\frac{\{x-\mu(\theta)\}^2}{2\sigma^2} + \kappa_1\cos(\theta-\mu_1) + \kappa_2\cos\{2(\theta-\mu_2)\}\right],$$
$$0 \le \theta < 2\pi, -\infty < x < \infty, \qquad (10.5)$$

with $\sigma > 0$, $\kappa_1, \kappa_2 \ge 0$, $0 \le \mu_1 < 2\pi$, $0 \le \mu_2 < \pi$, $\mu(\theta) = \mu' + \lambda\cos(\theta - \nu)$, $-\infty < \mu' < \infty$, $\lambda \ge 0$ and $0 < \nu \le 2\pi$, whose normalizing constant has an expression

$$C = (2\pi)^{3/2}\sigma\left[I_0(\kappa_1)I_0(\kappa_2) + 2\sum_{j=1}^{\infty} I_j(\kappa_2)I_{2j}(\kappa_1)\cos\{2j(\mu_1-\mu_2)\}\right].$$

The distribution is the maximum entropy distribution subject to $E(\cos\Theta)$, $E(\sin\Theta)$, $E\{\cos(2\Theta)\}$, $E\{\sin(2\Theta)\}$, $E(X)$, and $E(X\cos\Theta)$ taking specified values consistent with expectation with respect to the distribution (10.5). The model reduces to the one by Mardia and Sutton (1978) if $\kappa_2 = 0$. It is possible to have a lognormal model using the transformation $X = \log Y$.

The marginal distribution of Θ is a generalized von Mises distribution $\mathrm{GvM}(\mu_1, \mu_2, \kappa_1, \kappa_2)$, while the marginal distribution of X has a complex form. The conditional distribution of Θ given $X = x$ is also a generalized von Mises distribution $\mathrm{GvM}(\lambda_1, \lambda_2, \nu_1, \nu_2)$ with

$$\lambda_1\cos\nu_1 = \frac{\lambda}{\sigma^2}(x-\mu')\cos\nu + \kappa_1\cos\mu_1,$$
$$\lambda_1\sin\nu_1 = \frac{\lambda}{\sigma^2}(x-\mu')\sin\nu + \kappa_1\sin\mu_1,$$
$$\lambda_2\cos(2\nu_2) = -\frac{\lambda^2}{4\sigma^2}\cos(2\nu) + \kappa_2\cos(2\mu_2)$$

and

$$\lambda_2\sin(2\nu_2) = -\frac{\lambda^2}{4\sigma^2}\sin(2\nu) + \kappa_2\sin(2\mu_2),$$

and the conditional distribution of X given $\Theta = \theta$ is a normal distribution with mean $\mu(\theta)$ and variance σ^2.

Note that the distribution (10.5) is also obtainable as the conditional distribution of a trivariate normal distribution. Indeed, let $\boldsymbol{X} = (X, X_1, X_2)'$ be a random vector that follows the trivariate normal distribution with mean vector $\boldsymbol{\eta} = (\eta_1, \eta_2, \eta_3)'$ and covariance matrix

$$\Sigma = \begin{pmatrix} \sigma_1^2 & \rho_{12}\sigma_1\sigma_2 & \rho_{13}\sigma_1\sigma_3 \\ \rho_{12}\sigma_1\sigma_2 & \sigma_2^2 & \rho_{23}\sigma_2\sigma_3 \\ \rho_{13}\sigma_1\sigma_3 & \rho_{23}\sigma_2\sigma_3 & \sigma_3^2 \end{pmatrix},$$

where $-\infty < \eta_j < \infty$, $\sigma_j > 0$ $(j = 1, 2, 3)$, $-1 < \rho_{23} < 1$ and $1 - \rho_{12}^2 - \rho_{13}^2 - \rho_{23}^2 + 2\rho_{12}\rho_{13}\rho_{23} > 0$. Transform the trivariate random vector into a cylindrical coordinate system: $\boldsymbol{X} = (X, R\cos\Theta, R\sin\Theta)'$, where $R > 0$ and $0 \leq \Theta < 2\pi$. Then the conditional distribution of $(\Theta, X)'$ given $R = r$ has the form (10.5) after reparametrization.

10.3.4 Specified Cylindrical Regressions

In studying the dependency between a set of variables, the primary focus is to elucidate the relationships between them. The fundamental tool in this aspect is the regression function, specifically the conditional expectation of one (or more) variable(s) given the values of the others. Maximum entropy distributions are usually constructed under specified marginal or joint moments, and not conditional moments. Joint distributions, with or without conditional specifications, in general will force the regression to be the already built in ones. For example, the distribution in Example 1 of Section 10.3.5 forces the conditional mean direction of Θ, i.e. the regression of Θ on $X, E[\Theta|(X = x)] = \mu_{\theta|x}$, to be linear in x. This motivates the need of deriving cylindrical distributions with the specified forms of their regression functions and is addressed next following SenGupta et al. (2021).

Model 1

Consider first the case where regression of x on θ is specified. As an initial example, specify the regression function of X on θ as

$$\mu_{X|\theta} = E[X|(\Theta = \theta)] = b_0 + b_1\cos\theta + b_2\sin\theta.$$

This regression function is linear in terms of the first harmonics of θ and is also symmetric around $\arctan(b_2/b_1)$. It can be generalized to accommodate very rich classes, e.g. higher order harmonics as well as non-linear, asymmetric, highly peaked, flat-topped, heavy tailed, etc. functions—see Equations (3)–(5) of SenGupta and Ugwuowo (2006).

Then, let the conditional distribution of $X|(\Theta = \theta)$ be some popular, say normal, distribution, i.e. $N(\mu_{X|\theta}, \sigma^2)$. Next, let the marginal distribution of Θ be a popular circular, say von-Mises or circular normal distribution, $\mathrm{VM}(\mu, \kappa)$.

Then the cylindrical distribution of $(X, \Theta)'$ is given by

$$\begin{aligned} f(x, \theta) = &\frac{1}{2\pi I_0(\kappa)}\exp\{\kappa\cos(\theta - \mu)\}\frac{1}{\sqrt{2\pi}\sigma} \\ &\times \exp\left[-\frac{1}{2\sigma^2}\left\{x - (b_0 + b_1\cos\theta + b_2\sin\theta)\right\}^2\right]. \end{aligned}$$

It can be shown that the aforementioned distribution is the maximum entropy distribution subject to

$$E(X) = \frac{b_0}{\sigma^2}, \quad E(\cos\Theta) = -\frac{b_0 b_1}{\sigma^2} + \kappa\cos\mu, \quad E(\sin\Theta) = -\frac{b_0 b_2}{\sigma^2} + \kappa\sin\mu,$$

$$E(X\cos\Theta) = \frac{b_1}{\sigma^2}, \quad E(X\sin\Theta) = \frac{b_2}{\sigma^2}, \quad E(\cos^2\Theta) = \frac{b_1^2}{2\sigma^2},$$

$$E(\sin^2\Theta) = \frac{b_2^2}{2\sigma^2}, \quad E(X^2) = \frac{1}{2\sigma^2} \quad \text{and} \quad E(\cos\Theta\sin\Theta) = -\frac{b_1 b_2}{\sigma^2}$$

taking specified values consistent with these expectations with respect to that distribution. This distribution also results by specifying the conditional mean in the model of Mardia and Sutton (1978) as $\mu_{X|\theta}$ earlier.

Model 2

Consider now the case where regression of Θ on x is specified. As an initial example, specify the regression function of Θ on x as

$$E[\Theta|(X = x)] = \mu_{\Theta|x} = \alpha + 2\arctan(\beta x) \pmod{2\pi}.$$

As in Model 1 earlier, here also more general regression functions can be specified. Then, let the conditional distribution of $\Theta|(X = x)$ be some popular, say, the von Mises distribution, i.e. $\text{VM}(\mu_{\Theta|x}, \kappa)$. Next, let the marginal distribution of X be a popular, e.g. the normal distribution, $N(\mu, \sigma^2)$.

Then the cylindrical distribution of $(X, \Theta)'$ is given by

$$f(x,\theta) = \frac{1}{2\pi I_0(\kappa)}\exp[\kappa\cos\{\theta - \alpha - 2\arctan(\beta x)\}]\frac{1}{\sqrt{2\pi}\sigma}\exp\left\{-\frac{1}{2\sigma^2}(x-\mu)^2\right\}.$$

Using this method with different choices of the linear and circular distributions, a large variety of cylindrical distributions can be derived with the cylindrical regression functions as specified.

10.3.5 Conditionally Wrapped Distributions

In contrast to the foregoing methods, we now present a method where the dependency between Θ and X is already built in, as is the circular-linear regression function also. Simulation from these distributions for cylindrical random variables are quite convenient, especially when speed and volume, as in Big Data analysis, become imperative.

Let the joint pdf of $(X, Y)' \in \mathbb{R}^2$ be given by $f(x, y)$, to be also termed as the generating distribution, with $f_1(x)$ and $f_{2.1}(y; x)$ being the marginal density of X and the conditional density of $Y|(X = x)$ respectively. Then this method obtains the joint distribution of $(\Theta, X)'$ by wrapping $f_{2.1}(y; x)$, as

$$g(\theta, x) = f_1(x) \sum_{k=-\infty}^{\infty} f_{2.1}(\theta + 2k\pi; x). \tag{10.6}$$

Example 1

Seal and SenGupta (2012) proposed wrapping the conditional component of a bivariate linear distribution. Let the generating density be given by the bivariate normal density $N_2(x_1, x_2; \mu_1, \mu_2, \sigma_1^2, \sigma_2^2, \rho)$, or equivalently $N_2(\boldsymbol{x}; \boldsymbol{\mu}, \Sigma = (\sigma_{ij}))$ of $\boldsymbol{X} = (X_1, X_2)'$. Let, as per usual notations for conditional regression coefficients, means, and variances of a bivariate normal distribution, $\beta_{i.j} = \rho\sigma_i/\sigma_j$, $\mu_{i.j} = \mu_i + \beta_{i.j}(x_j - \mu_j)$, $\sigma_{ii.j} = \sigma_i^2(1 - \rho^2)$, $i, j = 1, 2$. Also denote $X_1 = X, X_{2k} = \Theta + 2\pi k$. Let $\phi(x; \cdot)$ and $\phi_{2.1}(x_2|x_1; \cdot)$ denote generically the "functions" corresponding to the marginal and conditional normal pdfs of X and $X_2|(X_1 = x_1)$, respectively. Then, by the aforementioned method, we get the cylindrical density (10.6) of $(\Theta, X)'$ as,

$$\begin{aligned} g(\theta, x) &= \phi(x; \mu_1, \sigma_1^2) \sum_{k=-\infty}^{\infty} \phi_{2.1}(x_{2k}|x; \mu_{2.1}, \sigma_{22.1}) \\ &= \frac{1}{(\sqrt{2\pi})^3\sigma_1} \exp\left[\frac{-(x-\mu_1)^2}{\sigma_1^2}\right] \\ &\quad \times \left[1 + 2\sum_{k=1}^{\infty} \exp\left(\frac{-\eta^2 k^2}{2}\right) \cos\left[k\left\{\theta - \mu_2 - \frac{\rho\sigma_2(x-\mu_1)}{\sigma_1}\right\}\right]\right], \end{aligned}$$

where $\eta^2 = \sigma_2^2(1 - \rho^2)$. This gives the marginal density of X as a normal and the conditional density of $\Theta|(X = x)$ as a wrapped normal density with mean direction $\mu_{\Theta|x} = [\mu_2 + (\rho\sigma_2/\sigma_1)(x - \mu_1)] \pmod{2\pi}$ and concentration parameter $\rho = \exp(-\eta^2/2)$.

We now derive the marginal density of Θ and the conditional density of $X|(\Theta = \theta)$. Rewrite the joint density $g(x, \theta)$ suitably exploiting the properties of $N_2(x_1, x_2)$. Then passing the integration inside the summation by the Fubini–Tonelli theorem, validated by the fact that the concerned functions are in exponential families, we get the marginal pdf of Θ as,

$$h(\theta) = \sum_{k=-\infty}^{\infty} \phi(x_{2k}; \mu_2, \sigma_2^2) \int_{-\infty}^{\infty} \phi_{1.2}(x|x_{2k}; \mu_{1.2}, \sigma_{11.2}) dx = \sum_{k=-\infty}^{\infty} \phi(x_{2k}; \mu_2, \sigma_2^2),$$

i.e. the marginal density of Θ is again a wrapped normal with mean direction $\mu_{\Theta|x} = \mu_2 \pmod{2\pi}$ and concentration parameter $\rho = \exp(-\sigma_2^2/2)$.

Finally, the conditional density of $X|(\Theta = \theta)$ is given by

$$f(x|\theta) = \sum_{k=-\infty}^{\infty} \phi(x_{2k}; \mu_2, \sigma_2^2)\phi_{1.2}(x|x_{2k}; \mu_{1.2}, \sigma_{11.2})/h(\theta)$$
$$= \sum_{k=-\infty}^{\infty} C_k(\theta)\phi_{1.2}(x|x_{2k}; \mu_{1.2}, \sigma_{11.2}),$$

where $C_k(\theta)$ is free of x. Thus the marginal pdf of X is that of an infinite sum of weighted normal variables.

By factorization theorem, X and Θ become independently distributed if and only if $\rho = 0$ (see Result 6.1 of Seal and SenGupta, 2012), under which they again follow normal and wrapped normal distributions respectively. Given the fact that both the marginal and the conditional densities of Θ are wrapped normal, the marginal of X is normal, and the conditional density of X is also in terms of normal variables, this simple normal-wrapped normal density can be a judicious choice for a cylindrical distribution.

Example 2

We consider now the case of the linear marginal with heavier tails than that of the normal in Example 1. Let the generating density be the bivariate Cauchy density $C(\boldsymbol{x}; \boldsymbol{\mu}, \Sigma)$ of $\boldsymbol{X} = (X_1, X_2)'$ given by

$$f(\boldsymbol{x}) = \frac{1}{2\pi|\Sigma|^{1/2}}[1 + (\boldsymbol{x} - \boldsymbol{\mu})'|\Sigma|^{-1}(\boldsymbol{x} - \boldsymbol{\mu})]^{-3/2}.$$

This gives the marginal density of X_1 as Cauchy and the conditional density of $\Theta|(X_1 = x_1)$ as a wrapped t density (since the conditional density of this bivariate Cauchy density is the t density, see, e.g. Ding, 2016), obtained by wrapping the t density with 2 degrees of freedom, $t(\mu_{2.1}, ((1 + d_1)/2)\Sigma_{22.1}; 2)$, where $d_1 = (x - \mu_1)^2/\sigma_{11}$, is the squared Mahalanobis distance of x_1, with scale parameter σ_{11}, from μ_1.

Semi-wrapped Distributions

The previous method of conditional wrapping has sometimes (see, e.g. Roy et al., 2016) been referred to as semi-wrapping.

The aforementioned method directly extends to yield distributions on the hyper-cylinder by wrapping as many components of the multivariate density of a vector random variable on $\mathbb{R}^n$ as desired to be circular.

10.3.6 Specified Marginals Through Normalizing Constants

We introduce now a novel and elegant method. The normalizing constants for some linear distributions are given simply in terms of their scale parameters. This fact can be exploited to derive a cylindrical distribution with a specified circular marginal. Write the normalizing constant of the "generating" linear distribution as the inverse of the kernel of some specified circular density of Θ. This yields, after proper normalization for the joint density, the cylindrical distribution with the marginal of Θ as that specified circular density. This method readily applies to some extremal distributions as is shown next.

Consider, as an example, the generating linear distribution of X to be the Gumbel Max distribution GMax$(0, 1/b), b > 0$. Then for its normalizing constant it follows that

$$\int_{-\infty}^{\infty} \exp(-bx) \exp[-\exp(-bx)]dx = 1/b. \tag{10.7}$$

Now choose $b \equiv b(\theta)$ to be the inverse of the kernel of any specified circular distribution. This yields the (joint) cylindrical distribution and, on proper normalization, the marginal distribution of Θ as that specified circular distribution.

Example 1

Let $b(\theta) = \exp[-\kappa \cos(\theta - \mu)]$. The resulting cylindrical density is

$$\begin{aligned} f(\theta, x) =& \frac{\beta}{2\pi I_0(\kappa)} \{1 + \lambda \sin(\theta - \mu)\} \exp\{-(x - \alpha)\beta b(\theta)\} \\ & \times \exp[-\exp\{-(x - \alpha)\beta b(\theta)\}]. \end{aligned}$$

The marginal distribution of Θ is symmetric (sine-skewed) von Mises for $\lambda = (\neq)\ 0$. For $\kappa = 0$, the marginal distribution of X becomes Gumbel Max distribution (see (10.7) with $b = \beta$ and $x = (x - a)$). With $\alpha = 0$, the conditional distribution of $X|(\Theta = \theta)$ is Gumbel Max distribution GMax$(0, 1/\beta b(\theta))$, given by

$$f_{X|\Theta}(x|\theta) = b(\theta) \exp\{-\beta b(\theta)x\} \exp\left[-\exp\{-\beta b(\theta)\}\right].$$

Example 2

Let $b(\theta) = 1 - \tanh(\kappa) \cos(\theta - \mu)$, be the wrapped Cauchy kernel. The resulting cylindrical density in its general form, the Gumbel sine-skewed von Mises distribution (GSSvM) of Kalaylioglu (2022), is given by

$$\begin{aligned} f(\theta, x) =& \frac{\beta}{2\pi \cosh(\kappa)} \{1 + \lambda \sin(\theta - \mu)\} \exp\{-(x - \alpha)\beta b(\theta)\} \\ & \times \exp\left[-\exp\{-(x - \alpha)\beta b(\theta)\}\right]. \end{aligned}$$

Then the marginal distribution of Θ is symmetric (sine-skewed) wrapped Cauchy for $\lambda = (\neq)\ 0$. For $\kappa = 0$, the marginal distribution of X becomes Gumbel Max distribution. The conditional distribution of $X|(\Theta = \theta)$ is Gumbel Max distribution GMax$(0, 1/\{\beta b(\theta)\})$.

It is evident that other choices of $b(\theta)$, e.g. inverses of the kernels of the symmetric (cardioid, power cardioid), asymmetric (generalized von Mises of

Yfantis and Borgman, 1982), asymmetric multimodal (Kim and SenGupta, 2021), flat-topped or sharply peaked (Batschelet, 1981), etc. circular densities, will lead to these corresponding marginal specified circular densities in the joint extremal family GMax$(0, 1/\{\beta b(\theta)\})$.

Other choices of generating functions can also be made, e.g. exponential, Erlang, gamma, and Weibull distributions with scale parameters. Using Weibull as the generating linear distribution and choosing $b(\theta)$ as the inverse wrapped Cauchy kernel, we obtain the WeiSSvM distribution derived, though from different motivation and approach as discussed under Model 1, by Abe and Ley (2017).

10.3.7 Method of Mixing Parameters

Let X and Θ be initially two independent linear and circular rvs with pdfs given by $f_i(x; \alpha_i, \beta_i), i = 1, 2$, respectively. Let the parameter spaces of α_i be such that they admit the representation $\alpha_i = z\delta_i$ with appropriately chosen support spaces for z and $\delta_i, i = 1, 2$. We will treat z as a rv Z with pdf $f_3(z; \tau)$. This introduces the dependency between X and Θ, and their cylindrical distribution is given by

$$f(\theta, x) = \int f_1(x; z\delta_1, \beta_1) f_2(\theta; z\delta_2, \beta_2) f_3(z; \tau) dz.$$

We illustrate the aforementioned method by a simple example.

Example

Let X and Θ be independently distributed as Exp(λ) and Cardioid(μ, ρ) respectively. Also let, $\lambda = z\eta, \rho = z\delta$, where $\eta \in (0, \infty), \delta \in (-1/2, 1/2)$. z is treated as a random variable Z with the pdf of an exponential random variable Exp(τ), $\tau > 0$, truncated at 1, i.e. with support $(0, 1)$. Then as given earlier, on integrating out z, we get the cylindrical density of $(\Theta, X)'$ (see SenGupta and Ong, 2014) as,

$$f(\theta, x) = \frac{\eta\tau}{2\pi\lambda^2(1 - e^{-\eta})} \left[\gamma(2, \lambda) + 2\delta \cos(\theta - \mu)\gamma(3, \lambda)/\lambda\right],$$

$$\gamma(2, \lambda) = 1 - (\lambda + 1)e^{-\lambda}, \quad \gamma(3, \lambda) = 2 - (\lambda^2 + 2\lambda + 2)e^{-\lambda}, \quad \lambda = \eta x + \tau.$$

Thus, the marginal and conditional densities of Θ are both Cardioid densities, Cardioid $(\rho_1(\delta, \eta, \tau))$ and Cardioid$(\rho_2(\delta, \eta, \tau, x))$ respectively where $\rho_i(\zeta), i = 1, 2$, are somewhat complicated functions of ζ. The marginal density of X is also explicitly available as

$$g(x) = \eta\tau\gamma(2, \lambda)/[\lambda^2(1 - e^{-\eta})].$$

It is interesting to note that X and Θ are independent if and only if $\delta = 0$, in which case Θ has a circular uniform distribution, but X has the same pdf as that for its marginal given by $g(x)$.

10.3.8 A Cylindrical Distribution Related to Toroidal Distributions

The distribution $\mathrm{BS}_2(\rho Q)$ of a pair of random vectors $(\boldsymbol{U}, \boldsymbol{V})$ on the torus (Kato, 2009), as defined in Section 9.12, has pdf

$$c(\boldsymbol{u}, \boldsymbol{v}) = \frac{1}{4\pi^2} \frac{1-\rho^2}{1-2\rho \boldsymbol{u}' Q \boldsymbol{v} + \rho^2}, \qquad \boldsymbol{u} = (u_1, u_2)', \ \boldsymbol{v} = (v_1, v_2)' \in S^1.$$

Transforming $(Z_U, Z_V) = (U_1 + iU_2, V_1 + iV_2)$, where $i = \sqrt{-1}$, and $\psi = \rho e^{i\theta}$ with θ satisfying

$$Q = \begin{pmatrix} \cos\theta & -\det Q \sin\theta \\ \sin\theta & \det Q \cos\theta \end{pmatrix}, \qquad 0 \le \theta < 2\pi,$$

we obtain the pdf of (Z_U, Z_V) as

$$c(\boldsymbol{u}, \boldsymbol{v}) = \frac{1}{4\pi^2} \frac{1-|\psi|^2}{|1-\psi z_v z_u^{-\det Q}|^2}, \qquad z_u, z_v \in \Omega,$$

where $|\psi| < 1$ and $\Omega = \{z \in \mathbb{C} |\ |z| = 1\}$.

Consider the case when $\det Q = 1$ and define a random vector $(Z_\Theta, X) = (Z_U, i(1-Z_V)/(1+Z_V))$. Then the marginal distribution of Z_Θ is a circular uniform and that of X is a standard Cauchy. The conditional distribution of Z_Θ given $X = x$ is a wrapped Cauchy $C^*(\{(1+ix)/(1-ix)\}\psi)$, whose notation appears in Section 4.4.1.3, and that of X given $Z_\Theta = z_\theta$ is a Cauchy distribution $C(-i(1-\overline{z_\theta}\psi)/(1+\overline{z_\theta}\psi))$ on the real line.

10.4 Distributions on a Discrete Cylinder

A random vector $(X, \Theta)'$ will be said to have the support as a discrete cylinder, if either or both of X and Θ are discrete random variables. Discrete circular distributions have been considered in some details in Chapter 6. We consider two approaches: conditional specifications and normalizing constants as functions of known kernels, for constructing discrete cylindrical distributions. Other approaches as given earlier for usual (continuous) cylindrical distributions may be generalized in the same lines.

For X discrete, the support will be taken as $\mathbb{N}$, the set of natural numbers or $\mathbb{Z}$, the set of integers. For Θ discrete, the support $\mathbb{S}_n^*$ will be taken on the n (chosen) equidistant angular points $2\pi j/n, j = 0, \dots, n-1$.

10.4.1 Conditionally Specified Distributions

Consider the approach of obtaining the pmf of a discrete distribution as the normalized pmf of a continuous distribution as discussed in Chapter 6. For simplicity let us choose the discrete normal and discrete circular normal or von Mises distributions. Observe that these two discrete distributions are still in the exponential family. The discrete von Mises distribution has been discussed in some details in Chapter 6. The pmfs of the discrete normal and discrete circular normal distributions are respectively given by,

$$p_X(k) = \frac{\exp\{-(k-\mu)^2/2\sigma^2\}}{\sum_{t=-\infty}^{\infty}\exp\{-(t-\mu)^2/2\sigma^2\}}, \quad k \in \mathbb{Z};$$

$$p_\Theta(r) = \frac{\exp\{\kappa\cos(2\pi r/n-\mu)\}}{\sum_{j=0}^{n-1}\exp\{\kappa\cos(2\pi j/n-\mu)\}}, \quad r \in \mathbb{S}_n^*.$$

Thus, the functions $p(\theta)$ and $q(x)$, with $\theta = 2\pi t/n$ and $x = k$, remain the same as in the von Mises-normal conditionals distribution of Section 10.3.2. But, the supports of Θ and X will respectively be $\mathbb{S}^1$ and $\mathbb{R}$ as before for the continuous case and $\mathbb{S}_n^*$ and $\mathbb{Z}$ as stated previously for the discrete case.

Using Result 10.1, the joint probability function of $(\Theta, X)'$ on a discrete cylinder is given as

$$p_{\Theta,X}(\theta, x) = \exp[(1 \ \cos\theta \ \sin\theta)' M (1 \ x \ x^2)],$$

where the supports of $(\Theta, X)' \in (\mathbb{S}_n^* \times \mathbb{R}), (\mathbb{S}^1 \times \mathbb{Z}), (\mathbb{S}_n^* \times \mathbb{Z})$, define discrete von Mises-(continuous) normal, (continuous) von Mises-discrete normal and discrete von Mises-discrete normal conditionals distributions respectively. As before, independence between Θ and X results if and only if $\tilde{M} = 0$.

The aforementioned approach can be easily adopted for discrete cylinders with, e.g. von Mises-Poisson, von Mises-binomial, generalized von Mises-Poisson, etc. conditionals distributions.

10.4.2 Generalization of the Method of Normalizing Constant

To obtain more flexible distributions, the method of normalizing constant may be generalized by suitably choosing some function of the kernel beyond its inverse. This also enriches the marginal distributions. We illustrate this method by an example in the following text.

Example

Consider a generalization where the cardioid kernel is replaced by the power cardioid kernel to produce a Poisson-power cardioid discrete cylindrical probability function (SenGupta et al., 2022) as

$$f_{\Theta,X}(\theta, x) = e^{-\beta y}\frac{(\beta y)^x}{x!} C_r(\rho) y^r, \quad x = 0, 1, 2, \ldots, \ \theta \in [0, 2\pi), \tag{10.8}$$

where $y = 1 + \rho\cos(\theta - \mu)$, $\beta > 0$, $\mu \in [0, 2\pi)$, $-1 < \rho < 1$, $r \in \mathbb{R}$, and

$$C_r^{-1}(\rho) = \int_0^{2\pi} [1 + \rho \cos\theta]^r d\theta = 2\pi\ {}_2F_1(-r/2, (1-r)/2; 1; \rho^2),$$

${}_2F_1$ denoting the Gauss hypergeometric function, wherein the hypergeometric series terminates if r is equal to a positive integer. Since $0 < y < 2$, the role of β is to extend the support of the parameter βy of the conditional Poisson distribution to the full range of $(0, \infty)$.

The marginal pdf of Θ is given by

$$f_\Theta(\theta) = C_r(\rho)y^r, \quad \theta \in [0, 2\pi), \ |\rho| < 1, \ r \in \mathbb{R}, \tag{10.9}$$

Observe that $f_\Theta(\theta)$ with (10.9) is a power Cardioid density which has been studied in details in Chapter 4 and includes as special cases the circular uniform, Cardioid, wrapped Cauchy, finite trigonometric series, and von Mises distributions.

The marginal pmf of X can be written down explicitly, though not in any well-known form.

The conditional pdf of $\Theta|(X = x)$ is obtained as,

$$f_{\Theta|x}(\theta|x) = C(\beta, \rho, r, x)e^{-\beta y}(\beta y)^{x+r},$$

where $C^{-1}(\cdot)$ is the normalizing constant. The conditional pmf of $X|(\Theta = \theta)$ is obtained as Poisson(βy),

$$f_X(x|\theta) = e^{-\beta y}(\beta y)^x/x!.$$

X and Θ become independent if and only if $\rho = 0$. Independence however forces Θ to be distributed as circular uniform and then X as Poisson(β). The conditional mean of X varies non-linearly with θ, i.e. increases (decreases) with $\cos\theta$ for $\rho > (<)\ 0$.

The aforementioned family $f_{\Theta,X}(\theta, x)$ in (10.8) can be extended in several ways. Truncated and inflated Poisson pmfs may replace the conditional Poisson pmf $f_{X|\Theta}(x|\theta)$. Multiplying the joint density by $(1 + \lambda\sin\theta)$, we can extend the conditional Cardioid pdf $f_{\Theta|X}(\theta|x)$ easily to the sine-skewed power Cardioid family. Further details are available from SenGupta et al. (2022).

10.5 Distributions on the Hyper-Cylinder

Distributions on the hyper-cylinder may be derived by generalizing some of the methods given earlier for the cylinder as well as ones specific to this higher-dimensional case. We present some such approaches and their resulting distributions in the following text.

10.5.1 Maximum Entropy Distribution

Johnson and Wehrly (1978) gave a maximum entropy distribution for several angular and linear variables. Denote an angular vector by $\boldsymbol{\Theta} = (\Theta_1, \dots, \Theta_p)'$ and put

$$H(\boldsymbol{\Theta}) = \begin{pmatrix} \cos\Theta_1 & \cdots & \cos(n\Theta_1) & \sin\Theta_1 & \cdots & \sin(n\Theta_1) \\ \vdots & & \vdots & \vdots & & \vdots \\ \cos\Theta_p & \cdots & \cos(n\Theta_p) & \sin\Theta_p & \cdots & \sin(n\Theta_p) \end{pmatrix}.$$

Let $\boldsymbol{X} = (X_1, \dots, X_q)'$ be a linear vector and suppose that $(\boldsymbol{\Theta}', \boldsymbol{X}')'$ has a joint pdf

$$f(\theta, \boldsymbol{x}) = c^{-1} \exp\left\{ -\frac{1}{2}\, \boldsymbol{x}'\Sigma^{-1}\boldsymbol{x} + \lambda'\Sigma^{-1}\boldsymbol{x} + a(\theta)'\Sigma^{-1}\boldsymbol{x} \right\}, \tag{10.10}$$

whose distribution is a member of an exponential family, where c is a normalizing constant, $\boldsymbol{\theta} \in [0, 2\pi)^p$, $\boldsymbol{x} \in \mathbb{R}^q$, $\lambda = (\lambda_1, \dots, \lambda_q)' \in \mathbb{R}^q$, $\boldsymbol{a}(\boldsymbol{\theta}) = (a_1(\theta), \dots, a_q(\theta))' \in \mathbb{R}^q$ and

$$\begin{aligned} a_m(\theta) &= \sum_{j=1}^{p}\sum_{k=1}^{n} a_{mjk} \cos\{k(\theta_j - \mu_{mjk})\} \\ &= \sum_{j=1}^{p}\sum_{k=1}^{n} \{\alpha_{mjk} \cos(k\theta_j) + \beta_{mjk} \sin(k\theta_j)\}, \quad m = 1, \dots, q. \end{aligned}$$

Here Σ^{-1} is a $q \times q$ positive definite matrix. The distribution with joint pdf (10.10) is the maximum entropy multivariate angular-linear distribution subject to $E(\boldsymbol{X}\boldsymbol{X}')$, $E(\boldsymbol{X})$ and $E\{\boldsymbol{X} \bigotimes H(\boldsymbol{\Theta})\}$ taking specified values consistent with expectation with respect to the distribution (10.10), where $\bigotimes$ denotes the Kronecker product.

If $\boldsymbol{X}$ is partitioned as $\boldsymbol{X} = (\boldsymbol{X}_1'; \boldsymbol{X}_2')'$, where $\boldsymbol{X}_1$ is $r \times 1$ and $\boldsymbol{X}_2$ is $(q-r) \times 1$, and correspondingly

$$\lambda = (\lambda_1'; \lambda_2')', \ \boldsymbol{a}(\theta) = (\boldsymbol{a}_1(\theta)'; \boldsymbol{a}_2(\theta)')', \ \Sigma = \begin{pmatrix} \Sigma_{11} & \Sigma_{12} \\ \Sigma_{21} & \Sigma_{22} \end{pmatrix},$$

then the conditional distribution of $\boldsymbol{X}_1$ given $\boldsymbol{X}_2 = \boldsymbol{x}_2$ and $\boldsymbol{\Theta} = \theta$ is an r-dimensional normal distribution with mean vector $\lambda_1 + \boldsymbol{a}_1(\theta) + \Sigma_{12}\Sigma_{22}^{-1}[\boldsymbol{x}_2 - \{\lambda_2 + \boldsymbol{a}_2(\theta)\}]$ and variance-covariance matrix $\Sigma_{11} - \Sigma_{12}\Sigma_{22}^{-1}\Sigma_{21}$. In particular, the mean of X_m $(m = 1, \dots, r)$ is given by

$$\nu_0 + \sum_{m=r+1}^{q} \nu_i x_i + \sum_{m=r+1}^{q}\sum_{j=1}^{p}\sum_{k=1}^{n} \{\gamma_{mjk} \cos(k\theta_j) + \delta_{mjk} \sin(k\theta_j)\}.$$

The theory provides a model for the regression of linear variables on other linear and angular variables. Johnson and Wehrly (1978) give a regression equation $x_1 = a + bx_2 + c_1 \cos\theta + c_2 \sin\theta$ for an air quality index x_1, temperature x_2, and wind direction θ as an example.

10.5.2 A Distribution Motivated by Uncorrelated Bivariate Normal

An extension of Mardia and Sutton's (1978) model to the hyper-cylinder appears in Mardia et al. (2004), whose pdf is of the form

$$f(\theta, z) = C^{-1} \exp\left\{\kappa \cos(\theta - \mu) - \frac{1}{2\sigma^2}\left|z - \mu - c\, e^{i\theta}\right|^2\right\}, \quad 0 \le \theta < 2\pi,\ z \in \mathbb{C} \tag{10.11}$$

with $C = 2\pi I_0(\kappa)2\pi\sigma^2$ for $\kappa \ge 0$, $0 \le \mu < 2\pi$ and $c \in \mathbb{C}$. If we put $\mu = a_1 + ia_2$, $z = x + iy$ and $c = b_1 + ib_2$, Equation (10.11) is represented by

$$f(\theta, x, y) = C^{-1} \exp\left[\kappa \cos(\theta - \mu) - \frac{1}{2\sigma^2}\left\{(x - \mu_{x|\theta})^2 + (y - \mu_{y|\theta})^2\right\}\right],$$

where $\mu_{x|\theta} = a_1 + b_1 \cos\theta - b_2 \sin\theta$ and $\mu_{y|\theta} = a_2 + b_1 \sin\theta + b_2 \cos\theta$.

The marginal distribution of Θ is a von Mises distribution, but that of Z is not so straightforward. The conditional distribution of Θ given $Z = z$ is a von Mises distribution, and that of Z given $\Theta = \theta$ is a normal distribution.

As mentioned earlier, hyper-cylindrical distributions may be obtained as semi-wrapped distributions also. The case of two linear variables and one circular variable forming a hyper-cylinder was considered by Fujita (2017) where the three variables were assumed to be independent with the linear distributions being normal and the circular distribution being von Mises.

10.6 Statistical Inference

For the parameter estimation of cylindrical distribution, the usual maximum likelihood estimation under the model is applicable when observations are available. Closed-form expressions for the solutions are not given in most cases, mainly because of complicated normalizing constants. One needs to use numerical optimization, including constrained, algorithms such as Nelder–Mead method, evolution algorithm, random search, etc. However, when the model contains many parameters, a local maximum might be chosen by the used optimization method and one should select the solution with care.

Some of the models given earlier, however, are amenable to simple estimation procedures, e.g. the likelihood equations become separable, in sets of parameters. This fact can lead to closed form simple estimators for these models as illustrated next.

In Model 4 with pdf (10.5), (Kato and Shimizu, 2008), the linear and circular parts have two different sets of parameters, $(\mu', \lambda, \nu, \sigma)$ and $(\kappa_1, \mu_1, \kappa_2, \mu_2)$, leading to two separate sets of separable likelihood equations. Further for this model, the regression of x on $(\cos\theta, \sin\theta)$ is multiple linear regression in a normal model.

Thus the MLEs of the corresponding partial regression coefficients ($\lambda \cos \nu, \lambda \sin \nu$) are readily available as their sample counterparts, from which the MLEs of λ and ν (and of μ') easily follow. The maximum likelihood estimates of the other parameters $\kappa_1, \kappa_2, \mu_1$ and μ_2 are obtained by essentially the same method as in the generalized von Mises distribution GvM$(\kappa_1, \kappa_2, \mu_1, \mu_2)$.

It is obvious that Θ and X are independent if $\lambda = 0$ in (10.5), and in this case the marginal distribution of X is normal. If one wants to investigate whether there is a dependence between Θ and X or not, the likelihood ratio test for H_0: $\lambda = 0$ vs. H_1: $\lambda > 0$ is available and its test statistic is given by $T = -n \log(1 - r_{1.23}^2)$. We do not have a chi-square distribution as a limiting distribution of T because λ is on the boundary under the null hypothesis H_0. A half-half mixture of zero and a chi-square distribution $Z^2\, I[Z > 0]$ is its limiting distribution, where Z has a standard normal distribution and I is an indicator function. The null hypothesis is rejected when the value of T is large. Model 4 reduces to the Mardia and Sutton (1978) model if $\kappa_2 = 0$. And the likelihood ratio test for H_0: $\kappa_2 = 0$ vs. H_1: $\kappa_2 > 0$ is given by the statistic $T = -2 \log(\max L_0 / \max L_1)$, where L_0 is the likelihood under H_0 and L_1 is the likelihood under H_1. The parameter κ_2 is on the boundary under H_0, and again a half-half mixture of zero and a chi-square distribution $Z^2\, I[Z > 0]$ is used as a limiting distribution.

The separability of the likelihood equations also occurs with several of the models constructed earlier with specified conditionals and specified cylindrical regressions. Thus, similar simplifications as given earlier for (10.5) hold for the MLEs of their model parameters also.

Another difficulty for estimation of the parameters in cylindrical models stems from the fact that the parameters are often arguments of trigonometric functions. However, here we can appeal to the method of profile likelihood, which has been exploited conveniently, e.g. by SenGupta et al. (2021) for estimating the parameter β of their specified cylindrical regression Model 2 in Section 10.3.4.

A striking fact for the discrete cylindrical model of SenGupta et al. (2022) that needs to be mentioned is that independence of Θ and X for this distribution is equivalent to the nullity of the circular-linear correlation coefficient, $R_{x\theta}^2$ defined in Section 10.1.2. Thus, test for independence can be easily constructed based on this correlation coefficient, which is facilitated by its known asymptotic normal distribution under the null hypothesis.

10.7 Applications

Kato and Shimizu (2008) give two illustrative examples. As Example 1, a cylindrical dataset on movements of blue periwinkles (Fisher, 1993, Table B.20) was analyzed using model (10.5) and the Mardia and Sutton (1978) submodel. The observations are directions and distances moved by small blue periwinkles after they had been transplanted downshore from the height at which they normally live. This same data set was also modeled by Seal and SenGupta (2012) using the conditionally wrapped cylindrical distribution, and it was observed that the regression curve from that model gave a good fit also. It has been shown (Mardia and Sutton, 1978) that the cylindrical model with asymmetric circular marginal is more suitable for fitting to this dataset. Moreover, Abe and Ley (2017) report that the WeiSSVM model gives a better fit for the periwinkle dataset. For this dataset, the WeiSSVM model is more parsimonious than the heavy-tailed distribution proposed by Imoto et al. (2019). As Example 2 in Kato and Shimizu (2008), observations of January surface wind direction and temperature at 12 h GMT at Kew for the years 1956–1960, the data from Table 1 of Mardia and Sutton (1978), were reanalyzed. In this case, it is concluded that the Mardia and Sutton model is a better model than (10.5) to thiss dataset.

In the area of oceanography and marine science, wave direction and wave height are thought of important measurement as well as meteorological factors such as wind direction and wind speed. Hokimoto and Shimizu (2008) proposed an angular-linear time series model for waveheight prediction using waveheight, wind speed and wind direction, and Hokimoto and Shimizu (2014) constructed a non-homogeneous hidden Markov model for predicting the distribution of sea surface elevation using wind direction, wind speed and sea surface elevation observations. On the other hand, Wang et al. (2015) used the projected normal distribution to model spatial and spatio-temporal data on significant wave heights and outgoing wave directions in the Adriatic Sea, off the coast of Italy, during the time of a storm. See also Wang and Gelfand (2014) for modeling space-time wave direction data. Wave directions and heights are also analyzed by Lagona et al. (2015) using a hidden Markov model based on the WeiSSVM model in Abe and Ley (2017).

Extremal events may also need to be modeled using cylindrical distributions. Kalaylioglu (2022) applied her GSSvM distribution to model two meteorological data sets: (i) fastest two-minute wind speed (mph) per hour and their directions in the three winter months of 1 December 2018 through 28 February 2019, recorded at the land station in Greenville Maine US Forestry Service and (ii) daily maximum air temperature and direction of the wind at 3 pm during May 1 2019 to September 30 2019 in Sydney. It was claimed empirically that the distribution seemed to fit the data sets well.

Zucchini and MacDonald (2009) proposed hidden Markov models for modeling time series of wind direction, while Ailliot et al. (2015) (see also references therein) provided a non-homogeneous hidden Markov-switching model and applied the model to a wind time series for a location off the French Brittany coast. In the study by Mastrantonio et al. (2015), wind direction and speed are analyzed by a Bayesian hidden Markov model using circular-linear general projected normal distribution. Fernández-Durán (2007) used the Johnson and Wehrly (1978) model with nonnegative trigonometric sums g and f_1 and Weibull f_2 to apply it to wind direction and ground-level ozone concentration data.

The conditional expectation of X given $\Theta = \theta$ for a cylindrical random vector $(\Theta, X)'$ gives a regression. As a simple example of (10.10), a regression function of temperature and wind direction was used to predict an air quality index in Johnson and Wehrly (1978). Jammalamadaka and Lund (2006) studied the relationship between the wind direction and the ozone level in a suburb of Houston. SenGupta and Ugwuowo (2006) applied possibly asymmetric regression models to renewable energy in environmental sciences. The same dataset as in SenGupta and Ugwuowo (2006) was studied by Liu et al. (2017) in the view point of influence diagnostics. Concerning the Mardia and Sutton (1978) model, Anderson-Cook (1997) gave an extension to the model and re-analyzed the relationship between wind direction and temperature at Kew analyzed by Mardia and Sutton. Later, Anderson-Cook (2000) proposed the second order regression model $z = b_0 + b_1 \cos\theta + b_2 \sin\theta + b_{11} \cos(2\theta) + b_{22} \sin(2\theta) + \varepsilon$ or $z = b_0 + r_1 \cos(\theta - \theta_{01}) + r_2 \cos(2\theta - \theta_{02}) + \varepsilon$ to study more the same dataset. Maruotti (2016), see also references therein such as Nuñez-Antonio and Gutiérrez-Peña (2014), studied longitudinal data modeling techniques with application to animal orientation. The biased-correlated random walk model is known as a flexible and powerful model for the analysis of data on animal movement in ecology. However, Rivest et al. (2016) have proposed a multivariate angular regression model for both angular and linear predictors with analysis of a dataset on bison trails.

In the arena of Statistical Machine Learning, Agiomyrgiannakis and Stylianou (2009) used semi-wrapped independent components normal mixture model for modeling and high-rate quantization of phase data of speech. Roy et al. (2016) applied the corresponding dependent model to color image segmentation data. Chraiti et al. (2019) also used the independent components model to perform Beamforming (unsupervised) Learning for mmWave Communication Theory. SenGupta et al. (2021) used their models based on specified regression for clustering cylindrical data on the surface wind direction and temperature at Kew mentioned earlier. These models may be used for unsupervised learning of data in ecology, oceanography, marine science, bioinformatics, environmental science, etc.

The discrete cylindrical distributions mentioned earlier have been used and have potential for many innovative real-life applications. For example, the Poisson regression model has been used by Jarvis et al. (2021) to assess Lunar phase as a cue for migrations to two species of explosive breeding amphibians and their implications for conservation. The distribution by SenGupta et al. (2022) has been used to study variation of occurrences of traffic accidents over the 24 hour period in a busy city in India.

There are also interesting applications, with recently emerging ones, of the hyper-cylindrical distributions. For example, for the case of two linear and one circular variables, Johnson and Wehrly (1978), give a regression equation $x_1 = a + bx_2 + c_1 \cos\theta + c_2 \sin\theta$ for an air quality index x_1, temperature x_2 and wind direction θ, which stems from their distributional model. Recently, Fujita (2017) also considered this case in the context of color image quantization and clustering in the Hue Saturation Value (HSV) color space, where Hue is an angular, and Saturation and Value are linear variables. SenGupta and Ugwuowo (2011) studied classification problem by the use of bootstrap likelihood ratio method for hyper-cylindrical and toroidal data and applied the theory to the measurements on the skull of people from two tribes in Nigeria in the area of anthropology and archeology. The features measured are two angular and one linear variable, i.e. the front angle (radian), the side angle (radian), and the circumference (cm) of the skulls.

References

Abe, T. and Ley, C. (2017). A tractable, parsimonious and flexible model for cylindrical data, with applications, *Econometrics and Statistics*, 4, 91–104.

Abe, T., Imoto, T., Shiohama, T. and Miyata, Y. (2022). On some flexible models for circular, toroidal, and cylindrical data, In *Directional Statistics for Innovative Applications: A Bicentennial Tribute to Florence Nightingale*, SenGupta, A. and Arnold, B. C. (Eds.), Springer Nature, Singapore, 229–243.

Agiomyrgiannakis, Y. and Stylianou, Y. (2009). Wrapped Gaussian mixture models for modeling and high-rate quantization of phase data of speech. *IEEE Transactions on Audio, Speech, and Language Processing*. 17, 775–786.

Ailliot, P., Bessac, J., Monbet, V. and Pène, F. (2015). Non-homogeneous hidden Markov-switching models for wind time series, *Journal of Statistical Planning and Inference*, 160, 75–88.

Anderson-Cook, C. M. (1997). An extension to modeling cylindrical variables, *Statistics & Probability Letters*, 35, 215–223.

Anderson-Cook, C. M. (2000). A second order model for cylindrical data, *Journal of Statistical Computation and Simulation*, 66, 51–65.

Batschelet, E. (1981). *Circular Statistics in Biology*, Academic Press, London.

Chraiti, M., Chizhik, D., Du, J., Valenzuela, R. A., Ghrayeb, A. and Assi, C. (2019). Beamforming learning for mmWave communication: Theory and experimental validation, https://www.researchgate.net/publication/%338208543_Beamforming_

Learning_for_mmWave_Communication_Theory_and_Experimental_Validation/link/5e06df104585159aa49f8d88/download.

Ding, P. (2016). On the conditional distribution of the multivariate *t* distribution, *The American Statistician*, 70, 293–295.

Fernández-Durán, J. J. (2007). Models for circular-linear and circular-circular data constructed from circular distributions based on nonnegative trigonometric sums, *Biometrics*, 63, 579–585.

Fisher, N. I. (1993). *Statistical Analysis of Circular Data*, Cambridge University Press, Cambridge.

Fujita, K. (2017). A clustering method for data in cylindrical coordinates, *Mathematical Problems in Engineering*. 2017, Article ID 3696850, 11 pages, https://doi.org/10.1155/2017/3696850.

Hokimoto, T. and Shimizu, K. (2008). An angular-linear time series model for wave-height prediction, *Annals of the Institute of Statistical Mathematics*, 60, 781–800.

Hokimoto, T. and Shimizu, K. (2014). A non-homogeneous hidden Markov model for predicting the distribution of sea surface elevation, *Journal of Applied Statistics*, 41, 294–319.

Imoto, T., Shimizu, K. and Abe, T. (2019). A cylindrical distribution with heavy-tailed linear part, *Japanese Journal of Statistics and Data Science*, 2, 129–154.

Jammalamadaka, S. R. and Lund, U. J. (2006). The effect of wind direction on ozone levels: A case study, *Environmental and Ecological Statistics*, 13, 287–298.

Jammalamadaka, S. R. and SenGupta, A. (2001). *Topics in Circular Statistics*, World Scientific, Singapore.

Jarvis, L. E., Grant, R. A. and SenGupta, A. (2021). Lunar phase as a cue for migrations to two species of explosive breeding amphibians–implications for conservation, *European Journal of Wildlife Research*, 67, https://doi.org/10.1007/s10344-020-01453-3.

Johnson, R. A. and Wehrly, T. E. (1977). Measures and models for angular correlation and angular-linear correlation, *Journal of the Royal Statistical Society*, B39, 222–229.

Johnson, R. A. and Wehrly, T. E. (1978). Some angular-linear distributions and related regression models, *Journal of the American Statistical Association*, 73, 602–606.

Kalaylioglu, Z. (2022). Analysis of correlated circular and extremal data with a flexible cylindrical distribution, *Environmental and Ecological Statistics*, 29, 207–222.

Kato, S. (2009). A distribution for a pair of unit vectors generated by Brownian motion. *Bernoulli*, 15, 898–921.

Kato, S. and Shimizu, K. (2008). Dependent models for observations which include angular ones, *Journal of Statistical Planning and Inference*, 138, 3538–3549.

Kim, S. and SenGupta, A. (2021). Multimodal exponential families of circular distributions with application to daily peak hours of PM2.5 level in a large city, *Journal of Applied Statistics*, 48, 3193–3207.

Lagona, F., Picone, M. and Maruotti, A. (2015). A hidden Markov model for the analysis of cylindrical time series, *Environmetrics*, 26, 534–544.

Liu, S., Ma, T., SenGupta, A., Shimizu, K. and Wang, M. Z. (2017). Influence diagnostics in possibly asymmetric circular-linear multivariate regression models, *Sankhyā B*, 79, 76–93.

Maksimov, V. M. (1967). Necessary and sufficient statistics for a family of shifts of probability distributions on continuous bicompact groups, *Teoriya Verojatnosteĭ i ee Primeneniya*, 12, 307–321 (in Russian). English translation: *Theory of Probability and its Applications*, 12, 267–280.

Mardia, K. V. (1975). Characterizations of directional distributions, In *A Modern Course on Statistical Distributions in Scientific Work*, **3**, Patil, G. P., Kotz, S. and Ord, J. K. (Eds.), Reidel, Dordrecht, 365–385.

Mardia, K. V. (1976). Linear-circular correlation coefficients and rhythmometry, *Biometrika*, 63, 403–405.

Mardia, K. V. and Jupp, P. E. (2000). *Directional Statistics*, Wiley, Chichester.

Mardia, K. V. and Sutton, T. W. (1978). A model for cylindrical variables with applications, *Journal of the Royal Statistical Society*, B40, 229–233.

Mardia, K. V., Kirkbride, J. and Bookstein, F. L. (2004). Statistics of shape, direction and cylindrical variables, *Journal of Applied Statistics*, 31, 465–479.

Maruotti, A. (2016). Analyzing longitudinal circular data by projected normal models: A semi-parametric approach based on finite mixture models, *Environmental and Ecological Statistics*, 23, 257–277.

Mastrantonio, G., Maruotti, A. and Jona-Lasinio, G. (2015). Bayesian hidden Markov modelling using circular-linear general projected normal distribution, *Environmetrics*, 26, 145–158.

Nuñez-Antonio, B. and Gutiérrez-Peña, E. (2014). A Bayesian model for longitudinal circular data based on the projected normal distribution, *Computational Statistics & Data Analysis*, 71, 506–519.

Rivest, L. P., Duchesne, T., Nicosia, A. and Fortin, D. (2016). A general angular regression model for the analysis of data on animal movement in ecology, *Journal of the Royal Statistical Society*, C65, 445–463.

Roy, A., Parui, S. K. and Roy, U. (2016). SWGMM: A semi-wrapped Gaussian mixture model for clustering of circular-linear data, *Pattern Analysis and Applications*, 19, 631–645.

Rukhin, A. L. (1972). Some statistical decisions about distributions on a circle for large samples, *Sankhyā A*, 34, 243–250.

Seal, B. and SenGupta, A. (2012). On the foundations of dependency models and parameters for random variables on cylinder R^3. *Calcutta Statistical Association Bulletin*, 64, 151–165.

SenGupta, A. (2004). On the constructions of probability distributions for directional data, *Bulletin of the Calcutta Mathematical Society*, 96, 139–154.

SenGupta, A. and Ong, S. H. (2014). A unified approach for construction of probability models for bivariate linear and directional data, *Communications in Statistics–Theory and Methods*, 43, 2563–2569.

SenGupta, A. and Ugwuowo, F. I. (2006). Asymmetric circular-linear multivariate regression models with applications to environmental data, *Environmental and Ecological Statistics*, 13, 299–309.

SenGupta, A. and Ugwuowo, F. I. (2011). A classification method for directional data with application to the human skull, *Communications in Statistics–Theory and Methods*, 40, 457–466.

SenGupta, A., Roy, M. and Chattopadhyay, A. K. (2021). Model-based clustering for cylindrical data, In *Advances in Statistics - Theory and Applications - Honoring the Contributions of Barry C. Arnold in Statistical Science*, Ghosh, I., Balakrishnan, N. and Ng, H. K. T. (Eds.), Springer, New York, 347–363.

SenGupta, A., Shimizu, K., Ong, S. H. and Das, R. (2022). A flexible family of mixed distributions for discrete linear and continuous circular random variables, In *Methodology and Applications of Statistics: A Volume in Honor of C.R. Rao on the Occasion of his 100th Birthday*, Arnold, Barry C., Balakrishnan, Narayanaswamy and Coelho, Carlos A. (Eds.), Springer International Publishing, 365–381.

Wang, F. and Gelfand, A. E. (2014). Modeling space and space-time directional data using projected Gaussian processes, *Journal of the American Statistical Association*, 109, 1565–1580.

Wang, F., Gelfand, A. E. and Jona-Lasinio, G. (2015). Joint spatio-temporal analysis of a linear and a directional variable: Space-time modeling of wave heights and wave directions in the Adriatic Sea, *Statistica Sinica*, 25, 25–39.

Yfantis, E. A. and Borgman, L. E. (1982). An extension of the von Mises distribution, *Communications in Statistics–Theory and Methods*, 11, 1695–1706.

Zucchini, W. and MacDonald, I. L. (2009). *Hidden Markov Models for Time Series: An Introduction Using R*, Chapman and Hall/CRC, Boca Raton, FL.

11

Distributions on the Disc and Hyper-Disc

In this chapter we describe probability distributions on a manifold different from circle, sphere, torus, and cylinder, i.e. on the disc and its higher dimensional extension, hyper-disc. A typical example of a dataset on the disc is daily observations in summer on highest humidity (converted from percentage to proportion) and its corresponding time. An example of a dataset on the hyper-disc is observations on the maximum infection rate from Covid 19 and its corresponding location in a given country. Note that for these data sets, both the linear and non-linear components are random and the linear variable is scaled down to the unit interval.

11.1 Preliminary Notions

11.1.1 Definitions and Properties

A disc or disk in this chapter means a set D bounded by the unit circle with center at the origin, i.e. $D = \{(x_1, x_2)' \in \mathbb{R}^2 \mid x_1^2 + x_2^2 \leq 1\}$. If we use the polar coordinates $x_1 = r\cos\theta$ and $x_2 = r\sin\theta$, where $0 < r \leq 1$ and $0 \leq \theta < 2\pi$, the region is represented by $D = \{(r, \theta)' \mid 0 < r \leq 1, 0 \leq \theta < 2\pi\}$. Thus, D represents the unit disc in the two-dimensional plane and r and θ represent the radial distance and the angular position of the concerned point in it from its center (origin). Figure 11.1 shows the region D. Generally, a p-dimensional ($p \geq 1$) disc is defined by $D_p = \{\boldsymbol{x} = (x_1, \dots, x_p)' \in \mathbb{R}^p \mid \|\boldsymbol{x}\|^2 = \sum_{j=1}^p x_j^2 \leq 1\}$. Hence, D_2 is equal to D. In terms of the polar coordinates $\boldsymbol{x} = r_x\boldsymbol{z}$, the region is represented by $D_p = \{(r_x, \theta_1, \dots, \theta_{p-1})'\}$ for $p \geq 2$ where $r_x \equiv \|\boldsymbol{x}\|$, $0 < r_x \leq 1$, $\boldsymbol{z}$, and θ_j, $j = 1, \dots, p-1$, are defined in (11.10).

11.1.2 Hyper-spherical–Linear Correlations

The correlation between the circular and linear random variables $\boldsymbol{\Theta}$ and R in the disc can be measured by the same multiple correlation coefficient $R^2_{r\theta}$ used

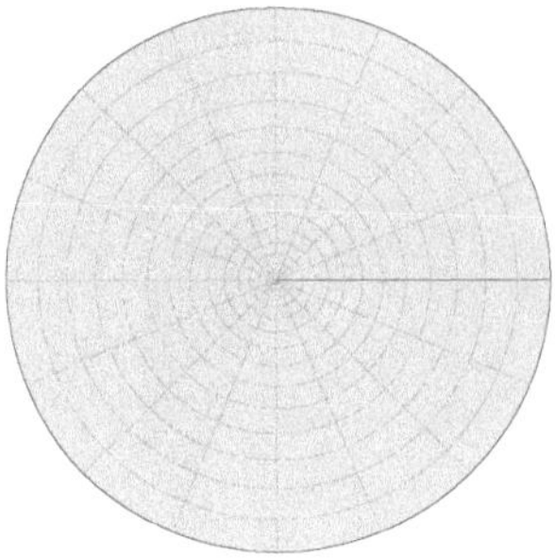

Figure 11.1 A figure of the disc.

in Chapter 10 for the cylinder. In the same spirit, the correlation between the spherical (hyper-spherical) and linear random variables $\boldsymbol{\Theta} = (\Theta_1, \dots, \Theta_{p-1})'$ and $R \equiv \|\boldsymbol{X}\|$ in the hyperdisc can be measured by $R^2_{r\boldsymbol{u}}$, the multiple correlation coefficient between R and $\boldsymbol{U}$.

11.1.3 Distributions on the Disc

A continuous distribution with support on D has a joint pdf $f(x_1, x_2)$, with a pair of random variables $(X_1, X_2)'$, which satisfies $f(x_1, x_2) \geq 0$ for all $(x_1, x_2)' \in D$ and $\int_D f(x_1, x_2) dx_1 dx_2 = 1$. Similarly, a discrete distribution has pmf that satisfies $f(x_1, x_2) \geq 0$ for all $(x_1, x_2)' \in D$ and for a countable number of points $(x_{1j}, x_{2k})' \in D$ with $j, k = 1, 2, 3, \dots, \sum_{j=1}^{\infty} \sum_{k=1}^{\infty} f(x_{1j}, x_{2k}) = 1$. A continuous (discrete) distribution on the p-dimensional unit disc D_p is similarly defined by pdf (pmf) $f(\boldsymbol{x})$ for $\boldsymbol{x} \in D_p$.

Remark 11.1 Observe that in terms of polar coordinates, the points considered for the cylinder with unit circle as base and height as unity also, and unit disc may seem to be represented similarly, i.e. by $(x \cos\theta, x \sin\theta)$. However, for the cylinder x represents the height of the point on the vertical axis on a 3-dimensional cylinder surface. Whereas, $x \equiv r$ for the disc represents the radial distance of the point inside a 2-dimensional disc plane. Since the base of the cylinder defined by the cylindrical coordinates (ρ, θ, x) here has been taken as the unit circle, the first coordinate ρ, corresponding to the radius of the base circle is always 1, and hence its specification was suppressed in our discussions.

Remark 11.2 Probability distributions for R need to be unit distributions, i.e. defined on the unit interval. Some such readily available distributions are beta and its variants, e.g. Kumaraswamy and its generalizations, unit logistic, unit gamma/Gompertz, etc. distributions. Another approach of obtaining such distributions is the method of truncation, e.g. truncated exponential on (0, 1).

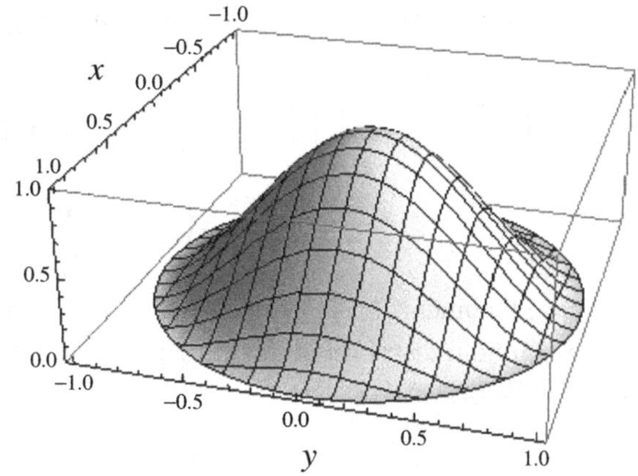

Figure 11.2 A distribution on the disc.

Circular beta models on the interval $(-1, 1)$ may be viewed as distributions on the one-dimensional unit disc D_1. An example of distribution on D is displayed in Figure 11.2. As in Section 11.1.1, the distribution with pdf $f(x_1, x_2)$ may be expressed by $f(r, \theta)$ when the polar transformation is used.

11.2 Uniform Distribution

Since the volume of the inside of the $(p-1)$-dimensional sphere $(p \geq 2)$ with radius a and center at the origin is given by $\pi^{p/2}a^p/\Gamma(p/2+1)$ (Section 8.4.1) it is obvious that the uniform distribution on the unit disc with random vector $\boldsymbol{X} = (X_1, \ldots, X_p)'$ has joint pdf

$$f(\boldsymbol{x}) = \begin{cases} \Gamma(p/2+1)/\pi^{p/2}, & \sum_{j=1}^p x_j^2 \leq 1, \\ 0, & \text{others,} \end{cases}$$

for $\boldsymbol{x} = (x_1, \ldots, x_p)'$. The pdf $f(\boldsymbol{x})$ is a function of $||\boldsymbol{x}||^2 = \sum_{j=1}^p x_j^2$ and is spherical. Hence, $\boldsymbol{X}$ has a stochastic representation $\boldsymbol{X} \stackrel{d}{=} R\boldsymbol{U}^{(p)}$, where $\boldsymbol{U}^{(p)} = (U_1, \ldots, U_p)'$ has the uniform distribution on the unit sphere S^{p-1}, and R is independent of $\boldsymbol{U}^{(p)}$ and distributed as a beta with pdf given by $f(r) = r^{p-1}/B(1,p) = pr^{p-1}$ for $0 \leq r \leq 1$. The characteristic function of $\boldsymbol{X}$ is given by

$$\begin{aligned} E\left(e^{i\,\boldsymbol{t}'\boldsymbol{X}}\right) &= E\left(e^{i\,||\boldsymbol{t}||X_1}\right) \\ &= \frac{2}{B(1/2,(p+1)/2)} \int_0^1 \cos(||\boldsymbol{t}||x_1)(1-x_1^2)^{(p-1)/2} dx_1 \\ &= {}_0F_1\left(\frac{p}{2}+1; -\frac{||\boldsymbol{t}||^2}{4}\right), \quad \boldsymbol{t} \in \mathbb{R}^p. \end{aligned}$$

For any nonnegative integers $m_1, \dots, m_p$ with $m = \sum_{j=1}^p m_j$, the mixed moments of $\boldsymbol{X}$ are

$$E\left(\prod_{j=1}^p X_j^{m_j}\right) = E\left(\prod_{j=1}^p (RU_j)^{m_j}\right) = E(R^m)E\left(\prod_{j=1}^p U_j^{m_j}\right)$$
$$= \frac{B(1, p+m)}{B(1,p)} E\left(\prod_{j=1}^p U_j^{m_j}\right) = \frac{p}{p+m} E\left(\prod_{j=1}^p U_j^{m_j}\right).$$

The mixed moments $E\left(\prod_{j=1}^p U_j^{m_j}\right)$ of $\boldsymbol{U}^{(p)}$ are given in Section 8.4.1.

The marginal distribution of $(X_1, \dots, X_k)'$ for $1 \le k < p$ is, from Theorem 2.10 (p. 36) in Fang et al. (1990),

$$\frac{\Gamma(p/2)}{\Gamma((p-k)/2)\pi^{k/2}} \int_{\sqrt{\sum_{j=1}^k x_j^2}}^1 \frac{1}{r^k} \left(1 - \frac{\sum_{j=1}^k x_j^2}{r^2}\right)^{(p-k)/2-1} pr^{p-1} dr$$
$$= \frac{\Gamma((p+2)/2)}{\Gamma((p-k+2)/2)\pi^{k/2}} \left(1 - \sum_{j=1}^k x_j^2\right)^{(p-k)/2}, \quad \sum_{j=1}^k x_j^2 \le 1, \tag{11.1}$$

and it is a distribution on the k-dimensional disc. If $k = 1$, then (11.1) reduces to a distribution on D_1

$$\frac{1}{B(1/2, (p+1)/2)} (1 - x_1^2)^{(p-1)/2} = \frac{1}{2^p B((p+1)/2, (p+1)/2)} (1 - x_1^2)^{(p-1)/2},$$
$$x_1^2 \le 1,$$

which is transformed into the pdf $u^{(p-1)/2}(1-u)^{(p-1)/2}/B((p+1)/2, (p+1)/2)$ of the corresponding symmetric beta distribution on the interval $[0, 1]$ by applying the transformation $U = (X_1 + 1)/2$.

11.3 General Methods of Constructing Distributions on the Disc

Several of the general methods presented for deriving cylindrical distributions may be adopted or modified to produce disc distributions also. We discuss some such methods next.

11.3.1 Distributions with Specified Marginals

When $p = 1$, it is possible to construct a distribution on the interval $(-1, 1)$ from a distribution on an interval (a, b) by using a transformation from (a, b) to $(-1, 1)$. For example, a beta distribution on $(0, 1)$ with random variable Y is transformed into the corresponding beta distribution on the one-dimensional disc D_1 with random variable X using $X = 2Y - 1$.

When $p = 2$, it is noted (Kato and Shimizu, 2008) that a Johnson–Wehrly (Johnson and Wehrly, 1977, 1978; Wehrly and Johnson, 1980) type distribution on the disc is given as follows. Let $f_1(r)$ be a specified linear pdf on the interval $[0, 1]$, $f_2(\theta)$ a specified circular pdf on $[-\pi, \pi)$ or $[0, 2\pi)$, $F_1(r)$ and $F_2(\theta)$ their distribution functions, respectively, and g is a pdf on the circle. Then

$$f(r, \theta) = 2\pi f_1(r) f_2(\theta) g[2\pi\{F_1(r) \pm F_2(\theta)\}]$$

is a pdf on the disc with marginal pdfs $f_1(r)$ and $f_2(\theta)$.

As an example, putting $f_1(r) = 2\gamma r(1 - r^2)^{\gamma-1}$, $\gamma > 0$, $f_2(\theta) = 1/(2\pi)$ and $g(\xi) = 1/(2\pi)$, we have the distribution,

$$f_\gamma(r, \theta) = 2\gamma r(1 - r^2)^{\gamma-1} \frac{1}{2\pi}, \quad 0 \leq r < 1, -\pi \leq \theta < \pi.$$

This distribution further appears as the bivariate spherically symmetric beta distribution on the unit disc as will be shown in the following text.

11.3.2 Distributions with Specified Conditionals

Refer to Result 10.1. For the disc, the same result can be applied restricting to conditional unit distribution for $R \in (0, 1)$ and circular distribution for $\Theta \in [0, 2\pi)$, as members of regular exponential families first. Then obtain the distribution of the points $\boldsymbol{X} = (X_1, X_2)'$ in the disc by the transformation $x_1 = r\cos\theta, x_2 = r\sin\theta$.

For example, take $p(\theta) = (1 \ \cos\theta \ \sin\theta)'$ and $q(r) = (1 \ r)'$, to get the $f(\theta, r)$ of Result 10.1. Then the von Mises-truncated exponential conditionals distribution $g(\boldsymbol{x})$ of the points $\boldsymbol{X}$ in the unit disc is given by

$$g(\boldsymbol{x}) = (1/r)\exp[(1 \ x_1/r \ x_2/r)M(1 \ r)'], \ r = (x_1^2 + x_2^2)^{1/2}, \ 0 < x_1^2 + x_2^2 < 1.$$

Observe that, as in general, independence between Θ and R results for $\tilde{M} = O$. Also, it is evident that $g(\boldsymbol{x})$ can be characterized as a maximum entropy distribution. Finally, it is obvious that this method can readily produce a large number of probability distributions on the disc.

11.3.3 Specified Disc Regression

The approach used for specified cylindrical regression in Chapter 10 may also be adopted here for regression with the variables defining the unit disc. Following the methods of Saw (1978) and Lai (1994), the density function of the circular beta distribution has been given in terms of rectangular coordinates in (2.2.23) of

Jammalamadaka and SenGupta (2001). In terms of polar coordinates it can be written as

$$f(\theta) = \frac{1}{2^{\alpha+\beta}B(\alpha,\beta)}\{1+\cos(\theta-\mu)\}^{\alpha-1/2}\{1-\cos(\theta-\mu)\}^{\beta-1/2},$$
$$\theta \in [0, 2\pi), \mu \in [0, 2\pi),$$

which is a Beta(α, β) density. It is clear that the mean direction parameter of Θ is μ. As for cylindrical regression detailed in Chapter 10, we can take $f(\theta)$ as the conditional density of Θ given $X = x$ and model the $E(\Theta|X = x) \equiv \mu_x$ as $\mu_x = 2\arctan(\delta + \eta x)$. The marginal density of X may be taken as any unit distribution, e.g. a linear beta distribution. This gives, as an example, the conditional-circular-beta - marginal-linear-beta distribution on the disc with specified circular–linear regression.

11.3.4 Möbius Transformations

As an alternative method, we can use the Möbius transformation on the disc, which is essentially the only conformal mapping from the unit disc in the plane to itself. And, using this mapping, distributions on the unit disc can be transformed into distributions on itself. For example, the Möbius distribution in the following is constructed by applying the Möbius transformation to a bivariate spherically symmetric beta distribution.

Jones (2004) constructed a skew distribution on the disc as follows. Consider the Möbius transformation

$$z = \frac{w - c}{1 - \bar{c}w} \tag{11.2}$$

from a complex number $w = re^{i\theta}$ $(0 \le r \le 1, -\pi \le \theta < \pi,\ i = \sqrt{-1})$ on the unit disc to $z = se^{i\phi}$ on the unit disc, where $c = ae^{i\mu}$ for $0 \le a < 1$ and $-\pi \le \mu < \pi$, and $\bar{c} = ae^{-i\mu}$, the complex conjugate of the complex number c. For a random vector $(S, \Phi)'$ with the bivariate spherically symmetric beta pdf $f_\gamma(s,\phi) = (\gamma/\pi)s(1-s^2)^{\gamma-1}$, $\gamma > 0$, we apply the Möbius inverse transformation $w = (z + c)/(1 + \bar{c}z)$ of (11.2). The resulting distribution for a random vector $(R, \Theta)'$ has pdf

$$f_{\gamma,a,\mu}(r,\theta) = \frac{\gamma(1-a^2)^{\gamma+1}r(1-r^2)^{\gamma-1}}{\pi\{1 - 2ar\cos(\theta-\mu) + a^2r^2\}^{\gamma+1}} \tag{11.3}$$

with parameters $\gamma > 0, 0 \le a < 1$ and $-\pi \le \mu < \pi$ on $0 < r < 1$ and $-\pi \le \theta < \pi$ in polar coordinates. The distribution is called the Möbius distribution (Jones, 2004), generally skew (see Figure 11.3). The pdf in Cartesian coordinates is given by

$$f_{\gamma,a,\mu}(u,v) = \frac{\gamma(1-a^2)^{\gamma+1}(1-u^2-v^2)^{\gamma-1}}{\pi\{1 - 2au\cos\mu - 2av\sin\mu + a^2(u^2+v^2)\}^{\gamma+1}}, \quad 0 \le u^2 + v^2 < 1. \tag{11.4}$$

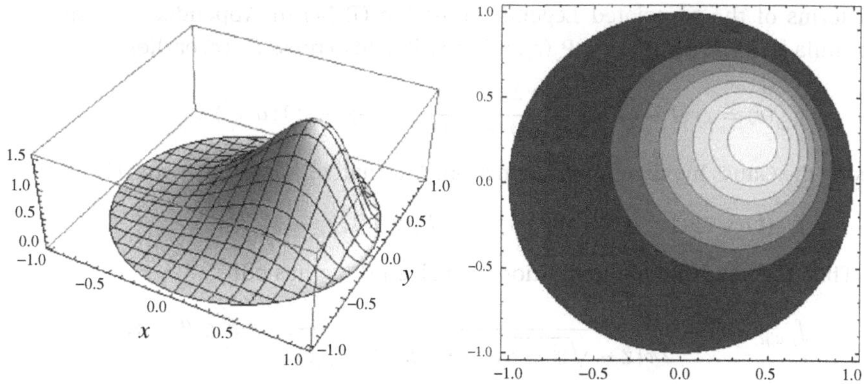

Figure 11.3 A Möbius distribution on the disc with parameters $\gamma = 3.5$, $a = 0.3$, and $\mu = \pi/6$ and its contour plot.

The parameter γ is a concentration parameter and μ controls the orientation of the distribution. When $a = 0$, the bivariate spherically symmetric beta distribution is recovered. The parameter a controls the skewness, more precisely the off-centeredness, of the distribution. When $a \neq 0$, the non-diagonal terms of the expected Fisher information matrix are all zero and the parameters γ, a, and μ are globally orthogonal.

Jones (2004) also studies unimodality, conditionals and marginals of the Möbius distribution, which are summarized next.

Unimodality

Since the Möbius distribution reduces to the bivariate spherically symmetric beta distribution when $a = 0$ and the unimodality of the beta distribution is mentioned in Section 11.4.2, we mainly deal with the case $0 < a < 1$. However, the result also contains the case $a = 0$.

Modality of (11.4) is studied in terms of polar coordinates $u = r\cos\theta$ and $v = r\sin\theta$ $(0 < r < 1, -\pi \leq \theta < \pi)$. Set

$$r_{\pm} = \frac{\sqrt{(\gamma-1)^2 + 8a^2(\gamma+1)} \pm (\gamma-1)}{4a} = \frac{2a(\gamma+1)}{\sqrt{(\gamma-1)^2 + 8a^2(\gamma+1)} \mp (\gamma-1)}.$$

For $\gamma \neq 1$, (11.4) is unimodal with its mode at $(\theta, r) = (\mu, r_-)$ when $\gamma > 1$ and is uniantimodal with its antimode at $(\theta, r) = ((\mu + \pi) \pmod{2\pi}, r_+)$ when $\gamma < 1$. For $\gamma = 1$, (11.4) has no modal point in the interior of the disc; for $a > 0$, its maximum takes at $(r, \theta) = (1, \mu)$ on the boundary of the disc.

Marginals and Conditionals

If a random vector $(R, \Theta)'$ has pdf (11.3), then the marginal distribution of R is given by

$$f_{\gamma,a}(r) = 2\gamma(1-a^2)^{\gamma+1} \frac{r(1-r^2)^{\gamma-1}}{(1-a^2r^2)^{\gamma+1}} P_\gamma\left(\frac{1+a^2r^2}{1-a^2r^2}\right), \quad 0 < r < 1$$

in terms of the associated Legendre function (B.34) in Appendix B. From the formula (B.36) with $P_\nu^0(z) = P_\nu(z)$, it is possible to express $f_{\gamma,a}(r)$ earlier as

$$f_{\gamma,a}(r) = 2\gamma(1-a^2)^{\gamma+1}\frac{r(1-r^2)^{\gamma-1}}{(1-a^2r^2)^{2\gamma+1}}\ {}_2F_1(-\gamma,-\gamma;1;a^2r^2)$$

using the Gauss hypergeometric function. Alternatively, it follows from (B.14) that

$$f_{\gamma,a}(r) = 2\gamma(1-a^2)^{\gamma+1}r(1-r^2)^{\gamma-1}\ {}_2F_1(\gamma+1,\gamma+1;1;a^2r^2).$$

Then the conditional distribution of Θ given $R = r$ has pdf

$$f_{\gamma,a,\mu}(\theta|r) = \frac{1}{2\pi\{z-\sqrt{z^2-1}\cos(\theta-\mu)\}^{\gamma+1}P_\gamma(z)}, \qquad -\pi \le \theta < \pi,$$

where $z = (1+a^2r^2)/(1-a^2r^2)$. The resulting distribution is a special case of the symmetric power cardioid family and the t-distribution on the circle. Concerning the conditional distribution, we have an alternative expression

$$f_{\gamma,a,\mu}(\theta|r) = \frac{1}{2\pi\{1+a^2r^2-2ar\cos(\theta-\mu)\}^{\gamma+1}\ {}_2F_1(\gamma+1,\gamma+1;1;a^2r^2)}$$

in terms of the Gauss hypergeometric function.

As an extension of the Möbius distribution, Bekker et al. (2019) have proposed a very flexible model

$$f(r,\theta) = \frac{C(1-a^2)^2r}{\{1-2ar\cos(\theta-\mu)+a^2r^2\}^2}\, g\left(\frac{a^2-2ar\cos(\theta-\mu)+r^2}{1-2ar\cos(\theta-\mu)+a^2r^2}\right) \tag{11.5}$$

for $0 \le a < 1$, $0 \le \mu < 2\pi$ and $0 \le r \le 1$, $0 \le \theta < 2\pi$. Here $g(\cdot)$ is a continuous function and $C^{-1} = \pi\int_0^1 g(u)du$. If $g(w) = (1-w)^{\gamma-1}$ with $\gamma > 0$ as a special case, then the distribution with density (11.5) reduces to the Möbius distribution. Examples of $g(\cdot)$ in Bekker et al. (2019) include $g(w) = w^{\beta-1}(1-w)^{\gamma-1}e^{-\lambda w}$ for $\gamma,\beta,\lambda > 0$ with $C^{-1} = \pi B(\beta,\gamma)_1F_1(\beta;\beta+\gamma;-\lambda)$ and $g(w) = w^{\beta-1}(1-w)^{\gamma-1}(1+w)^{-(\beta+\gamma)}$ for $\beta,\gamma > 0$ with $C^{-1} = \pi B(\beta,\gamma)/2^\beta$.

11.3.5 Independence of Components on the Disc

We concentrate on the case $p = 2$. It was noted earlier that the conditionally specified family of distributions in general readily admits independence of the components R and Θ on the disc by letting $\tilde{M} = O$. Another family admitting such independence is presented next.

As already stated, the case when $a = 0$ for the Möbius distribution gives an independent model for the variables R and Θ. It is noted that the Möbius distribution is possibly asymmetric about r, but has symmetric property around a line, i.e. for a fixed radius r, the density is symmetric about $\theta = \mu$ (see Figure 11.3). We show next that it is possible to construct a distribution which has asymmetric property about both r and θ, and study independence between the variables (cf. Wang (2013)).

Consider a modified Möbius transformation

$$M_c(z) = |z|\frac{z/|z| - c_0}{1 - \overline{c_0}z/|z|}, \quad c_0 = a_0 e^{i\mu_0},\ 0 \leq a_0 < 1,\ 0 \leq \mu_0 < 2\pi,$$

which preserves the length $|M_c(z)| = |z|$ for a point z (complex number) in the two-dimensional unit disc D_2. Applying the inverse modified Möbius transformation $M_c(\tau) = |\tau|(\tau/|\tau| + c_0)/(1 + \overline{c_0}\ \tau/|\tau|)$ to the Möbius distribution with density (11.3), we have

$$f_1(r,\theta) = \frac{\gamma(1-a^2)^{\gamma+1}(1-a_0^2)r(1-r^2)^{\gamma-1}}{\pi\left\{1+a_0^2-2a_0\cos(\theta-\mu_0)\right\}\left\{B_1(r,\theta)\right\}^{\gamma+1}}.$$

Here

$$B_1(r,\theta) = 1 + a^2r^2 - 2ar\,\frac{\cos(\theta-\mu) + a_0^2\cos(\theta+\mu-2\mu_0) - 2a_0\cos(\mu-\mu_0)}{1+a_0^2-2a_0\cos(\theta-\mu_0)}.$$

The marginal pdf of R is the same as that of the Möbius distribution. This modified Möbius distribution with density $f_1(r,\theta)$ obviously reduces to the Möbius distribution when $a_0 = 0$. If $a = 0$, the modified Möbius density reduces to an independent model

$$f_1(r,\theta) = 2\gamma r(1-r^2)^{\gamma-1} \times \frac{1-a_0^2}{2\pi\left\{1+a_0^2-2a_0\cos(\theta-\mu_0)\right\}},$$

since $B_1(r,\theta) = 1$. The distribution of Θ is a wrapped Cauchy $\mathrm{WC}(\mu_0, a_0)$. The aforementioned argument means that the modified Möbius distribution provides a family admitting independence of components. The conditional distribution of Θ given $R = r$ has density

$$f_1(\theta|r) = \frac{1}{2\pi\{1+a_0^2-2a_0\cos(\theta-\mu_0)\}\{B_1(r,\theta)\}^{\gamma+1}\ {}_2F_1(\gamma+1,\gamma+1;1;a^2r^2)},$$

but its detailed property is not yet investigated. However, if $a_0 = a$ and $\mu_0 = \mu$, the conditional density is expressed by

$$f_1(\theta|r) = C\,\frac{[1-\{2a/(1+a^2)\}\cos(\theta-\mu)\}]^{\gamma}}{\left(1-\left[2a\{1+a^2r+(1+a^2)r\}/\{(1+a^2)(1+a^2r^2)+4a^2r\}\right]\times\cos(\theta-\mu)\right)^{\gamma+1}}, \tag{11.6}$$

where

$$C = \frac{(1-a^2)(1+a^2)^{\gamma}(1-a^2r^2)^{2\gamma+1}}{2\pi\{(1+a^2)(1+a^2r^2)+4a^2r\}^{\gamma+1}{}_2F_1(\gamma+1,\gamma+1;1;a^2r^2)}.$$

The conditional distribution is related to the family (2.1) in Abe et al. (2010), though the interpretation of the parameters in (11.6) is not easy. It is obvious that the conditional distribution with (11.6) is uniform if $a = 0$.

An alternative model is constructed as follows. Consider $c_0 = a_0 r\, e^{i\mu_0}$, which depends on r, instead of $c_0 = a_0 e^{i\mu_0}$. Then we have

$$f_2(r,\theta) = \frac{\gamma(1-a^2)^{\gamma+1}(1-a_0^2r^2)r(1-r^2)^{\gamma-1}}{\pi\left\{1+a_0^2r^2-2a_0r\cos(\theta-\mu_0)\right\}\left\{B_2(r,\theta)\right\}^{\gamma+1}}.$$

Here $B_2(r,\theta)$ is obtained by replacing a_0 in $B_1(r,\theta)$ with $a_0 r$. The marginal pdf of R is the same as that of the Möbius distribution. The modified Möbius distribution with $B_2(r,\theta)$ reduces to the Möbius distribution if $a_0 = 0$, and R is not independent of Θ even if $a = 0$. Indeed, its density is given by

$$f_2(r,\theta) = 2\gamma r(1-r^2)^{\gamma-1} \times \frac{1-a_0^2r^2}{2\pi\left\{1+a_0^2r^2-2a_0r\cos(\theta-\mu_0)\right\}}.$$

The conditional distribution of Θ given $R = r$ is a wrapped Cauchy $\mathrm{WC}(\mu_0, a_0 r)$.

11.4 Some More Distributions on the Disc

11.4.1 Uniform Distribution in the Sphere

Since the volume of a unit disc is $4\pi/3$ as a special case of $p = 3$ in Section 11.2, the uniform distribution in the unit sphere has density $f(x_1,x_2,x_3) = 3/(4\pi)$, $x_1^2 + x_2^2 + x_3^2 \le 1$. Its marginal distribution in the unit circle has pdf

$$f(x_1,x_2) = \frac{3}{2\pi}\,(1-x_1^2-x_2^2)^{1/2}, \quad x_1^2+x_2^2 \le 1,$$

which is not a uniform distribution. The marginal distribution on the interval $[-1,1]$ has pdf

$$f(x_1) = \frac{3}{4}\,(1-x_1^2), \quad -1 \le x_1 \le 1,$$

whose transformation $U = (X_1+1)/2$ results in $f(u) = 6u(1-u)$ on $[0,1]$, which is a symmetric beta distribution with parameter 2.

11.4.2 Bivariate Spherically Symmetric Beta Distribution

From Section 8.4.1, if $\boldsymbol{X} = (X_1,\dots,X_n)'$ is distributed as an n-dimensional normal distribution with mean vector $\boldsymbol{0}$ and variance–covariance matrix I_n, $N_n(\boldsymbol{0}, I_n)$, then $(Y_1,\dots,Y_k)' = (X_1/\parallel \boldsymbol{X} \parallel, \dots, X_k/\parallel \boldsymbol{X} \parallel)'$ has pdf

$$\frac{\Gamma(n/2)}{\Gamma((n-k)/2)\pi^{k/2}}\left(1-\sum_{j=1}^{k} y_j^2\right)^{(n-k)/2-1}, \quad \sum_{j=1}^{k} y_j^2 \le 1,$$

which is identical with (11.1) with $p = n-2$ in Section 11.2.

A related distribution with support on the unit disc in the plane ($\gamma = p/2,\ k = 2$ in (11.1)) is the bivariate spherically symmetric beta (or Pearson type II) distribution with joint pdf

$$f_\gamma(x,y) = \frac{\gamma}{\pi}\,(1 - x^2 - y^2)^{\gamma-1}, \quad 0 \le x^2 + y^2 < 1,\ \gamma > 0. \tag{11.7}$$

The distribution has mode at the origin $(x,y)' = (0,0)'$ when $\gamma > 1$ (see Figure 11.2) and antimode at the origin when $\gamma < 1$, and it includes the uniform distribution on the disc when $\gamma = 1$. Thus, the parameter γ controls the concentration of the distribution. Marginals are symmetric beta distributions on the interval $(-1,1)$ with parameter $\gamma + 1/2$ whose pdfs are given by (8.3) with $p = 2(\gamma+1)$. The distribution with pdf (11.7) may be expressed as

$$f_\gamma(r,\theta) = 2\gamma r(1-r^2)^{\gamma-1}\frac{1}{2\pi}, \quad 0 < r < 1, -\pi \le \theta < \pi \tag{11.8}$$

in terms of polar coordinates r and θ, which means that its random variable R is independent of a circular uniform random variable Θ. It is possible to generate this distribution as a Johnson–Wehrly type distribution as shown earlier. An extension of (11.8) is given by

$$f(r,\theta) = \alpha\gamma r^{\alpha-1}(1-r^\alpha)^{\gamma-1}\frac{1}{2\pi}, \quad 0 < r < 1, -\pi \le \theta < \pi$$

using $f(r) = \alpha\gamma r^{\alpha-1}(1-r^\alpha)^{\gamma-1}$ with $\alpha > 0$ and $\gamma > 0$, the pdf of Kumaraswamy's (1980) distribution.

11.4.3 An Alternative Skew Distribution

As a model that can also exhibit skewness, using the marginal replacement approach, a p-dimensional beta distribution has been proposed by Jones (2002), with density

$$\frac{\Gamma(a+c)\Gamma(b)}{\Gamma(a)\Gamma(c)\Gamma(b-(p-1)/2)2^{a+c-1}\pi^{(p-1)/2}}\,(1+x_1)^{a-b}(1-x_1)^{c-b}(1-r_x^2)^{b-(p+1)/2}$$

for $0 \le r_x^2 = \sum_{j=1}^{p} x_j^2 < 1$ with $a > 0,\ b > (p-1)/2$ and $c > 0$.

When $p = 1$ in the aforementioned equation, we have (see Section 4.7 and p. 51 of Jammalamadaka and SenGupta (2001))

$$f(x_1) = \frac{1}{2^{a+c-1}B(a,c)}\,(1+x_1)^{a-1}(1-x_1)^{c-1}, -1 < x_1 < 1,$$

which is a Beta(a,c) density with parameter (a,c) on the one-dimensional disc, where B denotes the beta function. When $p = 2$, replacing b with $\gamma + 1/2$, we have a skew distribution on the disc whose pdf is given by

$$\frac{(1+x_1)^{a-\gamma-1/2}(1-x_1)^{c-\gamma-1/2}(1-x_1^2-x_2^2)^{\gamma-1}}{2^{a+c-1}B(a,c)B(1/2,\gamma)}, \quad 0 \le x_1^2 + x_2^2 < 1$$

with parameters $a, c, \gamma > 0$. The marginal distribution of X_1 is Beta(a, c) and the conditional distribution of X_2 given $X_1 = x_1$ is a beta distribution with parameter γ rescaled to the interval $(-\sqrt{1-x_1^2}, \sqrt{1-x_1^2})$. Interpretation of parameters is not simple in comparison with the Möbius distribution.

11.5 Distributions on the Hyper-Disc

Some of the general methods for deriving probability distributions on the disc presented earlier may be generalized to the case of hyperdiscs also. We first consider these in the following text.

11.5.1 Distributions with Specified Conditionals

Refer to Result 10.1. For the hyper-disc with $p = 3$, the same result can be applied restricting to conditional unit distribution for $R \in (0, 1)$ and spherical distribution for (Θ_1, Θ_2), $\Theta_1 \in [0, \pi)$, $\Theta_2 \in [0, 2\pi)$, as members of regular exponential families first. Then obtain the distribution of the points $\boldsymbol{X} = (X_1, X_2, X_3)'$ in the disc by the transformation as discussed in Section 11.3.2.

For example, take $p(\theta) = (1\ \cos\theta_1\ \sin\theta_1\cos\theta_2\ \sin\theta_2)'$ and $q(r) = (1\ r)'$, to get the $f(\theta_1, \theta_2, r)$ of Result 10.1. Then the Fisher–von Mises - truncated-exponential conditionals distribution $g(\boldsymbol{x})$ of the points $\boldsymbol{X}$ in the unit hyper-disc is given by

$$g(\boldsymbol{x}) = (1/r)\exp[(1\ x_1/r\ x_2/r\ x_3/r)M(1\ r)'],$$
$$r = (x_1^2 + x_2^2 + x_3^2)^{1/2},\ 0 < x_1^2 + x_2^2 + x_3^2 < 1.$$

Observe that, as in general, independence between (Θ_1, Θ_2) and R results for $\tilde{M} = 0$. It is also evident that $g(\boldsymbol{x})$ can be characterized as a maximum entropy distribution. Finally, it is obvious that this method can readily produce a large number of probability distributions on the hyper-disc.

11.5.2 Multivariate Möbius Transformation

For a general p (≥ 1), a multivariate Möbius transformation may be defined using the stereographic projection (cf. Ratcliffe, 2006, p. 122) from the p-dimensional unit disc $D^p = \{(x_0, x_1, \ldots, x_p)' | x_0 = 0, \sum_{j=1}^p x_j^2 < 1\}$ to the hyperbolic p-space H^p in Lorentzian $(p+1)$-space, which is the inner product space consisting of $(p+1)$-dimensional real vector space $\mathbb{R}^{p+1}$ together with the Lorentzian inner product defined by $\boldsymbol{u} \circ \boldsymbol{v} = -st + \sum_{j=1}^p u_j v_j$ for $\boldsymbol{u} = (s, u_1, \ldots, u_p)'$ and

$v = (t, v_1, \ldots, v_p)'$. Identify $\mathbb{R}^p$ with $\mathbb{R}^p \times \{0\}$ in $\mathbb{R}^{p+1}$. Then the stereographic projection is defined as

$$\varphi(x_1, \ldots, x_p) = \left(\frac{1 + r_x^2}{1 - r_x^2}, \frac{2x_1}{1 - r_x^2}, \ldots, \frac{2x_p}{1 - r_x^2} \right)', \quad r_x^2 = \sum_{j=1}^{p} x_j^2,$$

and the inverse of the stereographic projection from H^p with the coordinates $(s, u_1, \ldots, u_p)'$ to D^p as

$$\varphi^{-1}(s, u_1, \ldots, u_p) = \left(\frac{u_1}{1+s}, \ldots, \frac{u_p}{1+s} \right)'.$$

The Jacobian of the stereographic projection is given by

$$\frac{\partial(u_1, \ldots, u_p)}{\partial(x_1, \ldots, x_p)} = \frac{2^p(1 + r_x^2)}{(1 - r_x^2)^{p+1}}.$$

The Möbius transformation (Uesu et al., 2015) as a conformal mapping can be defined as the composite of the stereographic projection from the hyper-disc to the hyperbolic space, an isometry from the hyperbolic space to itself and the inverse stereographic projection from the hyperbolic space to the hyper-disc. See Figure 11.4 for the (inverse) stereographic projection.

Uesu et al. (2015) constructed a skew distribution on the hyper-disc by applying the multivariate Möbius transformation to a spherically symmetric distribution. For the case $p = 1$, the resulting distribution reduces to (4.42), the distribution studied by Seshadri (1991), and the case $p = 2$ corresponds to the Möbius distribution.

The Multivariate Möbius Transformation

The Möbius transformation we use is represented as a mapping $g(a_1, \ldots, a_p) = \varphi^{-1} \circ g_A \circ \varphi$ from the p-dimensional unit disc D^p to itself composed of the three mappings φ (stereographic projection) from D^p to the hyperbolic p-space H^p, g_A

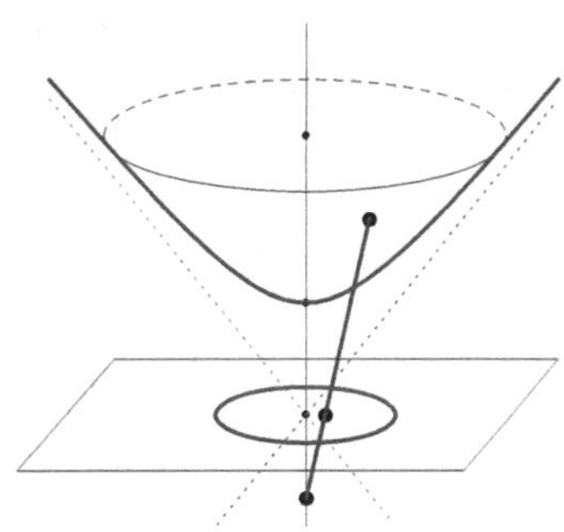

Figure 11.4 Stereographic projection from the unit disc to the hyperbolic space.

from H^p to itself, and φ^{-1} from H^p to D^p. Here the mapping g_A is as follows. We choose a $(p+1) \times (p+1)$ real matrix of the form

$$A = \begin{pmatrix} 0 & a_1 & \cdots & a_p \\ a_1 & & & \\ \vdots & & 0 & \\ a_p & & & \end{pmatrix}, \quad (a_1, \dots, a_p)' \neq (0, \dots, 0)'$$

and consider $\exp(A) = \sum_{k=0}^{\infty} A^k/k!$. Direct calculation yields

$$\exp(A) = \begin{pmatrix} \dfrac{1+\alpha^2}{1-\alpha^2} & \dfrac{2a_1\alpha}{r_a(1-\alpha^2)} & \cdots & \dfrac{2a_p\alpha}{r_a(1-\alpha^2)} \\ \dfrac{2a_1\alpha}{r_a(1-\alpha^2)} & & & \\ \vdots & & \delta_{jk} + \dfrac{2a_ja_k\alpha^2}{r_a^2(1-\alpha^2)} & \\ \dfrac{2a_p\alpha}{r_a(1-\alpha^2)} & & & \end{pmatrix},$$

which is a Lorentz transformation that maps from the hyperboloid model H^p onto itself. Here $r_a = \sqrt{\sum_{j=1}^{p} a_j^2}$, $\alpha = \tanh\left(r_a/2\right)$ and δ_{jk} is Kronecker's delta, i.e. $\delta_{jk} = 1$ for $j = k$ and 0 for $j \neq k$ $(j, k = 1, \dots, p)$.

Choose a point $\boldsymbol{u} = (s, u_1, \dots, u_p)'$ of H^p. Then $s = \sqrt{\sum_{j=1}^{p} u_j^2 + 1}$. And let $\exp(A)\boldsymbol{u} = \boldsymbol{v}$ with $\boldsymbol{v} = (t, v_1, \dots, v_p)'$. Then,

$$\begin{cases} t = \dfrac{2\alpha}{r_a(1-\alpha^2)} \displaystyle\sum_{j=1}^{p} a_j u_j + \dfrac{(1+\alpha^2)s}{1-\alpha^2}, \\ v_k = u_k + \dfrac{2\alpha^2 a_k}{r_a^2(1-\alpha^2)} \displaystyle\sum_{j=1}^{p} a_j u_j + \dfrac{2\alpha a_k s}{r_a(1-\alpha^2)}, \quad k = 1, \dots, p. \end{cases}$$

The mapping $(u_1, \dots, u_p)'$ to $(s, u_1, \dots, u_p)'$ gives a diffeomorphism from $\mathbb{R}^p$ to H^p. We adopt $(u_1, \dots, u_p)'$ as a coordinate system of the hypersurface H^p. Under this coordinate system, $\exp(A)$ determines a mapping from $\mathbb{R}^p$ to $\mathbb{R}^p$. This mapping is denoted by g_A. Then the Jacobian of the mapping g_A is

$$\frac{\partial(v_1, \dots, v_p)}{\partial(u_1, \dots, u_p)} = \frac{1}{(1-\alpha^2)s} \left\{ (1+\alpha^2)s + 2\alpha \sum_{j=1}^{p} \omega_j u_j \right\}$$

with $\omega_j = a_j/r_a$ $(j = 1, \dots, p)$.

The Jacobian of the mapping $g(a_1, \dots, a_p)(x_1, \dots, x_p)' = (y_1, \dots, y_p)'$ is the product of the Jacobians of the three mappings φ^{-1}, g_A and φ, and the Jacobian of the mapping $g(a_1, \dots, a_p)$ is expressed by

$$\frac{\partial(y_1, \dots, y_p)}{\partial(x_1, \dots, x_p)} = \left\{ \frac{1 - \alpha^2}{\sum_{j=1}^{p} (\alpha x_j + \omega_j)^2} \right\}^p$$

with $(a_1, \dots, a_p)' \neq (0, \dots, 0)'$.

A Possibly Asymmetric Distribution

Since $\{g(a_1, \dots, a_p\}^{-1} = g(-a_1, \dots, -a_p)$ from $\{\exp(A)\}^{-1} = \exp(-A)$, we generate the probability distribution induced by the mapping $g(-a_1, \dots, -a_p)$ from the multivariate spherically symmetric beta distribution with pdf

$$f_\gamma(y_1, \dots, y_p) = \frac{1}{C_{p,\gamma}} (1 - r_y^2)^{\gamma-1}, \quad 0 \leq r_y^2 = \sum_{j=1}^{p} y_j^2 < 1$$

with parameter $\gamma > 0$ and the normalizing constant $C_{p,\gamma} = \pi^{p/2}\Gamma(\gamma)/\Gamma(\gamma + p/2)$. The resulting distribution, called the multivariate Möbius distribution in this book, has pdf

$$\begin{aligned} f_{\gamma,a}(x_1, \dots, x_p) = \ & \frac{1}{C_{p,\gamma}} \left(\frac{1 - \alpha^2}{\sum_{j=1}^{p} (\alpha x_j - \omega_j)^2} \right)^{\gamma+p-1} (1 - r_x^2)^{\gamma-1}, \\ & 0 \leq r_x^2 = \sum_{j=1}^{p} x_j^2 < 1. \end{aligned} \tag{11.9}$$

The role of the parameters is as follows. The parameter r_a relates to symmetry or asymmetry. Actually, if $r_a = 0$, i.e. $(a_1, \dots, a_p)' = (0, \dots, 0)'$, then (11.9) reduces to the pdf of the multivariate spherically symmetric beta distribution. The parameter γ controls modality. In fact, (a) for $\gamma > 1$, the density (11.9) is unimodal with its mode at $(\omega_1 z_+, \dots, \omega_p z_+)'$, (b) for $0 < \gamma < 1$, the density (11.9) is unimodal with its antimode at $(\omega_1 z_-, \dots, \omega_p z_-)'$, and (c) for $\gamma = 1$, the distribution has no modal point in the interior of the disc, where

$$z_\pm = \frac{1}{2\alpha p} \left\{ -(\gamma - 1) \pm \sqrt{(\gamma - 1)^2 + 4\alpha^2 p(\gamma + p - 1)} \right\}.$$

In (11.9), a special case $p = 1$ gives a beta distribution on D_1 with density

$$f_{\gamma,a}(x) = \frac{(1 - \alpha^2)^\gamma (1 - x^2)^{\gamma-1}}{B(1/2, \gamma)(1 - \alpha x)^{2\gamma}}, \quad -1 < x < 1$$

for $\gamma > 0,\ \alpha = \tanh(r_a/2)$ and $r_a > 0$. This corresponds to (4.42) with $\lambda = 2\gamma$ and $z = 2\alpha/(1+\alpha)$. When $p = 2$ as another special case, we have the Möbius distribution on D with density

$$f_{\gamma,a}(u,v) = \frac{\gamma}{\pi}\left\{\frac{1-\alpha^2}{1+\alpha^2(u^2+v^2)-2\alpha(\omega_1 u+\omega_2 v)}\right\}^{\gamma+1}(1-u^2-v^2)^{\gamma-1},$$
$$0 \le u^2+v^2 < 1$$

for $\gamma > 0,\ (a_1,a_2)' \neq (0,0)',\ r_a = \sqrt{a_1^2+a_2^2},\ \alpha = \tanh(r_a/2)$ and $\omega_j = a_j/r_a$ $(j = 1,2)$, which is equivalent to (11.4) with $a = \alpha,\ \cos\mu = \omega_1$ and $\sin\mu = \omega_2$.

Angular Representation

We consider the following polar coordinate system of the p-dimensional disc with unit radius and center at the origin, denoted by $(r_x, \boldsymbol{z})'$, where $0 < r_x \le 1$, $\boldsymbol{z} = (z_1,\dots,z_p)'$ with

$$\left\{\begin{array}{rcl} z_p &=& \cos\theta_1, \\ z_{p-1} &=& \sin\theta_1\cos\theta_2, \\ z_{p-2} &=& \sin\theta_1\sin\theta_2\cos\theta_3, \\ &\vdots& \\ z_2 &=& \sin\theta_1\ \cdots\sin\theta_{p-2}\cos\theta_{p-1}, \\ z_1 &=& \sin\theta_1\ \cdots\sin\theta_{p-2}\sin\theta_{p-1}, \end{array}\right. \tag{11.10}$$

$0 \le \theta_1,\dots,\theta_{p-2} < \pi$ and $0 \le \theta_{p-1} < 2\pi$. The vector $\boldsymbol{z}$ is an element of S^{p-1}, the $(p-1)$-dimensional sphere with unit radius and center at the origin in $\mathbb{R}^p$. Letting $\boldsymbol{x} = r_x\boldsymbol{z}$, the Jacobian is given by

$$\frac{\partial(x_1,\dots,x_p)}{\partial(r_x,\theta_1,\dots,\theta_{p-1})} = r_x^{p-1}\prod_{j=1}^{p-2}\sin^{p-1-j}\theta_j,$$

and the angular representation of (11.9) is

$$\frac{1}{C_{p,\gamma}}\left\{\frac{1-\alpha^2}{\sum_{j=1}^p(\alpha x_j-\omega_j)^2}\right\}^{\gamma+p-1}(1-r_x^2)^{\gamma-1}r_x^{p-1}\prod_{j=1}^{p-1}\sin^{p-1-j}\theta_j$$

with θ_j in (11.10).

Conditional and Marginal Distributions

The density (11.9) has a representation with respect to the polar coordinate system $(r_x, \boldsymbol{z})'$ as

$$\frac{1}{C_{p,\gamma}}\left\{\frac{1-\alpha^2}{\sum_{j=1}^p(\alpha r_x z_j-\omega_j)^2}\right\}^{\gamma+p-1}(1-r_x^2)^{\gamma-1}r_x^{p-1}. \tag{11.11}$$

Thus, we have the conditional distribution of $\boldsymbol{z}$ for $p \geq 2$ given $r_x = \lambda$ $(0 < \lambda < 1)$ in (11.11) with density

$$\frac{1}{C_1}\left\{\sum_{j=1}^{p}(\alpha\lambda z_j - \omega_j)^2\right\}^{-(\gamma+p-1)} = \frac{1}{C_1}\left(1 + \alpha^2\lambda^2 - 2\alpha\lambda\sum_{j=1}^{p}\omega_j z_j\right)^{-(\gamma+p-1)}. \tag{11.12}$$

The normalizing constant C_1 is

$$C_1 = 2\pi^{p/2}(\alpha\lambda)^{1-p/2}(1-\alpha^2\lambda^2)^{-(\gamma+p/2)}P^{1-p/2}_{\gamma+p/2-1}(z), \quad 0 < \alpha < 1,$$

where P^{μ}_{ν} denotes the associated Legendre function and $z = (1+\alpha^2\lambda^2)/(1-\alpha^2\lambda^2)$. For $\alpha = 0$, we have $z = 1$ and $C_1 = \text{Vol}\,(S^{p-1}) = 2\pi^{p/2}/\Gamma(p/2)$. The conditional distribution with density (11.12) arises as the t-distribution on the sphere with $(2\gamma + p - 2)$ degrees of freedom. If $p = 2$, the density (11.12) reduces to

$$\begin{aligned} f(\theta|\lambda) &= \left[2\pi P_{\gamma}(z)\left\{z - \sqrt{z^2-1}\,(\omega_1 \sin\theta + \omega_2\cos\theta)\right\}^{\gamma+1}\right]^{-1} \\ &= \left[2\pi P_{\gamma}(z)\left\{z - \sqrt{z^2-1}\,\cos(\theta-\mu)\right\}^{\gamma+1}\right]^{-1}, \end{aligned}$$

where $\sin\mu = \omega_1$ and $\cos\mu = \omega_2$.

The conditional distribution of r_x given $\boldsymbol{z}$ has density

$$f(r_x|\boldsymbol{z}) \propto (1-r_x^2)^{\gamma-1}r_x^{p-1}\left(1 + \alpha^2 r_x^2 - 2\alpha r_x\sum_{j=1}^{p}\omega_j z_j\right)^{-(\gamma+p-1)},$$

for $0 < r_x < 1$.

The marginal distribution of r_x as a random variable has density

$$\begin{aligned} f(\lambda) &= \frac{2\Gamma(\gamma+p/2)}{\Gamma(\gamma)}(1-\alpha^2)^{\gamma+p-1}\alpha^{1-p/2} \\ &\times \frac{\lambda^{p/2}(1-\lambda^2)^{\gamma-1}}{(1-\alpha^2\lambda^2)^{\gamma+p/2}}P^{1-p/2}_{\gamma+p/2-1}\left(\frac{1+\alpha^2\lambda^2}{1-\alpha^2\lambda^2}\right), \quad 0 < \lambda < 1, \end{aligned} \tag{11.13}$$

for $0 < \alpha < 1$. From the formula (B.36), an alternative expression of the marginal distribution is given as

$$f(\lambda) = \frac{2(1-\alpha^2)^{\gamma+p-1}\lambda^{p-1}(1-\lambda^2)^{\gamma-1}}{B(p/2,\gamma)(1-\alpha^2\lambda^2)^{2\gamma+p-1}}\,{}_2F_1\left(1-\gamma-\frac{p}{2}, -\gamma; \frac{p}{2}; \alpha^2\lambda^2\right) \tag{11.14}$$

in terms of the Gauss hypergeometric function. Equation (11.14) is transformable into

$$f(\lambda) = \frac{2(1-\alpha^2)^{\gamma+p-1}}{B(p/2,\gamma)}\lambda^{p-1}(1-\lambda^2)^{\gamma-1}{}_2F_1\left(p+\gamma-1, \frac{p}{2}+\gamma; \frac{p}{2}; \alpha^2\lambda^2\right) \tag{11.15}$$

using (B.14). It can be confirmed, using equation 7.512.6, p. 813 of Gradshteyn and Ryzhik (2007) and ${}_1F_0(c;x) = (1-x)^{-c}$, that Equation (11.13) gives a density, i.e. $\int_0^1 f(\lambda)d\lambda = 1$. It is easy to see that Equation (11.14) or (11.15) tends to

$$f(\lambda) = \frac{2}{B(p/2,\gamma)} \lambda^{p-1}(1-\lambda^2)^{\gamma-1} \tag{11.16}$$

as α goes to zero, and (11.13) covers the case $\alpha = 0$. This is a special case of McDonald's (1984) generalized beta distribution of the first kind, and if $p = 2$, Equation (11.16) becomes a special case of Kumaraswamy's (1980) double-bounded density.

Fisher Information

The Fisher information matrix for the multivariate Möbius distribution with density $f_{\gamma,a} = f_{\gamma,a}(x_1, \dots, x_p)$ in (11.9) is given in the following text. Parametrization μ_j $(j = 1, \dots, p-1)$ using the polar coordinate (11.10) is used for ω_j.

The parameters γ and r_a are orthogonal to μ_j $(j = 1, \dots, p-1)$ and $\mu_1, \dots, \mu_{p-1}$ are mutually orthogonal in the sense that the non-diagonal elements of the Fisher information matrix are zeros. The diagonal elements of the Fisher information matrix are given by

$$E\left\{\left(\frac{\partial \log f_{\gamma,a}}{\partial \gamma}\right)^2\right\} = \Psi'(\gamma) - \Psi'(\gamma + p/2)$$

in terms of Ψ', known as the trigamma function, where $\Psi(\gamma)$ denotes the psi (digamma) function defined by $\Psi(\gamma) = d\{\log \Gamma(\gamma)\}/d\gamma$,

$$E\left\{\left(\frac{\partial \log f_{\gamma,a}}{\partial r_a}\right)^2\right\} = \frac{(\gamma + p - 1)^2}{2(\gamma + p/2)}$$

and

$$E\left\{\left(\frac{\partial \log f_{\gamma,a}}{\partial \mu_j}\right)^2\right\} = \frac{2(\gamma + p - 1)^2\alpha^2}{(\gamma + p/2)(1-\alpha^2)^2} \prod_{m=1}^{j-1} \sin^2 \mu_m, \quad j = 1, \dots, p-1$$

with interpretation $\prod_{m=1}^{j-1} \sin^2 \mu_m = 1$ for $j = 1$. The parameters $\omega_1, \dots, \omega_{p-1}$ are not mutually orthogonal:

$$-E\left(\frac{\partial^2 \log f_{\gamma,a}}{\partial \omega_j \partial \omega_k}\right) = \frac{2(\gamma + p - 1)^2\alpha^2}{(\gamma + p/2)(1-\alpha^2)^2}\left(\delta_{jk} + \frac{\omega_j\omega_k}{\omega_p^2}\right), \quad j, k = 1, \dots, p-1.$$

For the case $p = 2$, we have expressions $\iota_{\mu\mu} = 2a^2(\gamma+1)/(1-a^2)^2$, $\iota_{\gamma\gamma} = \Psi'(\gamma) - \Psi'(\gamma+1) = \gamma^{-2}$ since $\Psi(x+1) = \Psi(x) + 1/x$ and $\iota_{aa} = 2(\gamma+1)/(1-a^2)^2$ for the diagonal elements of the Fisher information matrix.

11.6 Statistical Inference

Maximum likelihood estimation for the multivariate Möbius distribution has been studied by Uesu et al. (2015). Suppose that a random sample of size $n, \boldsymbol{x}^{\{1\}}, \ldots, \boldsymbol{x}^{\{n\}}$ with $\boldsymbol{x}^{\{k\}} = (x_1^{\{k\}}, \ldots, x_p^{\{k\}})'$ for $k = 1, \ldots, n$, is drawn from a distribution on the hyper-disc with density (11.9) and unknown parameters γ, r_a and ω_j $(j = 1, \ldots, p-1)$. If $(a_1, \ldots, a_p)' \neq (0, \ldots, 0)'$, there exists $a_k \neq 0$ for some k. Without loss of generality we assume that $a_p \neq 0$, so that $\omega_p \neq 0$. If $(a_1, \ldots, a_{p-1})' = (0, \ldots, 0)'$ as a particular case, (11.9) has two parameters γ and $r_a = |a_p|$. Note that if $(a_1, \ldots, a_p)' = (0, \ldots, 0)'$ then (11.9) reduces to the p-dimensional spherically symmetric beta distribution, which has only one parameter γ.

Under the assumption that γ, r_a and ω_j $(j = 1, \ldots, p-1)$ are unknown, equating the first order differentials of the log-likelihood function with respect to γ, r_a and ω_j $(j = 1, \ldots, p-1)$ to zero, we obtain the estimating equations

$$n\left\{\Psi(\gamma) - \Psi\left(\gamma + \frac{p}{2}\right)\right\} = n\log(1-\alpha^2) - \sum_{k=1}^{n} \log\left\{\sum_{m=1}^{p} (\alpha x_m^{\{k\}} - \omega_m)^2\right\} + \sum_{k=1}^{n} \log(1 - \rho^2_{x^{\{k\}}}), \tag{11.17}$$

$$\frac{n\alpha}{1-\alpha^2} = -\sum_{k=1}^{n} \frac{\sum_{m=1}^{p} (\alpha x_m^{\{k\}} - \omega_m) x_m^{\{k\}}}{\sum_{m=1}^{p} (\alpha x_m^{\{k\}} - \omega_m)^2}$$

to be solved numerically.

The aforementioned discussion is for the case where the parameters of a_js are unknown. However, here is an interesting note for the case where a_js are known. The conditional distribution of r_x given $\boldsymbol{z}$, which is obviously obtained, has a closed form expression, though not of any standard known distribution. This conditional distribution becomes a member of the Curved Exponential Family (Amari, 1985), CEF(2, 1), in terms of the scalar parameter γ. This observation may be exploited to draw inferences based on conditional likelihoods.

There may be many three-dimensional applications. Maximum likelihood estimates of a_1, a_2 and a_3 are obtained by maximizing the function

$$L(a_1, a_2, a_3) = n\log(1-\alpha^2) - \sum_{k=1}^{n} \log\left\{\sum_{m=1}^{3} (\alpha x_m^{\{k\}} - \omega_m)^2\right\},$$

independent of γ. We can apply numerical optimization to this function using the methods (a) Nelder–Mead simplex, (b) simulated annealing, (c) differential evolution, and (d) random search. To obtain the maximum likelihood estimate of γ, the likelihood equation (11.17) is numerically solved. The estimated value is reliable because the left side of the equation is monotonically increasing. Approximate confidence intervals for the parameters can be numerically computed using the Fisher information.

11.7 Applications

The small dataset in Johnson and Wehrly (1977), which consists of ozone concentration and wind direction with size 19, was used for showing a good fit of the Möbius distribution on the disc ($p = 2$) in Jones (2004). Bekker et al. (2019) proposed the generalized Möbius distribution and analyzed wind speed and wind direction data observed at Marion Island. For the case of three-dimensional disc ($p = 3$), which can be also visualized, two examples, datasets on Gamma-Ray Bursts (GRBs) with Declination of source, Right ascension of source and BATFluence in astrophysics and on worldwide earthquakes with latitude, longitude, and magnitude, were given in Uesu et al. (2015).

References

Abe, T., Shimizu, K. and Pewsey, A. (2010). Symmetric unimodal models for directional data motivated by inverse stereographic projection, *Journal of the Japan Statistical Society*, 40, 45–61.

Amari, S. (1985). *Differential-Geometrical Methods in Statistics*, Lecture Notes in Statistics 28, Springer-Verlag, Berlin.

Bekker, A., Nagar, P., Arashi, M. and Rautenbach, H. (2019). From symmetry to asymmetry on the disc manifold: Modeling of Marion Island data, *Symmetry*, 11, 1030; https://doi.org/10.3390/sym11081030.

Fang, K. T., Kotz, S. and Ng, K. W. (1990). *Symmetric Multivariate and Related Distributions*, Chapman and Hall, London.

Gradshteyn, I. S. and Ryzhik, I. M. (2007). *Table of Integrals, Series, and Products*, 7th ed., Elsevier, Amsterdam.

Jammalamadaka, S. R. and SenGupta, A. (2001). *Topics in Circular Statistics*, World Scientific, Singapore.

Johnson, R. A. and Wehrly, T. E. (1977). Measures and models for angular correlation and angular-linear correlation, *Journal of the Royal Statistical Society*, B39, 222–229.

Johnson, R. A. and Wehrly, T. E. (1978). Some angular-linear distributions and related regression models, *Journal of the American Statistical Association*, 73, 602–606.

Jones, M. C. (2002). Marginal replacement in multivariate densities, with application to skewing spherically symmetric distributions, *Journal of Multivariate Analysis*, 81, 85–99.

Jones, M. C. (2004). The Möbius distribution on the disc, *Annals of the Institute of Statistical Mathematics*, 56, 733–742.

Kato, S. and Shimizu, K. (2008). Dependent models for observations which include angular ones, *Journal of Statistical Planning and Inference*, 138, 3538–3549.

Kumaraswamy, P. (1980). A generalized probability density function for double-bounded random processes, *Journal of Hydrology*, 46, 79–88.

Lai, M. K. (1994). *Some Results on the Statistical Analysis of Directional Data*, Master's Thesis, The University of Hong Kong.

McDonald, J. B. (1984). Some generalized functions for the size distribution of income, *Econometrica*, 52, 647–663.

Ratcliffe, J. G. (2006). *Foundations of Hyperbolic Manifolds*, 2nd ed., Springer, New York.

Saw, J. G. (1978). A family of distributions on the m-sphere and some hypothesis tests, *Biometrika*, 65, 69–73.

Seshadri, V. (1991). A family of distributions related to the McCullagh family, *Statistics & Probability Letters*, 12, 373–378.

Uesu, K., Shimizu, K. and SenGupta, A. (2015). A possibly asymmetric multivariate generalization of the Möbius distribution for directional data, *Journal of Multivariate Analysis*, 134, 146–162.

Wang, M. Z. (2013). *Extensions of Probability Distributions on Torus, Cylinder and Disc*, Ph.D. Thesis, Graduate School of Science and Technology, Keio University.

Wehrly, T. E. and Johnson, R. A. (1980). Bivariate models for dependence of angular observations and a related Markov process, *Biometrika*, 67, 255–256.

Appendix A

Some Mathematical Concepts for Directional Data

A.1 Fourier Series

The characteristic function or pth trigonometric moment of a circular random variable Θ with pdf $f(\theta)$ on the interval $[0, 2\pi)$ is defined by $\phi_p = E(e^{ip\Theta}) = \int_0^{2\pi} e^{ip\theta} f(\theta) d\theta$ with $i = \sqrt{-1}$ as in Section 2.4. Then, here is a problem whether it is possible to represent it as

$$f(\theta) \sim \frac{1}{2\pi} \sum_{p=-\infty}^{\infty} \phi_p e^{-ip\theta}, \tag{A.1}$$

for the function $f(\theta)$. The right side of (A.1) is called the (formal) Fourier series of f.

Let $f(\theta)$ be a square integrable function, denoted by $f \in L^2(0, 2\pi)$, which means that

$$\int_0^{2\pi} |f(\theta)|^2 d\theta < \infty.$$

Then the Fourier series of f converges to f in the following L^2 sense:

$$\lim_{M,N\to\infty} \int_0^{2\pi} \left| f(\theta) - \frac{1}{2\pi} \sum_{p=-M}^{N} \phi_p e^{-ip\theta} \right|^2 d\theta = 0.$$

If $f \in L^2(0, 2\pi)$, then we have Parseval's identity

$$\frac{1}{2\pi} \sum_{p=-\infty}^{\infty} |\phi_p|^2 = \int_0^{2\pi} |f(\theta)|^2 d\theta,$$

and for an arbitrary complex sequence $\{\phi_p\}_{p=0,\pm 1,\dots}$ which satisfies $\sum_{p=-\infty}^{\infty} |\phi_p|^2 < \infty$, there exists $f \in L^2(0, 2\pi)$ whose pth trigonometric moment is $\phi_p = \alpha_p + i\beta_p$. It turns out that a pdf f which satisfies $f \in L^2(0, 2\pi)$ has the Fourier series

$$f(\theta) = \frac{1}{2\pi} \sum_{p=-\infty}^{\infty} \phi_p e^{-ip\theta} = \frac{1}{2\pi} \left[1 + 2 \sum_{p=1}^{\infty} \left\{ \alpha_p \cos(p\theta) + \beta_p \sin(p\theta) \right\} \right]$$

in the sense of L^2 convergence.

If a continuous function f on $[0, 2\pi]$ satisfies the condition (a) $f(0) = f(2\pi)$, (b) continuously differentiable, and (c) $f' \in L^2(0, 2\pi)$, then the Fourier series of f is absolutely convergent as shown below. Let $a_n/(2\pi)$ and $b_n/(2\pi)$ be the Fourier coefficients of f and f', respectively, i.e.

$$a_n = \int_0^{2\pi} f(\theta) e^{in\theta} d\theta, \; b_n = \int_0^{2\pi} f'(\theta) e^{in\theta} d\theta.$$

Then integration by parts leads to

$$b_n = \left[f(\theta) e^{in\theta} \right]_0^{2\pi} - in \int_0^{2\pi} f(\theta) e^{in\theta} d\theta = -in\, a_n.$$

Hence, we have

$$\sum_{n=-\infty}^{\infty} |a_n| = |a_0| + \sum_{n \neq 0} \frac{|b_n|}{|n|} \leq |a_0| + \left(\sum_{n \neq 0} \frac{1}{n^2} \right)^{1/2} \left(\sum_{n \neq 0} |b_n|^2 \right)^{1/2} < \infty,$$

since $\int_0^{2\pi} |f'(\theta)|^2 d\theta = (2\pi)^{-1} \sum_{n=-\infty}^{\infty} |b_n|^2 < \infty$ from the condition (c).

We propose to extend this result to the multivariate case. In particular, there does exist a unique representation of a periodic function of two variables in terms of the corresponding bivariate Fourier series. Thus we have,

$$\begin{aligned} f(\theta, \phi) \sim \sum_{m=0}^{\infty} \sum_{n=0}^{\infty} \{ & a_{mn} \cos(m\theta) \cos(n\phi) + b_{mn} \cos(m\theta) \sin(n\phi) \\ & + c_{mn} \sin(m\theta) \cos(n\phi) + d_{mn} \sin(m\theta) \sin(n\phi) \}, \end{aligned}$$

where the constants $(\boldsymbol{a}, \boldsymbol{b}, \boldsymbol{c}, \boldsymbol{d})$ are well defined and can be obtained analytically. Constraining the periodicities of both θ and ϕ to 2π, we obtain distributions on the torus. A simple yet instructive example may be constructed to yield, termed here as a Bivariate Cardioid (BC) distribution, where all the constants are taken to be zeros whenever $m \geq 2$, $n \geq 2$. The marginals of this distribution are also cardioid distributions, but the conditionals are not. It is interesting to derive conditions for independence of θ and ϕ based on the coefficients $(\boldsymbol{a}, \boldsymbol{b}, \boldsymbol{c}, \boldsymbol{d})$. In particular, for the standard BC distribution where the mean directions of the marginals are taken to be zeros, the condition of independence reduces to simply $a_{11} = a_{10} \times a_{01}$. The significant advantage of this representation, as is also true for the corresponding univariate case, is that the (joint) trigonometric moments are directly expressible in terms of the above coefficients. However in general, this representation too lacks a multivariate exponential family structure, and hence is not conveniently amenable to standard procedures for statistical inference.

A.2 A Chebishev Inequality

As in Section 2.5 the circular variance of a circular random variable Θ is defined by $v = E\{1 - \cos(\Theta - \mu)\}$. We apply Chebishev's inequality to v (Mardia and Jupp, 2000, p. 31). Since

$$\cos(\theta - \mu) = 1 - 2\sin^2\left(\frac{\theta - \mu}{2}\right),$$

we have

$$v \geq \int_{|\sin\{(\theta-\mu)/2\}|\geq\varepsilon} 2\varepsilon^2 dF(t) = 2\varepsilon^2 P(|\sin\{(\Theta - \mu)/2\}| \geq \varepsilon),$$

for $0 < \varepsilon \leq 1$, and thus a Chebishev inequality

$$P\left(\left|\sin\frac{1}{2}(\Theta - \mu)\right| \geq \varepsilon\right) \leq \frac{v}{2\varepsilon^2}, \quad 0 < \varepsilon \leq 1. \tag{A.2}$$

The equality holds for a symmetric circular distribution given by

$$P\left(\sin\frac{1}{2}(\Theta - \mu) = \pm\varepsilon\right) = \frac{v}{2\varepsilon^2}, \; P\left(\sin\frac{1}{2}(\Theta - \mu) = 0\right) = 1 - \frac{v}{2\varepsilon^2},$$

and the Chebishev inequality (A.2) cannot be improved without any conditions.

A.3 Measures of Similarity or Dissimilarity

Let ϕ be a convex function in the sense that $\phi(\lambda x + (1 - \lambda)y) \leq \lambda\phi(x) + (1 - \lambda)\phi(y)$ for $0 < \lambda < 1$ and $x, y \in [a, b]$, an adequate interval. Suppose that $\phi(1) = 0$ and Θ is a circular random variable whose pdf is $f(\theta)$. The divergence or dissimilarity between f and another circular pdf g with respect to ϕ may be defined by

$$D_\phi(f\|g) = E_f\left[\phi\left\{\frac{g(\Theta)}{f(\Theta)}\right\}\right] = \int_0^{2\pi} \phi\left\{\frac{g(\theta)}{f(\theta)}\right\} f(\theta)d\theta.$$

It holds that $D_\phi(f\|g) \geq 0$ from the existence of a straight line which supports $\phi(x)$ at $x = E_f\{g(\Theta)/f(\Theta)\} = 1$ and Jensen's inequality, and, for a strictly convex function ϕ, $D_\phi(f\|g) = 0$ if and only if $f(\Theta) = g(\Theta)$ with probability 1. If $\phi(t) = (1 - \sqrt{t})^2$ or $\phi(t) = 2(1 - \sqrt{t})$, then $D_\phi(f\|g)$ is called the Hellinger divergence and denoted by $D_{\text{H}}(f\|g)$, and if $\phi(t) = -\log t$, then it is called the Kullback–Leibler divergence (Section 3.2.1) and denoted by $D_{\text{KL}}(f\|g)$.

Note that $0 \leq D_{\text{H}}(f\|g) \leq 2$ and $S(f\|g)$ defined by $S(f\|g) = 1 - D_{\text{H}}(f\|g)/2$ $(0 \leq S(f\|g) \leq 1)$ may be viewed as a measure of similarity between f and g. The Hellinger divergence is symmetric, i.e. $D_{\text{H}}(f\|g) = D_{\text{H}}(g\|f)$. Indeed, the square root of D_{H} satisfies the distance axiom, but the Kullback–Leibler divergence does not satisfy this property. It is possible to think of maximum and minimum divergences of $D_\phi(f(\theta)\|g(\theta + \xi))$ with respect to ξ $(0 \leq \xi < 2\pi)$ as in Section 4.3.3.5.

A.4 Central Limit Theorem

Suppose that $X_1, X_2, X_3, \ldots$ are independent and identically distributed (i.i.d.) real-valued random variables with mean $\mu\,(-\infty < \mu < \infty)$ and variance $\sigma^2\,(0 < \sigma < \infty)$, and let $S_n = \sum_{j=1}^{n} X_j$ denote their partial sum. Then the central limit theorem states that the standardized variable $(S_n - n\mu)/(\sigma\sqrt{n})$ converges to the standard normal random variable in distribution as n tends to infinity, i.e.

$$P\left(\frac{S_n - n\mu}{\sigma\sqrt{n}} \leq x\right) \to \int_{-\infty}^{x} \frac{1}{\sqrt{2\pi}} e^{-t^2/2} dt, \qquad n \to \infty,$$

for all $x \in \mathbb{R}$. The central limit theorem on the unit circle in the complex plane, however, has a difference from that on the real line. The partial sum of i.i.d. circular random variables converges to the circular uniform distribution (Section 4.2.1) under a mild condition. We refer to Mardia and Jupp (2000, pp. 62–63) and Lamperti (1996, p. 94) to explain this fact.

Let C denote a set of complex numbers with absolute value 1, and suppose that $X_1, X_2, X_3, \ldots$ are i.i.d. random variables with values in C. Then $X_j\,(j = 1, 2, 3, \ldots)$ are represented as $X_j = e^{i\Theta_j}$, where the Θ_j are i.i.d. angular random variables on the interval $[0, 2\pi)$ and $i = \sqrt{-1}$. We consider the limiting distribution of the partial product $\prod_{j=1}^{n} X_j$ or the partial sum $\sum_{j=1}^{n} \Theta_j$ in the sense of modulo 2π as $n \to \infty$. The continuity theorem states that if P_n are probability measures on C with corresponding characteristic sequences $\phi_n(m)$ (the analogue of the characteristic functions with integer m) and

$$\lim_{n\to\infty} \phi_n(m) = \phi(m),$$

exists for each integer m, then $\phi(m)$ is the characteristic sequence of a measure P on C, and a sequence of probability measures $\{P_n\}$ converges weakly to P. From this theorem, we have the following statement: if the distribution of the i.i.d. random variables X_j is not supported on a finite set of points that are equally spaced around the circle, the probability distribution of the product $\prod_{j=1}^{n} X_j$ converges weakly to the uniform distribution on C as n tends to infinity. This means that, under the condition, the sum $\sum_{j=1}^{n} \Theta_j$ converges weakly to the circular uniform distribution as $n \to \infty$.

Next consider the case where the distribution of the i.i.d. random variables Θ_j is supported on a finite set of equally spaced points, whose probability is assigned by p_r to the point $2\pi r$ for $r = 0, 1, \ldots, m - 1$, where $0 < p_0 < 1$. Suppose that m_0 is the smallest number of points such that the support is contained in these points and the characteristic function of Θ_j is

$$\phi_p = E(e^{ip\Theta_j}) = \sum_{r=0}^{m_0-1} p_r \omega^{rp},$$

where $\omega = \exp(2\pi i/m_0)$. Then $|\phi_p| < 1$ for values of s not divisible by m_0. Thus,

$$\lim_{n\to\infty} \phi_p^n = \begin{cases} 0 & \text{if } p \neq 0 \text{ mod } m_0, \\ 1 & \text{if } p = 0 \text{ mod } m_0. \end{cases}$$

This is the characteristic function of the discrete uniform distribution on m_0 points around the unit circle, because $\sum_{r=0}^{m_0-1} \omega^{rp} = 0$. This means that, under the assumption above, the sum $\sum_{j=1}^{n} \Theta_j$ converges weakly to the circular discrete uniform distribution as $n \to \infty$.

The case where m_0 points that support the distribution are not roots of unity occurs with $|\phi_{m_0}| = 1$ but $\phi_{m_0} \neq 1$. In this situation, convergence does not hold. An example is, as in Mardia and Jupp (2000, p. 63),

$$P(\Theta = c\pi) = q, \quad P(\Theta = c\pi + \pi) = 1 - q,$$

for an irrational number c and $0 < q < 1$. The characteristic function is

$$\phi_p = E(e^{ip\Theta}) = qe^{ipc\pi} + (1-q)e^{ip(c\pi+\pi)}, \quad p = 0, \pm 1, \dots,$$

and $\phi_{2p} = e^{2ipc\pi}$ for $p = 0, \pm 1, \dots$. In this example, $m_0 = 2$, $|\phi_2| = 1$ but $\phi_2 \neq 1$, and $\lim_{n\to\infty} \phi_{2p}^n$ does not exist for $p = \pm 1, \pm 2, \dots$.

Mardia (1972, p. 90) remarks that the limiting distribution of the sum with scaling factor $S_n^* = \sum_{j=1}^{n} \Theta_j/\sqrt{n} \,(\text{mod}\, 2\pi)$ tends to the wrapped normal (Section 4.4.2).

References

Lamperti, J. W. (1996). *Probability: A Survey of the Mathematical Theory*, Second Edition, Wiley, New York.

Mardia, K. V. (1972). *Statistics of Directional Data*, Academic Press, London.

Mardia, K. V. and Jupp, P. E. (2000). *Directional Statistics*, Wiley, Chichester.

Appendix B

Mathematical Formulae

1) Gamma function:

$$\Gamma(r) = \int_0^\infty t^{r-1}e^{-t}dt, \quad r > 0. \tag{B.1}$$

2) Duplication formula:

$$\Gamma(2z) = \frac{2^{2z-1}}{\sqrt{\pi}}\Gamma(z)\Gamma\left(z+\frac{1}{2}\right). \tag{B.2}$$

3) Stirling's formula:

$$\Gamma(z) \sim \sqrt{2\pi}z^{z-1/2}e^{-z}\left(1+\frac{1}{12z}+\cdots\right), \quad z \to \infty. \tag{B.3}$$

4) Incomplete gamma function:

$$\gamma(r;x) = \int_0^x t^{r-1}e^{-t}dt, \quad r > 0, x > 0. \tag{B.4}$$

5) Beta function:

$$B(a,b) = \int_0^1 x^{a-1}(1-x)^{b-1}dx = \frac{\Gamma(a)\Gamma(b)}{\Gamma(a+b)}, \quad a > 0, b > 0, \tag{B.5}$$

$$2^{2\nu}B\left(\nu+\frac{1}{2},\nu+\frac{1}{2}\right) = B\left(\nu+\frac{1}{2},\frac{1}{2}\right). \tag{B.6}$$

6) Incomplete beta function:

$$B_z(a,b) = \int_0^z x^{a-1}(1-x)^{b-1}\,dx, \quad a > 0, b > 0, 0 < z < 1. \tag{B.7}$$

7) Pochhammer's symbol:

$$(a)_k = \begin{cases} a(a+1)\cdots(a+k-1), & k \geq 1, \\ 1, & k = 0. \end{cases} \tag{B.8}$$

8) Kummer's or confluent hypergeometric function:

$$M(a,b,z) \;=\; {}_1F_1(a;b;z) = \sum_{k=0}^{\infty} \frac{(a)_k}{(b)_k} \frac{z^k}{k!}. \tag{B.9}$$

9) Integral representations for $0 < a < b$:

$${}_1F_1(a;b;z) = \frac{1}{B(a,b-a)} \int_0^1 e^{zt} t^{a-1} (1-t)^{b-a-1} dt, \tag{B.10}$$

$${}_1F_1(a;b;z) = \frac{2^{1-b} e^{z/2}}{B(a,b-a)} \int_{-1}^1 e^{-zt/2} (1+t)^{b-a-1} (1-t)^{a-1} dt. \tag{B.11}$$

10) Gauss hypergeometric function:

$${}_2F_1(a,b;c;z) = \sum_{k=0}^{\infty} \frac{(a)_k (b)_k}{(c)_k} \frac{z^k}{k!}. \tag{B.12}$$

11) (Gradshteyn and Ryzhik, 2007, 9.134.2, p. 1009):

$${}_2F_1(2\alpha, 2\alpha+1-\gamma; \gamma; z) = (1+z)^{-2\alpha} \, {}_2F_1\left(\alpha, \alpha+\frac{1}{2}; \gamma; \frac{4z}{(1+z)^2}\right). \tag{B.13}$$

12) (Gradshteyn and Ryzhik, 1994, 9.131.1, p. 1069; Gradshteyn and Ryzhik, 2007, 9.131.1, p. 1008):

$${}_2F_1(a,b;c;z) = (1-z)^{c-a-b} \, {}_2F_1(c-a, c-b; c; z). \tag{B.14}$$

13) (Abramowitz and Stegun, 1972, 15.3.21, p. 560; $a = \nu, c = 2\nu + 1$):

$${}_2F_1\left(\nu, \nu+\frac{1}{2}; 2\nu+1; -z\right) = 2^{2\nu} \left\{1 + (1+z)^{1/2}\right\}^{-2\nu}. \tag{B.15}$$

14) Generalized hypergeometric function:

$${}_pF_q(a_1, \dots, a_p; b_1, \dots, b_q; z) = \sum_{n=0}^{\infty} \frac{(a_1)_n \cdots (a_p)_n}{(b_1)_n \cdots (b_q)_n} \frac{z^n}{n!}, \tag{B.16}$$

where a_j, b_k are complex numbers and p, q are nonnegative integers. If $p < q+1$, the series converges absolutely at every point of the complex plane, if $p = q = 1$, it converges absolutely for $|z| < 1$, and if $p > q+1$, it diverges for all $z \neq 0$ unless it terminates.

15) Bessel function of the first kind:

$$J_n(z) = \frac{1}{2\pi} \int_{-\pi}^{\pi} e^{-ni\theta + iz\sin\theta} d\theta, \tag{B.17}$$

$$J_n(z) = \frac{1}{\pi} \int_0^{\pi} \cos(n\theta - z\sin\theta) d\theta, \quad n = 0, 1, 2, \dots \tag{B.18}$$

16) (Abramowitz and Stegun, 1972, 9.1.42, 9.1.43, p. 361):

$$\cos(z\sin\theta) = J_0(z) + 2\sum_{k=1}^{\infty} J_{2k}(z)\cos(2k\theta), \tag{B.19}$$

$$\sin(z\sin\theta) = 2\sum_{k=0}^{\infty} J_{2k+1}(z)\sin\{(2k+1)\theta\}. \tag{B.20}$$

17) (Abramowitz and Stegun, 1972, 9.1.27, p. 361):

$$J_{p-1}(x) + J_{p+1}(x) = 2pJ_p(x)/x, \tag{B.21}$$

$$J_{p-1}(x) - J_{p+1}(x) = 2J_p'(x). \tag{B.22}$$

18) Modified Bessel function of the first kind and order r as integral and infinite sum representations:

$$I_r(\kappa) = \frac{1}{2\pi}\int_0^{2\pi} \cos(r\theta)\, e^{\kappa\cos\theta} d\theta = \sum_{j=0}^{\infty} \frac{1}{\Gamma(r+j+1)j!}\left(\frac{\kappa}{2}\right)^{2j+r}. \tag{B.23}$$

19) (Abramowitz and Stegun, 1972, 9.6.18, p. 376):

$$I_\nu(z) = \frac{(z/2)^\nu}{\sqrt{\pi}\,\Gamma(\nu+1/2)}\int_{-1}^{1} (1-t^2)^{\nu-1/2} e^{\pm zt} dt, \quad \mathrm{Re}(\nu) > -1/2. \tag{B.24}$$

20) When z is large (Abramowitz and Stegun, 1972, 9.7.1, p. 377):

$$I_0(z) \sim e^z/\sqrt{2\pi z}. \tag{B.25}$$

21) (Abramowitz and Stegun, 1972, 10.2.13, p. 443):

$$\sqrt{\pi/(2z)}I_{1/2}(z) = (\sinh z)/z, \tag{B.26}$$

$$\sqrt{\pi/(2z)}\,I_{3/2}(z) = -(\sinh z)/z^2 + (\cosh z)/z. \tag{B.27}$$

22) (Abramowitz and Stegun, 1972, 9.6.26, p. 376):

$$pI_p(\kappa) + \kappa I_{p+1}(\kappa) = \kappa I_{p-1}(\kappa) - pI_p(\kappa). \tag{B.28}$$

23) For large $|z|$ (Abramowitz and Stegun, 1972, 9.7.1, p. 377):

$$I_\nu(z) \sim \frac{e^z}{\sqrt{2\pi z}}\left(1 - \frac{4\nu^2-1}{8z}\right), \quad |\arg z| < \frac{\pi}{2}. \tag{B.29}$$

24) Modified Bessel function of the third kind and order ν (Gradshteyn and Ryzhik, 1994, 8.432.6, p. 969; Gradshteyn and Ryzhik, 2007, 8.432.6, p. 917):

$$K_\nu(z) = \frac{1}{2}\left(\frac{z}{2}\right)^\nu \int_0^\infty \exp\left\{-\left(\tau + \frac{z^2}{4\tau}\right)\right\}\frac{d\tau}{\tau^{\nu+1}},$$
$$|\arg z| < \frac{\pi}{2}, \ \mathrm{Re}(z^2) > 0. \tag{B.30}$$

25) Integral representation (Abramowitz and Stegun, 1972, 9.6.25, p. 376; Gradshteyn and Ryzhik, 1994, 8.432.5, p. 969; Gradshteyn and Ryzhik, 2007, 8.432.5, p. 917):

$$K_\nu(xz) = \frac{\Gamma(\nu+1/2)(2z)^\nu}{\sqrt{\pi}x^\nu}\int_0^\infty \frac{\cos(xt)}{(t^2+z^2)^{\nu+1/2}}dt,$$
$$\mathrm{Re}(\nu+1/2)>0, \qquad x>0, \qquad |\arg z|<\pi/2. \tag{B.31}$$

26) (Abramowitz and Stegun, 1972, 9.6.6, p. 375):

$$K_{-\nu}(z) = K_\nu(z). \tag{B.32}$$

27) (Abramowitz and Stegun, 1972, 10.2.17, p. 444):

$$K_{\pm 1/2}(z) = \left(\frac{\pi}{2z}\right)^{1/2} e^{-z}. \tag{B.33}$$

28) Associated Legendre function:

$$\int_0^\pi (z+\sqrt{z^2-1}\cos x)^{1/\psi}dx = \pi P_{1/\psi}(z), \quad \psi>0,$$
$$\int_0^\pi \frac{1}{(z+\sqrt{z^2-1}\cos x)^{-1/\psi}}dx = \pi P_{-1/\psi-1}(z) = \pi P_{1/\psi}(z), \quad \psi<0. \tag{B.34}$$

29) Integral representation of the toroidal function, also called torus function (Gradshteyn and Ryzhik, 1994, 8.851.1, p. 1041; Gradshteyn and Ryzhik, 2007, 8.851.1, p. 981):

$$P^m_{n-1/2}(\cos\eta) = \frac{(-1)^m}{2\pi}\frac{\Gamma(n+1/2)}{\Gamma(n-m+1/2)}\int_0^{2\pi}\frac{\cos(m\varphi)d\varphi}{(\cosh\eta+\sinh\eta\cos\varphi)^{n+1/2}}. \tag{B.35}$$

30) (Gradshteyn and Ryzhik, 1994, 8.772.3, p. 1028; Gradshteyn and Ryzhik, 2007, 8.772.3, p. 970):

$$P^\mu_\nu(z) = \frac{1}{\Gamma(1-\mu)}\left(\frac{z-1}{z+1}\right)^{-\mu/2}\left(\frac{z+1}{2}\right)^\nu {}_2F_1\left(-\nu,-\nu-\mu;1-\mu;\frac{z-1}{z+1}\right),$$
$$\left|\frac{z-1}{z+1}\right|<1, \tag{B.36}$$

$$P^\mu_{-\nu-1}(z) = P^\mu_\nu(z). \tag{B.37}$$

31) Appell's function F_1 (Gradshteyn and Ryzhik, 2007, 9.180.1, p. 1018):

$$\begin{aligned} F_1(a;b_1,b_2;c;x,y) &= \frac{1}{B(c,c-a)}\int_0^1 \frac{t^{a-1}(1-t)^{c-a-1}}{(1-xt)^{b_1}(1-yt)^{b_2}}\,dt \\ &= \sum_{m=0}^{\infty}\sum_{n=0}^{\infty}\frac{(a)_{m+n}(b_1)_m(b_2)_n}{(c)_{m+n}m!n!}x^m y^n, \quad |x|<1,\ |y|<1. \end{aligned} \tag{B.38}$$

32) Relation between Appell's function F_1 and the Gauss hypergeometric function ${}_2F_1$ (Appell and Kampé de Fériet, 1926, Eq. (28), p. 24; Gradshteyn and Ryzhik, 2007, 9.182.1, p. 1019):

$$F_1(a;b_1,b_2;b_1+b_2;x,y) = (1-y)^{-a}\,{}_2F_1\left(a,b_1;b_1+b_2;\frac{x-y}{1-y}\right). \tag{B.39}$$

33) Appell's double hypergeometric function (Gradshteyn and Ryzhik, 2007, 9.180.4, p. 1018):

$$F_4(a,b;c,d;x,y) = \sum_{m=0}^{\infty}\sum_{n=0}^{\infty}\frac{(a)_{m+n}(b)_{m+n}}{(c)_m(d)_n}\frac{x^m y^n}{m!n!}, \quad \sqrt{x}+\sqrt{y}<1. \tag{B.40}$$

34) Confluent hypergeometric series of two variables (Gradshteyn and Ryzhik, 2007, 9.261.1, p. 1031):

$$\Phi_1(\alpha,\beta,\gamma,x,y) = \sum_{m,n=0}^{\infty}\frac{(\alpha)_{m+n}(\beta)_m}{(\gamma)_{m+n}m!n!}x^m y^n, \quad |x|<1. \tag{B.41}$$

35) Lerch function (Erdélyi et al., 1953, 1.11, p. 27; Gradshteyn and Ryzhik, 2007, 9.55, p. 1039):

$$\begin{aligned} \Phi(z,s,a) &= \sum_{k=0}^{\infty}\frac{z^k}{(a+k)^s} \quad (|z|<1) \\ &= \frac{1}{\Gamma(s)}\int_0^{\infty}\frac{t^{s-1}e^{-at}}{1-ze^{-t}}\,dt \\ &\quad (\mathrm{Re}(a)>0 \text{ and either } |z|\le 1, z\ne 1, \mathrm{Re}(s)>0 \text{ or } z=1, \mathrm{Re}(s)>1), \end{aligned} \tag{B.42}$$

where any term with $a+k=0$ is excluded.

36) Polylogarithm function:

$$\mathrm{PLog}(z,s) = \sum_{k=1}^{\infty}\frac{z^k}{k^s} = \frac{1}{\Gamma(s)}\int_0^{\infty}\frac{t^{s-1}}{1-ze^{-t}}\,dt = \Phi(z,s,0). \tag{B.43}$$

37) Theta function (Abramowitz and Stegun, 1972, 16.27.3, p. 576):

$$\vartheta_3(z,q) = 1 + 2\sum_{n=1}^{\infty} q^{n^2} \cos(2nz). \tag{B.44}$$

References

Abramowitz, M. and Stegun, I. A. Eds. (1972). *Handbook of Mathematical Functions with Formulas, Graphs, and Mathematical Tables*, Dover, New York.

Appell, P. and Kampé de Fériet, J. (1926). *Fonctions Hypergéométriques et Hypersphériques: Polynomes d'Hermite*, Gauthier-Villars, Paris.

Erdélyi, A., Magnus, W., Oberhettinger, F. and Tricomi, F. G. (1953). *Higher Transcendental Functions*, Volume 1, McGraw-Hill, New York.

Gradshteyn, I. S. and Ryzhik, I. M. (1994). *Table of Integrals, Series, and Products*, Fifth Edition, Academic Press, London.

Gradshteyn, I. S. and Ryzhik, I. M. (2007). *Table of Integrals, Series, and Products*, Seventh Edition, Elsevier, Amsterdam.

Index

C

d

e

f

g

h

t

W

Printed in the USA/Agawam, MA
August 5, 2026

908663.005